Statistical Methods in
Agriculture and
Experimental Biology

OTHER STATISTICS TEXTS FROM CHAPMAN & HALL

Practical Statistics for Medical Research
Douglas Altman
The Analysis of Time Series – An Introduction
C. Chatfield
Problem Solving: A Statistician's Guide
C. Chatfield
Statistics for Technology
C. Chatfield
Indroduction to Multivariate Analysis
C. Chatfield and A. J. Collins
Applied Statistics: Principles and Examples
D. R. Cox and E. J. Snell
An Introduction to Statistical Modelling
A. J. Dobson
Introduction to Optimization Methods and their Applications in Statistics
B. S. Everitt
Multivariate Statistics – A Practical Approach
B. Flury and H. Riedwyl
Readings in Decision Analysis
S. French
Multivariate Analysis of Variance and Repeated Measures
D. J. Hand and C. C. Taylor
Multivariate Statistical Methods – a primer
Bryan F. Manly
Elements of Simulation
B. J. T. Morgan

Probability Methods and Measurement
Anthony O'Hagan
Essential Statistics
D. G. Rees
Foundations of Statistics
D. G. Ress
Decision Analysis: A Bayesian Approach
J. Q. Smith
Applied Statistics: A Handbook of BMDP Analyses
E. J. Snell
Applied Statistics: A Handbook of GENSTAT Analyses
E. J. Snell and H. R. Simpson
Elementary Applications of Probability Theory
H. C. Tuckwell
Intermediate Statistical Methods
G. B. Wetherill
Statistical Process Control: Theory and practice
G. B. Wetherill and D. W. Brown
Statistics in Research and Development
Second edition
R. Caulcutt
Modelling Binary Data
D. Collett
Statistical Analysis of Reliability Data
M. J. Crowder, A. C. Kimber, T. J. Sweeting and R. L. Smith

Further information of the complete range of Chapman and Hall *statistics books is available from the publishers.*

Statistical Methods in Agriculture and Experimental Biology

SECOND EDITION

R. Mead

Professor, Department of Applied Statistics
University of Reading, UK

R. N. Curnow

Professor, Department of Applied Statistics
University of Reading, UK

and

A. M. Hasted

Q I Statistics Ruscombe
Reading, UK

CHAPMAN & HALL
London · Glasgow · New York · Tokyo · Melbourne · Madras

Published by Chapman & Hall, 2-6 Boundary Row, London SE1 8HN

Chapman & Hall, 2-6 Boundary Row, London SE1 8HN, UK

Blackie Academic & Professional, Wester Cleddens Road, Bishopbriggs, Glasgow G64 2NZ, UK

Chapman & Hall, 29 West 35th, Street, New York NY10003, USA

Chapman & Hall Japan, Thomson Publishing Japan, Hirakawacho Nemoto Building, 6F, 1-7-11 Hirakawa-cho, Chiyoda-ku, Tokyo 102, Japan

Chapman & Hall Australia, Thomas Nelson Australia, 102 Dodds Street, South Melbourne, Victoria 3205, Australia

Chapman & Hall India, R. Seshadri, 32 Second Main Road, CIT East, Madras 600 035, India

First edition 1983
Reprinted 1986
Second edition 1993

©1993 Chapman & Hall

Typeset in 10/12pt Times by Thomson Press (India) Ltd, New Delhi
Printed in Great Britain by T.J. Press Ltd, Padstow, Cornwall

ISBN 0 412 35470 5 (HB)   0 412 35480 2 (PB)

A catalogue record for this book is available from the British Library

Library of Congress Cataloging-in-Publication data available

∞ Printed on permanent acid-free text paper, manufactured in accordance with the proposed ANSI/NISO Z 39.48-199X and ANSI Z 39.48-1984

Contents

Preface

Our aim in this edition, as in the first edition of the book, has been to describe and explain those statistical ideas which we believe are an essential part of the intellectual equipment of a scientist working in agriculture or on the experimental side of biology. Much of the material in the book has grown out of our experience as advisory statisticians and from teaching introductory statistics courses for agricultural students, both undergraduates and postgraduates. The examples in the early chapters are taken mainly from agricultural experiments involving field crops or farm animals but later examples are concerned predominantly with laboratory experiments and with biological investigations.

In this second edition we have expanded the scope to include some additional topics not usually covered in introductory courses or textbooks on statistics. It is particularly important that, even in an introductory statistics course, students should develop an appreciation of the breadth of statistical methodology now available. The development of computing facilities, and particularly of statistical packages, means that there is a very large library of statistical methods available to any scientist with access to any substantial computing facilities. Therefore, although some of the topics in this book would not feature in most introductory courses, we hope that there is a sufficient discussion of more advanced topics for the student/scientist to have some understanding of what would be involved in using some of the available advanced methods. Some students in their first reading of the book may well omit some of the more advanced chapters or sections.

Experimental scientists should have a clear understanding of the principles of statistics governing the planning of experiments and the analysis and interpretation of experimental data. Therefore, while covering the details of methods through worked examples, our main aim has been to help our readers understand why and when the various methods should be used, or not used! We emphasize the importance of thinking carefully about the purpose of each experiment; of using the available experimental resources as efficiently as possible; and then of extracting all the relevant information from the data. We also stress the importance of checking any assumptions that need to be made about the data before it can be analysed.

The mathematical knowledge required has been deliberately kept at a low level even at the cost of omitting mathematical details which would be well within the understanding of some undergraduates in biological subjects. Omission of mathematical details sometimes leads to more lengthy verbal justification of the statistical ideas, but if this encourages the reader to think more deeply about the statistical philosophy this will be a positive benefit.

In this second edition we discuss calculations in terms of both manual calculation and the use of computer packages. Most data analysis is now performed using computers and we have tried to make clear what information will be required to be input to the computer and what forms of information may reasonably be expected from a sensible computer package. We have deliberately not linked our examples of computing to a particular computer package but have discussed input and output in terms of an ideal package (all the features we discuss do exist in one or more standard packages). However, we have retained many worked numerical examples and suggested exercises for the reader since we believe that only by working through examples of analyses is it possible to understand fully what is being assumed, and how the analysis provides estimates of the accuracy of the results of the experiment. We would expect that when the book is used as a textbook for an introductory course in statistics the lecturer will plan parallel practical sessions using the locally preferred statistical computer package.

Turning to the contents in more detail, the first seven chapters present the basic statistical ideas of estimation and hypothesis testing and develop the major ideas of experimental design and the associated methods of analysis for standard experimental designs. In Chapter 8 we examine the assumptions inherent in the analytic techniques of the previous chapters and discuss methods of detecting and dealing with those situations where the assumptions are suspected or demonstrated to be untenable. Some of the more difficult methods of analysis required for particular forms of data arising from these first eight chapters are dealt with further in this second edition in a set of three chapters concentrating on the general ideas of modelling. These three chapters which are rather more advanced are Chapters 11, 12 and 14.

Relationships between two or more variables are discussed in Chapters 9, 10, 11 and 12, starting with simple linear regression, progressing to multiple regression and combinations of regression and experimental design and concluding with a discussion of the use of non-linear models. Chapters 13 and 14 discuss the analysis and interpretation of frequency data for discrete variables and examine examples of generalized linear models which by the date of this second edition must be considered part of the standard repertory of statistical tools for agricultural and biological research.

In Chapter 15 we consider some of the methods for analysing multiple measurements made for each experimental unit. These ideas are important and relevant to the practical research scientist, but are rather more advanced. In Chapter 16 we bring together various important ideas about experimental design and look at some specific areas of application. Finally there is a very short

Chapter 17 on some aspects of sampling, principally related to sampling within experiments.

Our examples of data from experiments are drawn from a wide range of agricultural and biological research and have been garnered over a long period of time. We have been unable to trace some of the sources through the mists of time. We are very grateful to our various colleagues, past and present, particularly at the University of Reading but also at the University of Aberdeen, at the National Vegetable Research Station at Wellesbourne (as it was then) and elsewhere for the use of their data. We hope that when these colleagues recognize their data they will approve of the way they have been used and accept our apologies for not acknowledging the use of their data. We have, in all cases, used the units in which the data were originally measured. This may be attributed to laziness but we prefer to believe that using the data as they were recorded minimizes possibilities of misinterpretation. To transform all data into a single system of units is a cosmetic operation imposing an unnatural homogeneity upon the reality.

We have listed only a few references. These are mainly cited in the late chapters and are intended to provide suggestions for further reading on particular topics. The tables of basic statistical distributions included in the Appendix are in the form used in the Department of Applied Statistics at Reading.

We have benefited, over many years from the comments and suggestions of colleagues, past and present. We thank them for their interest influence and ask their understanding of those instances when, in our stubbornness, we stick to our own views.

For their assistance in converting our scribbled manuscript into a readable form, we are grateful to Mrs Audrey Wakefield who produced the final typescript, and the late Mrs Rosemary Stern, Mr Ian Dale and Mrs Yolanda Martinez who drew the figures.

R. Mead
R. N. Curnow
A. M. Hasted

Reading, UK

1

Introduction

1.1 THE NEED FOR STATISTICS

The statistical methods described in this book have been developed to help in the quantitative study of variation and in the analysis and interpretation of data from experiments which are subject to variation. In biology the amount of variation and the relative importance of the different causes of variation are often of interest in themselves. In experiments variation tends to obscure the effects of different treatments, making comparisons difficult. This in turn can lead to mistakes in interpreting the results of experiments and to mistaken judgements about the best treatments to recommend for commercial use.

It is becoming a cliché to assert that everything is variable. However, the concept of variation is so fundamental to scientific experimentation that it is vitally important that anyone who comes into contact with experimental results, either as a research worker or as a user, must appreciate the universality of variation.

The following are a few of the situations in which variation occurs, sometimes without our noticing it.

1. The time of arrival of a particular train at a station will vary from day to day, certainly in terms of minutes and possibly by quite large amounts of time.
2. The life of apparently identical electric light bulbs will vary considerably.
3. The height of the tide line on a beach will vary from day to day, partly in a systematic way, but also in an apparently irregular fashion.
4. In a vegetable show the marrows exhibited will all be magnificent but some will be more magnificent than others. There will in fact be considerable variation in weight (is it really credible that the two best marrows will weigh exactly the same, to the nearest gram?). They will also vary in length, colour and many other characteristics.
5. If a number of apparently similar daffodil bulbs are planted in similar situations in the garden, one will bloom first and one will have the longest stem. The greater the detail in which the plants are examined, the more forms of variation will be observed.
6. If a pair of twin pigs are examined, being given exactly the same diet in the same environment, one of them will increase its body weight faster than the

other, and in general any characteristic will be slightly different for the two animals. Think of any pair of identical human twins you know, their very likeness makes the slight differences both of physique and of character the more noticeable.

These examples come from very different situations and the forms of the variation are particular to the situation. What they have in common is that in each case the causes of the variation are numerous. Thus the variation in train arrival times may result from the particular driver, the number of passengers, the condition of the coaches, various meteorological factors and the temper of various station masters. Even if it were possible to standardize all these factors it would not be reasonable to expect the train to arrive at the same time to the nearest second each day. The removal of major factors causing variation would only serve to spotlight minor ones. In the same way if we consider the growth of a particular plant we can write down an extensive list of factors which might be expected to have an effect on the growth of the plant, e.g. differences in seed weight; soil variation, both in terms of microclimate and in large-scale variation; depth of planting and competition from other plants in the vicinity. If we consider the harvesting and weighing of a number of different plants then further variation between the weights of plants can arise from the different times of harvest, from the different machines used to weigh the plants and from the subjective recording of the weights. If only part of the plant is weighed then the extraction of that part might involve further variation. However many factors we attempt to control, we shall always find that the final weights of two plants will be different. Variation is found even in experiments with plants grown in controlled environment rooms.

The amount of variation is generally much greater with biological material than it is with inanimate material. The main reasons for this are the ways in which the various genetic systems of inheritance act so as to maintain variability in a population and the extreme sensitivity of living matter during its growth and development to the effects of environment. In the physical sciences processes are more often deterministic and the principal source of variation is error in measurement. The greater variability of living material emphasizes the particular need for methods for dealing with variation in the biological sciences, medicine and agriculture.

The object of an agricultural experimenter is generally to measure the effect of varying some factor, for example the level of protein in poultry diets. The variability of the birds with which he has to work means that if the protein were kept at a fixed level considerable variation would be observed in the 'yield' of interest, e.g. weight gain over a certain period. It is therefore logical to expect that if different levels of protein are applied to different birds, the variation in the weight gains observed would be due partly to the different levels of feeding and partly to the basic variation between birds fed at the same level. The problem for the experimenter is to disentangle these two parts of the variation, so as to obtain an estimate of the 'true' differences caused by his treatments, i.e. the feeding levels, where 'true' is to be understood as meaning the difference that would have been observed if the birds and environmental conditions used in the test had been identical for all the treatments. The experimenter also needs some idea as to how accurate this estimate is.

The purpose of statistics is to provide methods for analysing data from such experimental situations. Sometimes the need for statistics is questioned on the grounds that statistics can only emphasize what is already obvious in a set of results, and that what is not obvious is not important even if picked out by some statistical procedure. These doubts arise partly from an incomplete consideration of what is required in interpreting experimental results and partly from an ignorance of the power of statistical methods. To illustrate further the need for statistical methods, results from an experiment investigating the effect of irrigation of cabbage plants at four different plant spacings are shown in Table 1.1.

All eight combinations of frequent or rare irrigation at each spacing were tried on small plots in each of three fields. The reasons for choosing such a form of experiment will be considered more fully later. The experimental form is, however, reasonable; all eight combinations are tried because the effect of irrigation may be different for different spacings; also it seems sensible to have more than one plot of each combination, both to give some idea of how variable the experimental material is and also to obtain a more precise estimate of any particular treatment effect.

Obvious conclusions to be drawn from these results are that frequent irrigation produces larger cabbages and that closer spacing reduces the size of cabbages. It might possibly to claimed as obvious that the effect of irrigation is greater at the closer spacings, though this is more debatable, the results for field B being contrary to this conclusion. This form of effect could, of course be of

Table 1.1 Yields of cabbage (mean fresh weight per head in kg) for 24 plots

Irrigation	Spacing	Field A	Field B	Field C
Frequent	1 (21 in)	1.11	1.03	0.94
Frequent	2 (18 in)	1.00	0.82	1.00
Frequent	3 (15 in)	0.89	0.80	0.95
Frequent	4 (12 in)	0.87	0.65	0.85
Rare	1 (21 in)	0.97	0.86	0.92
Rare	2 (18 in)	0.80	0.91	0.68
Rare	3 (15 in)	0.57	0.72	0.77
Rare	4 (12 in)	0.60	0.69	0.51

considerable scientific and economic importance. The experimenter probably expects the first two conclusions from previous work, and is really interested in this more complex effect. However, even if it is accepted that all important effects can be picked out by eye, there are a number of questions still to be answered. Thus the experimenter would want to know how much of an improvement the frequent irrigations achieved at the wide spacing. Since the running cost of the irrigation equipment, and possibly the purchase of further equipment, must be taken into account, a minimum increase of, say, 100 g per head may be necessary for the more frequent irrigation to be economic. The actual improvements in the three fields are 140, 170 and 20 g, giving an average value of 110 g. Do these results give sufficient confidence that if the frequent irrigation is used generally, the average increase in head size will be greater then 100 g? Similarly, how much confidence can we place in the apparent dependence of the response to irrigation on spacing? Since the spacings used are four out of the many possible spacings that might have been chosen, can we deduce any pattern in the weight obtained for different spacings, and perhaps predict the weight obtained at a spacing not included in the experiment? Can we do the same for the response to frequent irrigation, at different spacing?

This is not a complex experiment—with only 24 observations it would hardly be so described—yet the number of questions implicit in the formulation of such an experiment is considerable and most are not capable of immediate answer on inspection of the results by eye. It is, we hope, clear that to obtain full information from such data a more refined quantitative approach must be used.

The development of such a quantitative approach, often termed 'statistical science', has inevitably been in the language of mathematics, and over the last fifty years a vast literature of mathematical theory has been amassed. The basic principles of statistical science are simple, logical and distinct from mathematical concepts, however, and the following chapters do not merely give a series of methods for dealing with particular forms of data but also seek to explain the fundamental logic of statistical methods. Without an understanding of the fundamental logic, the systematic application of methods will lead to useless or even ridiculous conclusions. The amount of mathematics involved in the expounding of this logic has been firmly kept to a minimum and does not assume more than 16-year-old school mathematics. The reader may sometimes feel that he is lost in the principles and we suggest that when this happens he may find it more useful to read on to the application of these principles to the analysis of data before returning to re-read the section on the logic of the methods. Throughout the book there are examples of calculations on data in which the methods are illustrated in considerable detail. We believe that a close following of these examples will be helpful to the reader and that he will derive particular benefit by working through the exercises, again on actual data, which usually follow the examples closely.

1.2 THE USE OF COMPUTERS IN STATISTICS

Most of the statistical ideas discussed in this book were developed before the advent of computers. Statistical methods might very probably have developed along different paths if computers had been available eighty years ago instead of only thirty years ago. However, the underlying principles on which the analysis of data from design experiments or of data on relationships are based would not have been different. Almost all statistical analysis of data will use a computer, and most probably will involve the use of a statistical package.

During the last twenty years many computer programs and packages for the calculation of statistical analyses have been written. These range from simple programs to calculate the analysis of variation for a particular simple experiment structure; through more general programs which can analyse data from any simple standard design, where the treatments may have structure; to powerful packages which can handle data from designed experiments with all kinds of block sizes and blocking and treatment structures. These more powerful packages can also analyse data from designs with several levels of variation and with any form of treatment structure, whether factorial, partially factorial, confounded or in patterns beyond the scope of the book. A computer program requires information to be presented in a form appropriate to that program. When we talk about a package we are referring to a very large program, for which a new language has been devised for communication, and which offers a wide range of facilities which can be used in ways specified by the user, provided he has learnt the language!

It is now probable that, wherever an agricultural or biological research worker is employed, there will be access to a computer and that either there are programs immediately available to produce the analysis of any data that he requires, or such programs do exist but have not yet been implemented on the available computer. The time has therefore passed when it is sensible for an experimenter or a statistician to write his own programs except occasionally for very specific and non-standard calculations. Not all the packages available are easy to use and some produce peculiar forms of output and even incorrect results. However, to ignore the available general programs and to write one's own is a waste of time. An analogy would be to ignore all information from the newspapers, radio and television and to try to set up a chain of letter writers around the world. By writing your own program, you may get the analysis printed in exactly the form you want but the program will be slow and inefficient and may include mistakes, and you will spend a long time checking that the program is correct.

The principle advantage which the computer offers is the sheer speed of its calculations. A large analysis will require at most a few seconds with most packages. The other great advantage is in repeated calculations. If you have results from a designed experiment and have many different variables for each experimental unit, then the initial instructions defining the form of analysis for the computer program can be used for all the variables without further definition. The speed, of course, relates to the time required after the data and analysis instructions have been typed in. There are undoubtedly very small sets of data for which analysis by hand using a small calculator is still more economical. It is of course also true that you may learn more about the principles of methods of analysis by working through the detailed calculations yourself. Once!

The standard manner of storing data on computers is in a two-dimensional matrix, or array, as shown in Fig. 1.1. Each row in the matrix corresponds to a single observational unit. Each column in the matrix records a single piece of information about each unit. These pieces of information are of two kinds. The first kind is categorical information about the units, identifying the particular characteristics of each unit. The second kind of information is of various measurements or variables recorded for all the observational units.

Most general packages include a range of facilities for deriving new column variables from existing column variables by simple arithmetical or functional operations. Examples include: weight gain = difference between final weight and initial weight; $\sqrt{\text{weight}}$; weight per unit area. The great advantage of being

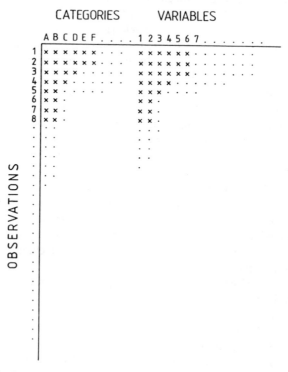

Fig. 1.1 Standard matrix format for data storage of experimental or survey data within a computer.

able to operate on complete columns is that only a single instruction is needed for an entire set of observations. There is therefore no need to calculate derived variables prior to the computer analysis of data, though when such derived variables are calculated within the computer program it is a very good general rule to have the values of the new variable output so that they can be checked visually for any values which appear suspicious.

This checking of values within each column is a facility which should usually be available within the package. A range of forms of checks are possible. Some programs simply provide the maximum and minimum values in a column. Others can identify all those values, and the corresponding units, lying outside a specified range. In addition to providing checks, simple summaries should be available for each column. These would normally include the mean and standard deviation and medians and other percentiles. Graphical output is also available with most packages and normally includes frequency distributions for single variables and scatter plots for pairs of variables.

Packages for the analysis of large quantities of data will include a much wider set of data handling techniques. One class of technique provides tabulation of subsets of the data by selecting those observations (rows) which satisfy conditions defined in terms of one or more variables (column). Thus, for example if the rows correspond to different farms, then one column may define the type of farm (arable, dairy, mixed, etc.) and it may be required to produce a subtable including only the results for the arable farms. Alternatively, a subtable could include only those farms of a certain size or in a particular geographical area.

In addition to the facilities for table manipulation, packages for large quantities of data should include a range of possible presentations of data through graphical outputs. The packages should also include facilities for calculating simple summary statistics and either the routines for more complex analyses or at least a method of storing subsets of the raw or derived data which can be subjected to more complex analyses using other packages. New packages are continually becoming available and improved facilities being developed. It is crucial therefore to have a sound understanding of the basic statistics methodology.

In this book, although we emphasize the principles underpinning the statistical analysis and interpretation of data, and illustrate methodology with simple examples, we also hope to show the computer can provide all the necessary calculations. We do not relate the computations to a particular computer package, nor do we attempt the almost impossible task of providing a comparative assessment of the available packages. Our discussion of the use of computer packages for statistical analysis is in an idealized framework of 'the form in which information will be required for presentation to the computer' and 'the form in which output results should be forthcoming from the computer'. Hopefully anyone using this book to learn about statistical methods will be able to use a statistical package in parallel to reading the book, and doing some simple calculations by hand.

2
Probability and distributions

2.1 PROBABILITY

The main problem, introduced in the first chapter, which statistics attempts to solve is the disentangling of the effects of different treatments from background variation due to a host of factors which we either cannot control or decide not to control because of the expense. We shall want to know how much reliance can be placed on the observed differences between the effects of different treatments. From some knowledge of the variability of the biological material we shall want to estimate how likely it is that the differences we have observed between the treatments could have arisen because rather better animals or plants have been subjected to one treatment compared to another. This brings us on to the idea of probability.

The word 'probable' is used in various senses in the English language: statisticians use the term probability in a rather narrow quantitative sense. Thus if a single unbiased coin is tossed it seems reasonable to suggest that the probability of a head is approximately half. There are two sides to the coin, apparently equally balanced, and there is no logical reason for supposing that the head (or tail) side has any particular preference for looking up at the sun, or down towards the earth. This is a very simple form of probability argument. For more complicated situations a more detailed approach is necessary. Suppose, for example, that we know that a certain farmer has a herd of 100 cows. If we were to select, in an arbitrary manner, one cow from this herd what is the probability that this particular cow will be brown? Or, to look at a rather more useful characteristic, what is the probability that its average weekly milk yield will be greater than 20 gallons? To answer these questions we would need to know what the colours and milk yields of all the cows in the herd are. In statistical terminology, we need to know the **distribution** of colour within the herd and the **distribution** of milk yields.

If we recorded the colour of each cow we would perhaps obtain the following

8 black cows
25 black and white cows
35 brown and white cows
32 brown cows

Table 2.1 Average weekly milk yields (in gallons) of a herd of 100 cows

13.2	12.6	16.2	14.3	13.4	17.8	18.8	15.6	18.0	7.6
11.1	21.3	16.8	17.1	23.4	16.9	16.3	13.7	19.7	16.7
9.6	9.1	14.7	16.9	19.1	11.1	17.0	16.7	18.2	13.5
17.9	17.6	14.7	18.8	14.3	18.4	15.8	22.0	20.0	19.4
15.5	12.7	18.3	10.4	13.4	12.4	19.0	21.0	14.7	11.7
6.1	23.5	13.9	14.6	23.2	21.9	16.8	17.6	15.5	25.6
16.3	20.0	25.9	26.2	24.1	21.0	20.5	17.3	22.3	15.6
19.5	13.7	10.7	12.5	13.2	16.6	13.6	20.5	15.0	19.7
10.8	12.1	25.9	15.2	10.5	20.0	10.0	13.3	29.7	19.0
19.8	18.0	16.5	15.9	16.2	19.8	22.5	20.1	20.1	12.3

Thus the proportion of brown cows in the herd is 32 out of 100 and we can reasonably say that if we select a cow from the herd in an arbitrary manner, unconnected with the colour of the cow, the probability that the selected cow will be brown is 32/100.

Consideration of the probability of the selected cow having an average milk yield greater than 20 gallons per week is slightly more complex but the general argument is the same. If we recorded the milk yields of all one hundred cows, we should obtain results as in Table 2.1.

By counting the number of yields of 20 gallons or more we obtain the proportion of 23 out of 100. Hence the probability that an arbitrarily selected cow will have an average milk yield greater than or equal to 20 gallons per week is 23/100 or roughly 1/4.

Data such as those in Table 2.1 are difficult to appreciate quickly and it is therefore desirable to have a simple method of presenting such data. A visual method is to construct a **histogram** of the data. For this we group the individual yields into a series of intervals on the yield scale: for the milk yield data we could use intervals: 6–8 gallons a week, 8–10, 10–12, etc., the first few observations in Table 2.1 then being grouped into the intervals 12–14, 12–14, 16–18, 14–16, 12–14, etc. In this way we obtain a summary of the original data in the form of a set of frequencies in these intervals. We can now represent these frequencies graphically using the horizontal axis for the yield scale and the vertical axis for frequency and, in each interval on the horizontal axis, drawing a column whose area is proportional to the frequency for the interval. Using the intervals of width 2 gallons/week suggested earlier we can construct a histogram for the milk yield data and this is shown in Fig. 2.1.

This histogram gives an immediate visual impression of the distribution of milk yield among the 100 cows of the herd. Obviously the effectiveness of a histogram depends on the number of observations and on the size of the intervals chosen for grouping the yields. A useful guide in choosing the size of the interval is to divide the range of observations into between 8 and 12 equal intervals though, if the number of observations is very large, the number of intervals

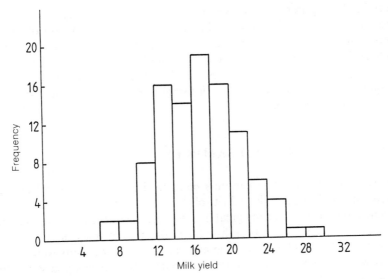

Fig. 2.1 Histogram of the data in Table 2.1. Grouping the yields into intervals of width 2 gallons per block.

may be increased. A histogram with too few observations or too many intervals will be irregular and unhelpful. We can see at once in Fig. 2.1 that most yields are close to 18 gallons per week and that all values lie in the range 6–30. We can also read off quite quickly the proportions in various ranges; thus 75 out of 100 yields are in the range 12–22 gallons per week, so that if we select a single cow, the probability of finding that its milk yield is outside this range is 1/4, a quarter.

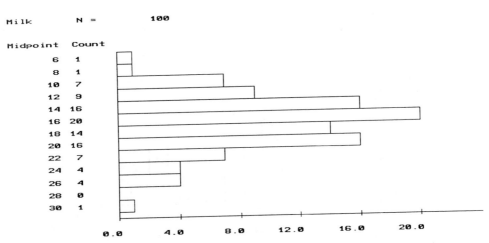

Fig. 2.2 Histogram of milk yield data, Table 2.1, produced by a computer package.

Forming histograms by hand is obviously very tedious. They can be quickly produced by a statistical package. Figure 2.2 shows a histogram of the cow data produced by a typical package. Although the histogram is presented on its side and the intervals are different the underlying shape of the distribution is obviously the same. Most packages will allow the user to specify the intervals he requires but choose by default a sensible number of intervals for the size and range of the data set.

2.2 POPULATIONS AND SAMPLES

In an experiment treatments can only be applied to a small number of plants, animals or field plots. The effect of the treatments on these particular experimental units is not of any great interest because by the time the experiment is over the plants will have been harvested, the animals may be dead and the field plots will be used for some other experiment. What we really want to know is what would happen if we applied the treatments to a much larger number

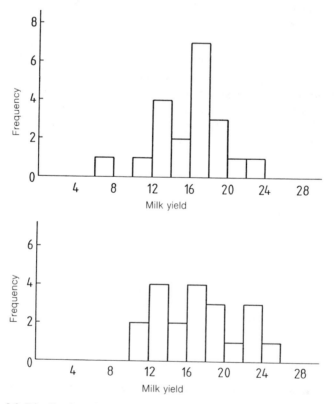

Fig. 2.3 Distributions for two samples of 20 from the data of Table 2.1.

of animals or plants. The material used in the experimental is called a sample and is taken to be representative of a larger population of material. We shall discuss later the ways in which samples of material should be chosen so that they are representative of a population. At the moment we need to consider what can be said about a population given only the results of a sample.

First we will consider the herd of 100 cows as the population and look at samples from this population. If we take the first two lines of Table 2.1 as a sample of 20 we can look at the distribution of milk yields in this sample. This distribution is shown in Fig. 2.3(a) which obviously has a strong resemblance to the population shown in Fig. 2.1. We can also calculate the proportion of the sample yields greater than 20 gallons/week, giving 2/20, and this gives us an estimate of the probability of such a yield in the population. Notice that this

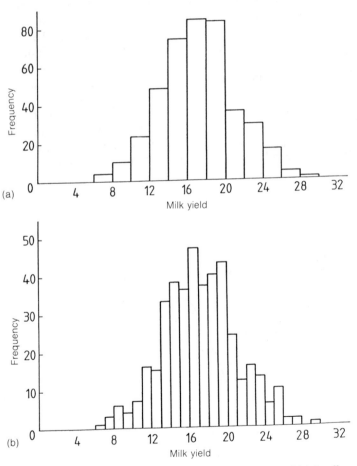

Fig. 2.4 Distributions for samples of 400 grouped in intervals, of (a) 2 gallons per week and (b) 1 gallon per week.

estimate 1/10 is quite a bit different from the true probability in the population 24/100. Nevertheless, if we had only measured milk yields for a sample of 20 the proportion 1/10 would be the only obvious **estimate** of the probability in the population. There are, of course, many other ways in which we could take a sample of 20 and each would give an estimate. Thus if we used the fifth and seventh columns we obtain the distribution shown in Fig. 2.3(b) and a proportion of yields greater than 20 gallons per week of 5/20, i.e. a probability estimate of 1/4 which is very close to the population value.

It is clear that the distributions of Fig. 2.3(a) and (b) approximate to the population distribution of Fig. 2.1. It is equally clear that conclusions drawn from the samples will be inaccurate to some degree as predictions for the population. It is also reasonable to assume that as the size of the sample increases the sample distribution will tend to get closer to the population distribution.

Now consider the herd of 100 cows not as a population but as a sample from a regional population which may have, perhaps, 100 000 cows. The sample distribution of Fig. 2.1 gives us an idea of the shape of the population distribution. If we increase our sample size to 400 we would get a distribution such as that of Fig.2.4(a) which is obviously more regular than Fig. 2.1. We now have quite high frequencies in each interval. By reducing the interval widths to 1 gallon/week we obtain a more continuous distribution but, as shown in Fig. 2.4(b), it becomes more irregular again. By gradually increasing the sample size and reducing the grouping intervals we shall obtain a smoother distribution which will approach closer and closer to the population distribution.

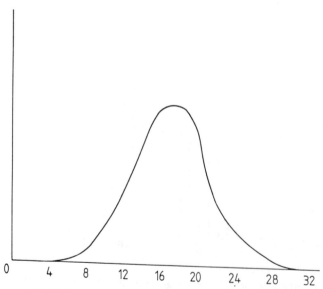

Fig. 2.5 Population distribution for milk yield data.

The variable which we are examining, milk yield, can theoretically take any value, i.e. it is not restricted to values of 16.1, or 13.3 but can be 16.1087 or 13.3786. We can therefore make the grouping interval extremely small, provided we have a large enough sample, and the population distribution can be effectively represented by a continuous curve, as shown in Fig. 2.5.

2.3 MEANS AND VARIANCES

We have used histograms to represent distributions pictorially and we can make qualitative comparisons of distributions by reference to their different histograms. For quantitative comparisons we shall obviously require a more concise description of the data than is afforded by a histogram. The most obvious attributes of a distribution are its position and spread.

There are two common statistical measures of position, the mean and the median. Of these the mean is defined to be the average of the values of the distribution and is the most commonly used measure because of its simplicity and other theoretical advantages. Thus for the first sample of 20 from the data of Table 2.1, the mean is

$$\text{mean} = (13.2 + 12.6 + 16.2 + 14.3 + \cdots + 13.7 + 19.7 + 16.7)/20$$
$$= 320.5/20 = 16.025$$

The mean of a population is denoted by μ, and the mean of a sample, containing n individuals, whose values are referred to as $x_1, x_2, x_3, \ldots, x_n$, by $\bar{x}$. The formal definition of the mean for a sample distribution is

$$\bar{x} = \frac{\Sigma(x)}{n}$$

where Σ is the symbol for summation and means add the quantity in brackets for all members of the sample.

The other measure of location, the median, is defined as that value for which exactly half the values in the population of the sample are greater, and half smaller. Again using the first sample of 20 from the data of Table 2.1 as an example, if we write the 20 values in order of increasing size we find that the tenth value is 16.3 and the eleventh 16.7. The median could be taken as any value between 16.3 and 16.7: for simplicity it is usually taken to be the average, i.e. 16.5. The median is a useful measure of position for distributions in which most of the observations are clustered together but there are a few very large or very small observations as well. As an example, the average income of people working in this country can be a slightly misleading figure because it is so much influenced by the small number of people earing a very large income. A more relevant measure of position for a discussion of many aspects of income would be the salary which divides the population into two equal groups, half with the salaries above this value and half below. This would be the median salary.

While the standard measure of position, the mean, is simply defined the usual measure of spread is not so simple. The most obvious measures of spread is the range of values. Unfortunately this is based on extreme values which may well be eccentric and also depends heavily on the size of sample, which makes a range observed for a sample of little use as an estimate of the range for the population. To illustrate this defect, consider again the population of 100 milk yields. The population range is $29.7 - 6.1 = 23.6$. The ranges for the two samples of 20 described earlier are (a) $23.4 - 7.6 = 15.8$, (b) $24.1 - 10.0 = 14.1$. If samples of 5 were taken these ranges would generally be further reduced.

The range being unsuitable as a measure of spread we would like to construct a measure from the deviations of individual values from the mean. The average deviation from the mean,

$$\frac{\Sigma(x - \bar{x})}{n}$$

is useless since the sum of deviations from the mean is zero! This is easily shown

$$\Sigma(x - \bar{x}) = \Sigma(x) - \Sigma(\bar{x}) = \Sigma(x) - n\left[\frac{\Sigma(x)}{n}\right] = 0$$

Hence the simplest measure of spread involving deviations from the mean is obtained by taking the average absolute deviation from the mean, i.e. ignoring the signals of the deviation. However, this measure is not easily handled mathematically and the average squared deviation is used instead. Most of the advantages of this are mathematical but some will become apparent later in the book. This measure of spread is termed the **variance** and is formally defined for a sample as

$$s^2 = \frac{\Sigma[(x - \bar{x})^2]}{(n - 1)}$$

where Σ again means sum over all members of the sample.

For the first sample of 20 from the milk yield data the variance is

$$s^2 = [(13.2 - 16.025)^2 + (12.6 - 16.025)^2 + (16.2 - 16.025)^2 \dots$$
$$+ (16.7 - 16.025)^2]/19$$
$$= 243.96/19$$
$$= 12.84$$

The variance of a sample is measured in the units of squared observations. To obtain a measure of spread in the same units as the mean we use the positive square root of the variance and this is called the **standard deviation**. For the example above, the sample standard deviation is $\sqrt{12.84} = 3.58$. Hence we can summarize the observations in the sample by the mean 16.02 and the standard deviation 3.58.

There are two further points to notice in the definition of the variance. First, it is computationally simpler to calculate the variance as

$$\{\Sigma(x^2) - [\Sigma(x)]^2/n\}$$

divided by $(n-1)$. This is equivalent to the previous form as can be seen from the following:

$$\begin{aligned}
\Sigma[(x-\bar{x})^2] &= \Sigma(x^2 - 2x\bar{x} + \bar{x}^2) \\
&= \Sigma(x^2) - \Sigma(2x\bar{x}) + \Sigma\bar{x}^2 \\
&= \Sigma(x^2) - 2\Sigma(x)\bar{x} + n(\bar{x}^2) \\
&= \Sigma(x^2) - 2\Sigma(x)\Sigma(x)/n + n[\Sigma(x/n)]^2 \\
&= \Sigma(x^2) - 2[\Sigma(x)]^2/n + [\Sigma(x)]^2/n \\
&= \Sigma(x^2) - [\Sigma(x)^2/n]
\end{aligned}$$

However, the latter form is considerably quicker to calculate. Whichever form is used, care must be taken to avoid errors due to rounding-off in the calculations.

The second point to notice is the division by $(n-1)$ rather than n which would seem to be the more obvious divisor in calculating an average squared deviation. By taking a large number of small samples from the large population of values it is not difficult to show that the use of n as the divisor will on average give a value that is lower than the variance of the whole population. This can also be proved mathematically. We call a value that is going to be right on average an unbiased estimate.

Although we only make the calculation for a single sample at any one time it would seem sensible to carry out the calculations in such a way that if they were repeated on different samples they would on average give a value that was neither too high nor too low. The divisor $(n-1)$ is called the **degrees of freedom** of the variance. It has this name because the variance is calculated from n deviations from the mean and only $(n-1)$ of these deviations can be chosen 'freely'. The other deviation is determined by the fact that the n deviations must add to zero.

As well as giving summary statistics, a good statistical package will offer various ways of displaying a data distribution. One of these is the box-and-whisker plot or boxplot for short. Figure 2.6 shows a boxplot of the milk yields given in Table 2.1. The box contains the middle 50% of data values, the median or mid value is marked with a line across the box. The whiskers or lines either side of the box show the range of the lower and upper 25% of data values. Any extreme values are highlighted, in this boxplot the extreme value is marked with the symbol *. Boxplots are very useful in showing whether a distribution

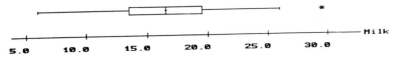

Fig. 2.6 Box and whisker plot of milk yield data, Table 2.1.

is skewed or symmetrical, for this data the median is in the middle of the box and the whiskers are of about the same length suggesting a symmetrical distribution as can also be seen in the histograms.

2.4 THE NORMAL DISTRIBUTION

The mean and variance to a considerable extent characterize a distribution, though it is clearly possible to construct situations where two distributions have the same mean and variance but are entirely different in other respects. There is, however, one particular form of population distribution which is completely defined by the mean and variance so that the probability of a single observation being in any particular range of values depends only on the mean and standard deviation of the distribution. This distribution, which is by far the most important in statistics, is called the **normal distribution**. It is bell-shaped and the general form of the distribution is shown in Fig. 2.7. Apart from some very useful mathematical properties the normal distribution is important because a remarkably large number of observed variables are approximately normally distributed. As an example of this Fig. 2.8 shows the distribution of heights in 1940 of men born in the UK in 1883, the frequencies being tabulated in Table 2.2.

The normal distribution is symmetrical and clustered about the mean. In fact approximately 68% of the distribution is within a standard deviation, σ, of the mean, μ, i.e. range $\mu - \sigma$ to $\mu + \sigma$. Only 2.3% of the values are greater than

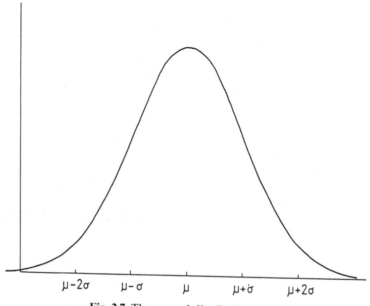

Fig. 2.7 The normal distribution curve.

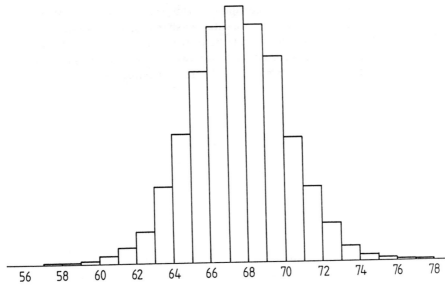

Fig. 2.8 Distribution of heights (inches) of men in the United Kingdom.

Table 2.2 Frequency distribution of heights in 1940 of adult males born in the UK in 1883

Height (in)	Number of men	Height (in)	Number of men
57–58	2	68–69	1230
58–59	4	69–70	1063
59–60	14	70–71	646
60–61	41	71–72	392
61–62	83	72–73	202
62–63	169	73–74	79
63–64	394	74–75	32
64–65	669	75–76	16
65–66	990	76–77	5
66–67	1223	77–78	2
67–68	1329		

$\mu + 2\sigma$ (with 2.3% less than $\mu - 2\sigma$). To tabulate the proportions of a normal distribution in various ranges we only have to know the position and spread, i.e. μ and σ.

However, it would obviously be better to have a single tabulation which does not depend on μ or σ. To do this, instead of considering the distribution of x which has mean μ and standard deviation σ, we consider the distribution of $z = (x - \mu)/\sigma$. Now z is still normally distributed. To demonstrate this consider the form of z. First, we subtract a constant, μ, from x; this moves the distribution along the horizontal axis, it does not alter its shape. Second, we divide by σ; this contracts or expands the size of the distribution but does not alter the shape, since each point of the distribution is moved towards the mean a constant fraction of its distance away from the mean. The three distributions of x, $x - \mu$,

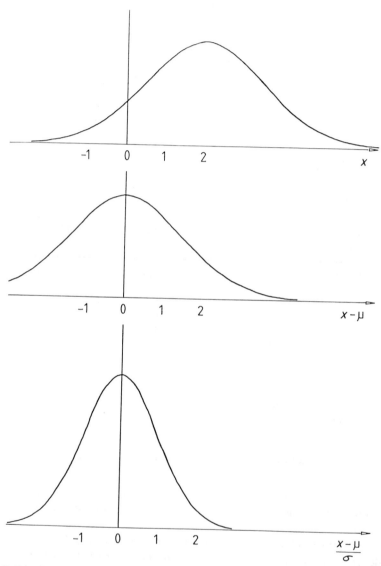

Fig. 2.9 Distributions of x, $x - \mu$, $(x - \mu)/\sigma$ where x is normally distributed with mean μ and variance σ^2.

and $z = (x - \mu)/\sigma$ are shown in Fig. 2.9. The mean of the distribution of z is clearly 0 and the standard deviation is 1 (the spread measured by σ has been reduced by a factor $1/\sigma$). Thus z is in a standard form and is called a **standardized normal deviate**. The distribution of z is given in Table A.1, the distribution being given in the form of the proportion of the distribution less than a particular value of z. Thus the proportion less than $z = -0.5$ is 0.3085 and the proportion less than $z = +0.5$ is 0.6915, so that the proportion between $z = -0.5$ and $+0.5$ is 0.3830. In terms of the distribution of x this means that the proportion of the distribution between $\mu - 0.5\sigma$ and $\mu + 0.5\sigma$ is 0.3830 and therefore the probability that a single observation, taken arbitrarily from a normally distributed variable, lies within 0.5σ of the mean value, μ, is 0.3830. To see just how closely the distribution of heights in Table 2.2 agrees with the normal distribution we will compare some of the frequencies of the two distributions. The mean and standard deviation of the height distribution are approximately 67.5 and 2.5 respectively. From the distribution of the standardized normal distribution we see that we would expect 68.36% of the distribution between 65 and 70 ($\mu - \sigma$ to $\mu + \sigma$), 92.8% between 63 and 72, and 99.72% between 60 and 75. A quick check shows that the observed proportions of this population of heights in these ranges are 68.07%, 92.4%, and 99.63%, showing a very good agreement with expectation.

2.5 SAMPLING DISTRIBUTIONS

Sample means are extremely important because in experiments we apply treatments to samples of material and use the mean results or yields as measures of the effects of the treatments. We shall now consider the distribution of sample means and how this distribution is related to the distribution of individual values.

As an example of the calculation of a sample mean it was found earlier that the mean of the sample of 20 from the herd of 100 cows was 16.025. If we calculate the mean for the second sample (columns 5 and 7 in Table 2.1) we have

$$\bar{x} = (13.4 + 23.4 + 19.1 + \cdots + 10.0 + 22.5)/20 = 341.1/20 = 17.055$$

If we take other samples we could get more values of the sample mean, in general all different, and by taking sufficient samples we could obtain the distribution of the values of the sample mean for a given size of sample. What properties would we expect the distribution of the sample mean to have? First, we would expect that the mean value of the distribution of the sample mean would be the same as the mean value of the distribution of the original observations (since there is no reason for expecting it to be either greater of smaller). Second, we would expect that the mean of a number of observations should be a better estimate of the population mean so that we would expect the spread of the distribution of the sample mean to be less than that of the distribution of the original observations.

At this point, for the first time, we shall quote without proof a mathematical result. This result says that if a variable x is distributed with mean μ and variance σ^2, then the variable $\bar{x}$ which is the mean of a sample of n observations of x, is distributed with mean μ and variance σ^2/n. The variance of the distribution of the sample mean is, as expected, less than the variance of an individual observation. It requires a methematical argument to show that it is less by the factor $1/n$. The square root of the variance of a sample mean is called the **standard error** of the mean rather than the standard deviation of the mean. This term is used to avoid confusion with the standard deviation of the distribution of individual observations.

Two other mathematical results concerning the distribution of the sample mean emphasize the importance of the normal distribution. First, if the distribution of the original observation, x, is normal, then the distribution of the sample mean, $\bar{x}$, is also normal. Second, for almost all forms of distribution of the original observation, x, the distribution of the sample mean, $\bar{x}$, for a sample of n observations, tends to the normal distribution as n increases. In fact the tendency for distributions of sample means to become normal is so strong that, for many forms of original distribution, if n is more than 5 the distribution of the sample mean is almost indistinguishable from the normal distribution. The result, which is known as the Central Limit Theorem, is extremely useful. It can be used to explain why so many biological quantities have an approximately normal distribution. Most measurements in biology are made on characteristics that have been influenced by many different genetical and environmental factors. For example, the milk yield of a cow is determined by the effects of different genes at many loci on its chromosomes and by a vast number of environmental influences such as feeding, climate and disease. This means that the milk yield of a cow is really the compound effect of a very large number of factors each contributing a small stimulus towards a higher or a lower yield. This averaging process tends to produce a normal distribution for many biological variables. We shall discuss later the problems of analysing observations that are not normally distributed.

To illustrate these results about the distribution of the sample mean we shall consider some data on weights of individual carrots grown in rows. The weights of 144 carrots in an arbitrary order are given in Table 2.3 and the distribution is shown in Fig. 2.10(a); obviously the distribution is not normal.

To find the mean and variance of this sample we calculate

$$\Sigma(x) = 43473 \qquad \Sigma(x^2) = 20128387$$
$$\bar{x} = 43473/144 = 302$$
$$s^2 = [20128387 - (43473)^2/144]/143$$
$$= 48979$$

Thus the mean and standard deviation of the individual observations are 302 and 221 respectively.

Table 2.3 Weights of 144 carrots (grams) in an arbitrary order

405	109	549	72	32	248	221	234	287	15
388	166	106	257	62	325	673	610	709	494
498	194	499	144	93	203	179	70	717	185
373	380	353	640	509	74	238	149	36	95
529	147	376	662	86	217	380	589	113	496
78	172	318	120	296	1017	197	202	116	129
66	201	292	59	341	606	97	284	915	512
290	87	39	429	643	191	305	258	252	385
216	66	288	670	390	100	579	106	245	426
599	874	597	656	80	175	106	543	340	134
325	55	274	18	71	211	331	254	87	77
749	21	446	88	283	212	31	286	343	91
556	93	132	54	657	223	325	121	436	124
771	505	251	173	258	484	28	813	663	318
393	157	325	87						

Now, we can use a statistical package to take random samples of four from the data in Table 2.3 and to calculate the mean for each sample of four carrots. The first sample of four chosen randomly consisted of (reading along the rows) the 11th value 388, the 57th value 197, the 90th value 426 and the 115th value 283 giving a mean of $388 + 197 + 426 + 283/4 = 323.5$. Repeating this sampling procedure 50 times produced samples whose means are given in Table 2.4.

The distribution of these means is shown in Fig. 2.10(b) and clearly it is much closer to a normal distribution than the distribution of the original data.

The mean and variance of this set of 50 samples means are: $\bar{x} = 308.8$ and $s^2 = 12611$. Thus the mean and standard error of the means of samples of four observations are 309 and 112 respectively.

If the process is repeated taking 50 samples of 16 carrots and calculating the mean for each sample, the means for each sample are given in Table 2.5 and

Table 2.4 Mean weight of 50 samples of four carrots

323.5	232.3	452.8	226.5	254.3	416.5
241.8	331.5	138.8	428.3	250.0	181.5
626.5	166.8	528.8	367.5	469.5	219.5
371.0	449.5	298.3	218.3	270.5	430.5
314.8	381.0	309.5	287.0	387.8	436.5
85.0	307.0	451.3	228.0	191.8	389.0
416.8	197.8	158.8	251.0	432.5	271.0
222.0	317.8	335.5	164.8	333.0	213.0
186.0	278.3				

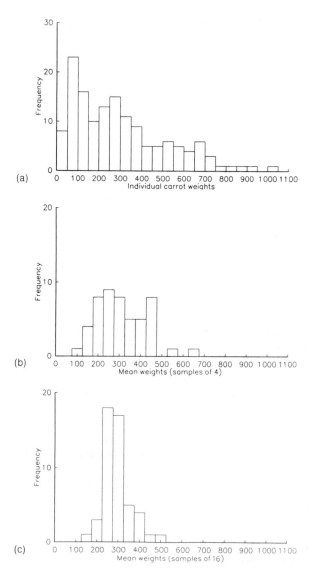

Fig. 2.10 Distributions of (a) 144 individual carrot weights, (b) 50 sample means for samples of size 4, (c) 50 sample means for samples of size 16.

the distribution is shown in Fig. 2.10(c). This distribution is even less variable than that for samples of four carrots.

The mean and standard error of the means of 16 observations are 292 and 61 respectively.

The means of our set of samples are of course not equal to the mean of the 144 individual observations but from our mathematical result that the variance

Table 2.5 Mean weight of 50 samples of 16 carrots

200.6	377.8	276.3	301.8	291.6	400.2
296.8	207.3	273.4	371.4	228.2	319.5
266.4	263.8	359.9	279.3	318.7	242.3
287.9	261.8	301.1	174.6	228.1	335.3
243.5	308.2	494.9	275.7	303.8	237.4
250.8	276.0	298.8	246.8	266.6	262.4
378.9	306.4	224.9	315.3	237.3	274.9
339.2	236.9	378.9	436.3	346.4	268.8
259.5	291.0				

of the distribution of means of samples of n observations is σ^2/n, we would expect the variances of the three distributions to be σ^2, $\sigma^2/4$ and $\sigma^2/16$, so that the standard deviations should be σ, $\sigma/2$ and $\sigma/4$ respectively. The estimated values of 221, 112 and 61 agree tolerably well with this expectation.

EXERCISE 2.1 *A frequency distribution*

The gains in weight (lb) of 40 pigs during a 20-day period and written in random order were

33	2	39	17
53	1	34	22
34	11	33	20
29	30	33	19
39	19	33	53
57	24	39	21
12	53	36	53
24	44	32	25
39	19	32	40
36	30	30	21

(a) Construct a histogram to demonstrate the distribution of these weight gains.
(b) Calculate the mean, variance and standard deviation of the distribution.
(c) Calculate the mean for each of the four columns of data. Calculate the mean, variance and standard deviation of these four means. Compare them with the values to be expected from the mean, variance and standard deviation of the individual values.

3

Estimation and hypothesis testing

3.1 ESTIMATION OF THE POPULATION MEAN

The main object of an experiment is usually to estimate certain quantities. These may be the yields of a particular variety of wheat grown under different specified conditions; or a single quantity such as the increase in yield of a variety when grown at a high fertiliser level as compared with the yield at a low level; or even the difference between the increase due to higher fertiliser application for one variety and the corresponding increase for a second variety. Usually there is a perfectly obvious estimate for any quantity in which the experimenter is interested. Thus if the experimenter wants to estimate the yields per hectare of a particular wheat variety and he has an experimental plot the area of which is 1/1000 hectare, with a surrounding area of wheat sufficient to ensure that the growth of the plants in the experimental plot is typical of normal field conditions, then the estimate of yield per hectare is the yield from the experimental plot multiplied by 1000. Similarly if the experimenter has a number of experimental plots of this variety and these give yields (in tonnes per hectare) of 2.0, 2.5, 2.8, 2.4, 2.3 then the obvious estimate of yield is the sample mean, $\bar{x} = (2.0 + 2.5 + 2.8 + 2.4 + 2.3)/5 = 2.4$.

An estimate of this kind is called a **point estimate**, and we can be sure of two things about the estimate. First, if we included more and more plots in the mean we would eventually get the population mean. Second, that with a small number of plots the estimate is bound to deviate from the true population value we are trying to estimate. Rather than a point estimate, it would be vastly preferable to have an **interval** estimate so that we could say, for example, that we were 95% confident that the population mean value lies somewhere between two values. Suppose we can assume that the plot yields of a wheat variety are normally distributed with standard deviation, $\sigma = 0.2$ tonnes per hectare. This means that, if we took a large number of observations of plot yield and drew a histogram to represent the yield, then as the number of observations increased the shape of the histogram would tend to the shape of the normal distribution, shown in Fig. 2.7. Further, whatever the mean value of the distribution, the standard deviation is 0.2 tonnes per hectare, so that approximately 68% of the observed values lie within 0.2 tonnes per hectare of the population mean value and only 5% are more than 0.4 tonnes per hectare away from the population mean value. Now, given a single plot yield, x, we know that the probability that this value is within two standard deviations of the population mean is 95%. Hence we can argue that the probability that the population mean is within two standard deviations ($= 0.4$ tonnes per hectare) of the observed value, x, is also 95%. Thus we have an interval estimate for the population mean, between $x - 0.4$ and $x + 0.4$, and we have 95% confidence that the population mean lies in this interval. In other words if we use this form of interval estimate a large number of times then the population mean value will fail to be included in the interval only 5% of those times.

This form of argument is a little complex and we shall now approach the idea of confidence intervals in another way, by considering how we might test a particular hypothesis about the population mean in the light of some experimental results.

3.2 TESTING HYPOTHESES ABOUT THE POPULATION MEAN

Consider again the situation where we have a single observation, x, of the yield of a new wheat variety and where we know that the distribution of yields is normal and that the standard deviation of the distribution of yields is σ tonnes per hectare. Suppose we also know the yield of a standard variety in these conditions to be μ. We wish to test the hypothesis that this new variety does not differ in average yield from the standard variety. In other words, assuming that the observation comes from a normal distribution with standard deviation, σ, we wish to test the reasonableness of the hypothesis that the mean of the distribution is μ. To obtain a measure of the reasonableness of the hypothesis we assume the hypothesis to be true and calculate the probability of obtaining by chance a value of x at least as different from μ as that actually obtained. The reasons for using this particular probability is that any result more extreme than x provides stronger evidence against the hypothesis than does x. Consequently the probability of a result at least as extreme as x is the probability that the evidence against the hypothesis would have been as great as, or greater than, that provided by the actual value observed, x. We can find this probability by calculating $z = (x - \mu)/\sigma$ and consulting Table A.1 to find the probability of at least as extreme a value of the standardized normal distribution as the calculated value, z. The principle of the method is shown in Fig. 3.1.

Let us suppose that the value of z is 1.7. The proportion of the standardized normal distribution that is greater than 1.7 or less than -1.7 may easily be read from Table A.1 as $2 \times 0.0446 = 0.089$. Thus about 9% of the distribution is at least as extreme as $z = 1.7$, or in terms of the original scale of measurement, the probability of obtaining a yield at least 1.7σ different from the hypothetical

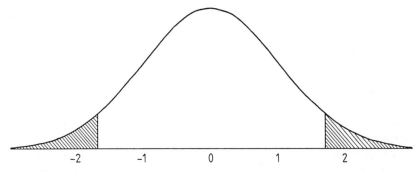

Fig. 3.1 The standardized normal distribution with the proportion of the distribution more extreme than 1.7 shaded.

mean value, μ, is 9%. We now either accept that an event has occurred which would only occur 9% of the time if our hypothesis is true, or we reject the hypothesis that the new variety has the same yield as the standard. It is customary to refer to the hypothesis being tested as the **null hypothesis** (generally the hypothesis of no effect) and to talk of tests of the null hypothesis in terms of **significance**. We say that if an event would occur less than, for example, 5% of the time under a certain hypothesis then that event provides evidence against the hypothesis at the 5% significance level, or more shortly that such a result is significant at the 5% level.

As an example suppose that the mean yield of the standard wheat variety is 2 tonnes per hectare, the standard deviation is known to be 0.2 tonnes per hectare, and a yield of 2.5 tonnes per hectare is observed for the new variety. Then $z = (x - \mu)/\sigma = (2.5 - 2.0)/0.2 = 2.5$. The probability of obtaining, by chance, a value of z greater than 2.5 or less than $- 2.5$ is $2 \times 0.00621 = 0.0124$ or about 1.25%. Thus we would say that the observed yield of 2.5 tonnes per hectare, is significantly different from the hypothetical yield of 2 tonnes per hectare at the 1.25% significance level.

If a test is performed using a statistical package the exact level of significance is usually presented, otherwise tables can be used. There are three commonly tabulated levels of significance, 5%, 1% and 0.1%. Results can then be classified as non-significant, significant at 5%, significant at 1%, or significant at 0.1%. Note that any result which is significant at the 1% level is in the most extreme 1% of the distribution and must therefore be significant to the 5% level. Thus significance at the 1% level provides stronger evidence against the hypothesis than significance to the 5% level. The choice of a threshold significant level is difficult. There are two mistakes than can be made. One is to demand too high a level of significance and thereby accept the null hypothesis on many occasions when it is in fact not true. On the other hand, too low a level of significance will mean that the null hypothesis is often rejected when in fact it is true. The balance between these two mistakes has to be struck by considering their likely consequences. If the null hypothesis is that irrigation has no effect on the yield of grassland then the consequence of too high a significance level may be that the opportunities for increased yields from irrigation are lost. On the other hand, too low a significance level may mean that expensive irrigation equipment is purchased without any compensating increase in yields. In experiments aimed at commercial decisions these ideas can be readily quantified. It is much more difficult to decide on an appropriate significance level when the purpose of the experiment is to add to our understanding of some biological process. The consequences of rejecting null hypotheses when they are true or of accepting null hypotheses when they are untrue are then much more difficult to quantify.

We shall return to these points later when we analyse some real experimental data and can discuss the appropriateness of the ideas of hypothesis testing when applied to the interpretation of real experimental results. All that should be remembered at this point is that the classification of a result as significant is not proof that the null hypothesis is false. The result is significant at 5% if the probability of obtaining this result or a more extreme one, when the null hypothesis is true, is less than 0.05. In this case the null hypothesis is rejected in favour of a more plausible hypothesis and this more plausible hypothesis tested in a further experiment. Using these procedures we shall occasionally reject a correct hypothesis. In fact, when the null hypothesis is true the use of a 5% significance level will incorrectly reject the null hypothesis in one test in twenty. Also the non-rejection of a null hypothesis is not proof of the truth of that hypothesis, merely a statement that the evidence against the hypothesis is insufficient to persuade us to reject the hypothesis. In fact it is possible to argue that no null hypothesis is exactly true. It is difficult to believe that any two varieties of potatoes are identical in yield or that irrigating a grass sward has no effect on yield. It is important to remember that the significance of an experimental result is determined not only by the truth or otherwise of the null hypothesis but also by the amount of information provided by the experiment.

An alternative approach which allows these two rather different determinants of significance to be separated is that of confidence intervals, which we have already discussed briefly in section 3.1. Confidence intervals and hypothesis testing are very closely related. If we consider all those hypothetical values of μ which would not be rejected as significantly different from the observed value, x, at the 5% level, these provide a range of values exactly the same as the confidence interval for μ. Consider again the single observation of a new wheat of yield of 2.5 tonnes per hectare, when the standard deviation of the distribution of yields is known to be 0.2 tonnes per hectare. A hypothetical population mean yield, μ, will be rejected if the difference between x and μ is greater than $1.96\sigma = 0.392$. Hence the acceptable range of values for μ is $2.5 - 0.392$ to $2.5 + 0.392$ or $(2.108$ to $2.892)$ which is the same as the confidence interval calculated earlier. The general argument is illustrated in Fig. 3.2. The three distributions shown, with means μ_1, x and μ_2 are identical in shape. When x is in the extreme 5% of the distribution with mean μ_1 then μ_1 itself is in the

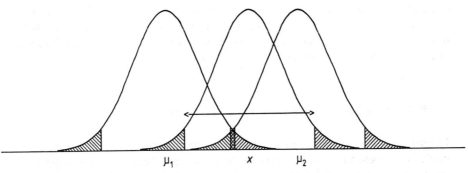

Fig. 3.2 Normal distributions with means μ_1, x, μ_2 and standard deviation σ and showing the extreme 5% of each distribution (shaded) and the 95% confidence interval centred at x.

extreme 5% of the distribution about x and therefore μ_1 is outside the 95% confidence interval centred at x. Similarly, for the distribution about μ_2, x is not in the extreme 5% of the distribution so that μ_2 is not in the extreme 5% of the distribution about x and therefore μ_2 is inside the 95% confidence interval centred at x.

So far we have considered the situation when we have only one observation of yield. The previous arguments can, however, be applied when we have a number of observations. In this situation the obvious estimate of the population mean is the sample mean x. We know that if the standard deviation of the original population distribution is σ then the standard error of the distribution of sample means for samples of size n is $\sigma/\sqrt{n}$. Therefore to test a hypothetical population mean of μ we calculate

$$z = \frac{\bar{x} - \mu}{\sigma/\sqrt{n}}$$

Suppose that in the wheat example, instead of having only a single observed yield of 2.5 tonnes per hectare, we had five further observations of 2.1, 2.4, 2.0, 2.6 and 2.2 tonnes per hectare. The sample mean is

$$\bar{x} = (2.5 + 2.1 + 2.4 + 2.0 + 2.6 + 2.2)/6 = 2.3$$

The standard error of the distribution of means of sample size is $\sigma/\sqrt{n} = 0.2/6 = 0.082$. We therefore calculate

$$z = \frac{2.3 - 2.0}{0.082} = 3.7$$

and on comparing this with the distribution in Table A.1 we find a probability of 0.0002 of such an extreme value occurring by chance, so that we have stronger evidence for the superiority of the new variety with more observations in our sample. It is, of course, not necessarily true that more observations will give a more significant result. Rather, more observations should result in a smaller standard error and hence a sample mean that will on average be nearer the true value. The sample mean should get nearer to the true value but this does not imply that it will be nearer the hypothesized mean unless the hypothesized mean is in fact the true value.

3.3 POPULATION VARIANCE UNKNOWN

In formulating significance tests and constructing confidence intervals we have made two assumptions in addition to the hypothesis being tested, namely that the yields we are considering are normally distributed and that we know the standard deviation of the distribution of yields. The latter assumption is usually unreasonable since the variation in yield depends on so many factors. Thus before we can carry out a significance test we need an estimate of σ and this must come from the variation of the sample of values obtained in the experiment. We have already pointed out (section 2.3) that the sample variance, s^2, is an estimate of σ^2 and we shall therefore estimate σ by s where

$$s^2 = \frac{\Sigma(x^2) - [\Sigma(x)]^2/n}{n-1}$$

The variance of the sample mean is σ^2/n, and so is estimated by s^2/n. Thus, in our test, we replace the standard error $\sigma/\sqrt{n}$ by the estimated standard error of the mean, $s/\sqrt{n}$. Hence the test statistic when the standard deviation of the distribution of yields is unknown, and has to be estimated, is

$$t = \frac{\bar{x} - \mu}{s/\sqrt{n}}$$

If we consider the distribution of values of t when the sample observations are taken from a normal distribution with mean μ (i.e. our hypothesis is correct) then it can be shown, mathematically, or empirically by taking a large number of samples from a normal distribution, that the distribution of t is similar to the normal distribution but that the distribution is flatter and less peaked. The distribution is know as 'Student's t' and its exact form depends on the precision with which s estimates σ. The name of this distribution is the result of the use of the pseudonym 'Student' by W. S. Gossett in the original publication on the distribution. Obviously if s is calculated from a sample of 50 observations it is likely to be more accurate than if it is calculated from a sample of 4 observations; and the better the estimate s is, the closer the distribution of t will be to the normal distribution. When we defined the sample variance we said that the variance from a sample of n was based on $(n-1)$ **degrees of freedom**. The degrees of freedom (abbreviated to d.f.) give a measure of the precision of s as an estimator of σ and so the different forms of the Student's t-distribution are characterized by the degrees of freedom on which s is based. Figure 3.3 shows

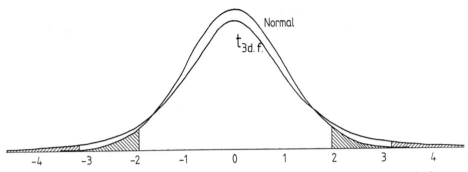

Fig. 3.3 The Student's *t*-distribution for three degrees of freedom and the equivalent normal distribution.

the form of the *t*-distribution for three degrees of freedom compared with the normal distribution.

The distribution of *t*, like the normal distribution, is symmetrical and the standard form of tabulation of the distribution is to record, for each value of the degrees of freedom, the value of *t* for which a given proportion, *P*, of the distribution is more extreme (including both tails of the distribution). This form of the tabulation of the *t*-distribution is given in Table A.2. To exemplify the method of tabulation, consider the distribution with seven degrees of freedom. For a proportion $P = 5\%$, Table A.2 shows that the value 2.36 cuts off 5% of the distribution, 2.5% at each end, so that 95% of the distribution of *t* lies between the values -2.36 and 2.36. Similarly 99% lies between -3.50 and $+3.50$.

To illustrate the use of the *t*-test consider the sample of six yields of a new variety of wheat discussed earlier: 2.5, 2.1, 2.4, 2.0, 2.6 and 2.2 tonnes per hectare. We wish to measure the extent to which these yields are compatible with a hypothetical mean yield of 2.0 tonnes per hectare.

First we calculate

$$s^2 = [(2.5^2 + 2.1^2 + 2.4^2 + 2.0^2 + 2.6^2 + 2.2^2)$$
$$- (2.5 + 2.1 + 2.4 + 2.0 + 2.6 + 2.2)^2/6]/5$$
$$= [32.02 - 190.44/6]/5$$
$$= 0.056$$

and

$$s = 0.237$$

The standard error of the sample mean is thus $0.237/\sqrt{6} = 0.097$ and the test statistic takes the value $t = (2.3 - 2.0)/0.097 = 3.10$. Table A.2 shows that 5% of the distribution of *t* with 5 d.f. is more extreme than -2.57 or $+2.57$. Hence the observed value of *t* is in the most extreme 5% of the distribution of *t* and the observed deviation of the sample mean from the hypothetical mean is significant at the 5% level. As with the normal distribution significance test we can construct 95% confidence intervals for the true mean yield of the new variety by considering those hypothetical values of μ which do not give a significant value of t at the 5% level, i.e. those values of μ such that

$$(2.3 - \mu)/0.097$$

lies in the range -2.57 to $+2.57$. This gives as a 95% confidence interval $2.3 \pm 0.097 \times 2.57$, i.e. an interval $(2.05, 2.55)$. In general terms this confidence interval for the true mean μ is given by:

sample mean $\pm t \times$ standard error of sample mean

where t is the 5% value for the t distribution with $(n-1)$ degrees of freedom.

EXERCISE 3.1 *Simple hypothesis testing and estimation*

(a) Farmer A tries a new form of artificial fertilizer for cereal crops. The yields for a large number of previous years for his particular variety of wheat with a standard fertilizer are normally distributed with mean 24 cwt per acre and standard deviation 3.5 cwt per acre. The yield in a single year with a new fertilizer is 29 cwt per acre. Assuming that the normality and the standard deviation are the same with the new as with the old fertilizer, how strong is the evidence that the new fertilizer will give yields higher on average than the standard fertilizer? Find the 95% confidence interval for the mean yield with the new fertilizer.

Farmer B has also tried the new fertilizer on his standard variety of corn, the yield of which he knows to be normally distributed with mean 96 bushels per acre and standard deviation 10 bushels per acre. He has obtained a yield of 112 bushels per acre. Ignoring questions of cost, which farmer has the stronger evidence for changing to the new fertilizer?

(b) Three farmers use different methods of fattening pigs to bacon weight. Using similar pigs they record the number of days from weaning to bacon weight for each pig as follows

Farmer C	Farmer D	Farmer E
105	107	100
112	108	107
99	104	100
97	112	113
104	101	103
107	103	115
	105	98
	108	110
		105

Calculate the mean and variance for each sample and use a t-test to compare each farmer's results with the average local figure of 110 days from weaning to bacon weight. Obtain 95% confidence intervals for the mean number of days for each farmer and comment on the results.

3.4 COMPARISON OF SAMPLES

So far we have been concerned with comparing the results of a single sample with a hypothetical mean value. Such a hypothetical value is rarely available since yields vary considerably from year to year and from site to site. Experiments are usually performed to compare yields obtained with different treatments, at the same site and in the same year, rather than to determine the actual level of yields for a single treatment. The standard situation in practice is that in which we wish to compare two or more different samples of plots, plants or animals, the samples having been subjected to different treatments.

We shall want to test the hypothesis that the difference between the true treatment means is a certain size; in particular that the difference is zero. We shall also want to obtain point and interval estimates for the true difference between treatments. Previously we have considered the value of a sample mean compared with the distribution of sample means under a given hypothesis. Now we shall be considering the value of the difference between two sample means by referring to the distribution of the difference between two sample means, again under some particular hypothesis.

The question therefore arises as to what form the distribution of a difference between two sample means takes. We shall assume that both populations from which the two samples are taken have normal distributions and write the mean and standard deviation μ_1 and σ_1 for the first population, and μ_2 and σ_2 for the second population. If the sample sizes are n_1 and n_2 then the means and variance of the distribution of the sample means are μ_1 and μ_2, and σ_1^2/n_1 and σ_2^2/n_2, respectively. Here we must again quote a mathematical result without proof, that the quantity $(\bar{x}_1 - \bar{x}_2)$ is normally distributed with mean $(\mu_1 - \mu_2)$ and variance $(\sigma_1^2/n_1 + \sigma_2^2/n_2)$. It is only reasonable to expect that the mean of the distribution of $(\bar{x}_1 - \bar{x}_2)$ should be $(\mu_1 - \mu_2)$. At the first sight, however, the addition of the variance is perhaps surprising. However, the variance of the difference between $\bar{x}_1$ and $\bar{x}_2$ could not possibly be the difference between the variances of $\bar{x}_1$ and $\bar{x}_2$ (if the variances of the means are equal the variance of their difference would be zero and this obviously cannot be). Also if we consider the variances of $(\bar{x}_1 - \bar{x}_2)$ and of $\bar{x}_1$ then clearly the former must be greater since in addition to the variability of $\bar{x}_1$ it includes also the variability of $\bar{x}_2$. Figure 3.4 illustrates this additivity of variances, showing the distributions of $\bar{x}_1$, $\bar{x}_2$ and $\bar{x}_1 - \bar{x}_2$.

In practice of course we rarely know σ_1^2 and σ_2^2 and have to estimate them by s_1^2, and s_2^2, the sample variances. Hence we would expect to calculate the

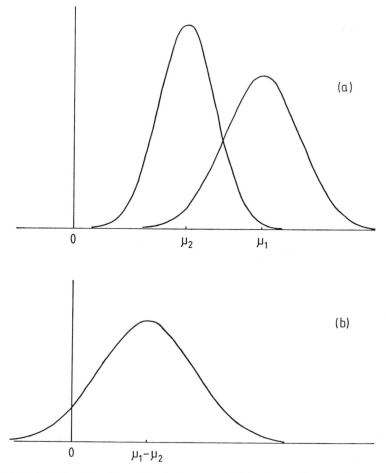

Fig. 3.4 Distribution (a) of two sample means and (b) of the difference between two sample means.

test statistic

$$t = \frac{(\bar{x}_1 - \bar{x}_2) - (\mu_1 - \mu_2)}{\sqrt{(s_1^2/n_1 + s_2^2/n_2)}}$$

and compare this with the appropriate t-distribution. The difficulty arises in deciding which is the appropriate t-distribution. In other words, what are the degrees of freedom of the estimated variance $(s_1^2/n_1 + s_2^2/n_2)$? We can say with confidence that the degrees of freedom which measure the precision of the estimate are greater for the combined variances than the lower of the degrees of freedom for the two separate variances and less than the sum of the two degrees of freedom. Being cautious we could use the t-distribution for the lower degrees of freedom of the two individual variances but it would be much better if we could find an alternative form of test.

3.5 A POOLED ESTIMATE OF VARIANCE

In order to obtain an alternative simpler form of test we need to make the assumption that the variances of the two populations are equal, i.e. that $\sigma_1^2 = \sigma_2^2 = \sigma^2$. This is often a reasonable assumption. When the two samples being compared are the results of different treatments being applied to initially similar material, this assumption is equivalent to assuming that each treatment has the effect of changing the yield of each experimental unit by a fixed amount. When $\sigma_1^2 = \sigma_2^2 = \sigma^2$, s_1^2 and s_2^2 are both estimates of σ^2 and hence we can calculate a **pooled** estimate of σ^2,

$$s^2 = \frac{(n_1 - 1)s_1^2 + (n_2 - 1)s_2^2}{(n_1 - 1) + (n_2 - 1)}$$

As stated earlier, we are assuming that s_1^2 and s_2^2 are estimating the same population value, σ^2. The reasonableness of this assumption should always be considered in terms of the biological situation. A purely statistical judgement can also be made by calculating the ratio of the two sample variances, s_1^2 and s_2^2, and deciding whether this ratio is too different from 1 to allow us to continue to assume that they are estimating the same quantity. The formal statistical test for such a comparison of variances is called the F-test. This test provides some guidance on the reasonableness of using the pooled estimate of variance.

Notice that s^2 is not a straightforward average of s_1^2 and s_2^2 (unless $n_1 = n_2$) but is a 'weighted' average taking account of the different sizes of samples and therefore the different amounts of information in each sample about the value of σ^2. If we consider the definitions of s_1^2 and s_2^2 we see that we can write the above formula as

$$s^2 = \frac{\Sigma(x_1 - \bar{x}_1)^2 + \Sigma(x_2 - \bar{x}_2)^2}{(n_1 - 1) + (n_2 - 1)}$$

where x_1 denotes members of sample 1 and x_2 members of sample 2, and Σ means sum over the appropriate sample. Thus our definition of s^2 treats all deviations from the mean equally. This estimate of s^2 has $(n_1 + n_2 - 2)$ degrees of freedom, the sum of the degrees of freedom from the two samples. We now reconsider the variance of the difference between the sample means, $\sigma_1^2/n_1 + \sigma_2^2/n_2$. Since $\sigma_1^2 = \sigma_2^2 = \sigma^2$, we can write this variance $\sigma^2/n_1 + \sigma^2/n_2$ or $\sigma^2(1/n_1 + 1/n_2)$, and can estimate it by $s^2(1/n_1 + 1/n_2)$. This gives an estimate of the standard error of the difference in means,

$$\sqrt{s^2\left(\frac{1}{n_1} + \frac{1}{n_2}\right)}$$

Now we calculate our test statistic

$$t = \frac{(\bar{x}_1 - \bar{x}_2) - (\mu_1 - \mu_2)}{\sqrt{[s^2(1/n_1 + 1/n_2)]}}$$

or, for the particular test of the hypothesis that the two samples come from populations with the same mean,

$$t = \frac{(\bar{x}_1 - \bar{x}_2)}{\text{standard error of difference}} = \frac{(\bar{x}_1 - \bar{x}_2)}{\sqrt{[s^2(1/n_1 + 1/n_2)]}}$$

and compare this with the significance points of the t-distribution with $(n_1 + n_2 - 2)$ degrees of freedom.

As usual we can construct confidence intervals for the true difference between the treatment effects, the 95% confidence interval being

$$\bar{x}_1 - \bar{x}_2 \pm t \times \text{standard error of difference}$$

i.e.

$$(x_1 - x_2) \pm t \sqrt{\left[s^2\left(\frac{1}{n_1} + \frac{1}{n_2}\right)\right]}$$

where t is the 5% value for the t-distribution with $(n_1 + n_2 - 2)$ d.f.

As an example of this method of comparing samples, consider again the comparison of wheat varieties. Suppose that in addition to his six estimates of yield with the new variety the farmer has ten similar areas with his standard variety so that his full set of yields are

New variety: 2.5, 2.1, 2.4, 2.0, 2.6, 2.3 tonnes per hectare

Standard variety: 2.2, 1.9, 1.8, 2.1, 2.1, 1.7, 2.3, 2.0, 1.7, 2.2 tonnes per hectare

We have already calculated the mean $\bar{x}_1 = 2.3$ and variance $s_1^2 = 0.056$ for the first sample. For the second sample

$$\bar{x}_2 = 2.0$$
$$s_2^2 = 0.047$$

In this instance the two sample estimates of σ^2, 0.056 and 0.047, seem reasonably close. Hence our pooled estimate of the common variance is

$$s^2 = \frac{5(0.056) + 9(0.047)}{6 + 10 - 2} = 0.050$$

and our estimate of the standard error of difference is

$$\sqrt{0.05\left(\frac{1}{6}+\frac{1}{10}\right)}=0.1155$$

the value of t for testing whether the new variety is different from the standard variety is

$$t = \frac{2.3-2.0}{0.1155}$$
$$= 2.60$$

Comparison of this value with the tabulated point of the t-distribution with $n_1 + n_2 - 2 = 14$ degrees of freedom shows that the observed value is greater than the 5% significance point (2.14) but less than the 1% significance point (2.98). Hence the observed difference is significant at the 5% level but not at the 1% level.

The 95% confidence interval for the true difference between the varieties is $(2.3 - 2.0) \pm 0.1155 \times 2.14$, i.e. $(0.053, 0.547)$.

All the results above, about the distributions of means and differences between means, are special cases of the following general result, which we shall need in our later discussions of factorial experiments and regression.

If the r independent quantities $x_1, x_2, \ldots, x_r$, are normally distributed with mean $\mu_1, \mu_2, \ldots, \mu_r$ and variances $\sigma_1^2, \sigma_2^2, \ldots, \sigma_r^2$ then the linear combination of them

$$X = \alpha_1 x_1 + \alpha_2 x_2 \ldots + \alpha_r x_r$$

where $\alpha_1, \alpha_2, \ldots, \alpha_r$ are constants, is normally distributed with mean

$$\alpha_1 \mu_1 + \alpha_2 \mu_2 + \ldots + \alpha_r \mu_r$$

and variance

$$\alpha_1^2 \sigma_1^2 + \alpha_2^2 \sigma_2^2 + \ldots + \alpha_r^2 \sigma_r^2$$

The distribution of the mean of a sample of r observations follows by taking the μs and σs all equal and $\alpha_1 = \alpha_2 = \ldots = \alpha_r = 1/r$. By taking $r = 2$, $a_1 = 1$, $a_2 = -1$ we obtain the result for the difference between the means of two samples (x_1 is $\bar{x}_1$, x_2 is $\bar{x}_2$, σ_1^2 is σ_1^2/n_1 and σ_2^2 is σ_2^2/n_2).

EXERCISE 3.2   *A comparison of two samples*

To test a fish-liver oil for its vitamin D potency, 20 day-old chicks were fed a mash to which was added 1% of the oil as the only source of vitamin D; 10 similar chicks were fed the same basic mash, but with the addition of 1% of a standard cod-liver oil of known vitamin D potency.

After 8 days the individual weights of the chicks in the two groups were as shown below

Chick weights at 8 days (g)

Type of oil		
Standard	Unknown	Unknown
69	68	65
66	69	63
71	66	66
69	64	68
68	67	66
71	69	72
74	65	69
69	71	67
73	68	66
70	70	61

(a) Calculate the mean, standard deviation and standard error of the mean for each group. (The uncorrected sum of squares of individual values is 49050 for the weights of chicks receiving the standard oil and 89918 for the unknown.)

(b) Find the difference between the two means and the standard error of this difference:

(i) When the variances of the two groups are not assumed to be the same.

(ii) When the variances of the two groups are assumed to be the same.

(c) Using the standard error based on the pooled variance (ii), carry out a t-test of the hypothesis that the difference between the mean weights of the two groups occurred by chance.

(d) Calculate a 95% confidence interval for the true difference between the two treatments.

4
A simple experiment

4.1 RANDOMIZATION AND REPLICATION

What considerations should guide us when we are planning an experiment? Taken at the simplest level there are two basic statistical requirements for a good experiments, namely **replication** and **randomization**.

The need for replicating, i.e. each treatment being applied to more than one experimental unit, should by now be clear. The material to be used in the experiment will always be variable. To have any idea of the 'value' of the results of an experiment and of the validity of the conclusions drawn from the results of an experiment we must obtain an estimate of this variability. This can only be done with a sample of results for each treatment. Also the more units used for each treatment the lower will be the standard errors of the estimates of treatment effects and therefore the more accurate the experiment.

Randomization must be used in the allocation of treatments to units. We know that the experimental units to be used differ, so that if we tried the same treatment on all units we should obtain a distribution of values. Before the experiments we do not know which units are 'better' and to make the comparison between treatments 'fair' we must allocate the treatments to units so that each treatment has the same chance of being allocated to the better units. In statistical terminology, we randomize the allocation of treatments to units in such a way that the probability of a particular treatment being allocated to a particular unit is the same for all treatments and units. Thus if we have three treatments and 15 units and each treatment is to be applied to five units, then the method of allocation must have the property that the probability of treatment A being applied to unit 7 is $1/3$, the probability of treatment B being applied to unit 10 is $1/3$, and so on.

The probabilities are not independent because if A is allocated to unit 1 then for the remaining 14 units we must include A four times only but B and C each five times. Therefore if A is allocated to unit 1 the allocation system must be so organized that A is less likely than B or C to be allocated to unit 2, the probabilities for allocation to unit 2 being $4/14$ for A and $5/14$ for B and for C.

How do we devise a practical scheme of random allocation? We can approach the allocation either from the viewpoint of allocating treatments to each unit in turn, or by allocating units to each treatment in turn. Suppose we try to allocate treatments to units and we consider the situation with only two treatments and 10 units, five to have treatment A, and five treatment B. To determine the allocation for the first unit we could toss a coin and allocate treatment A to unit 1 if the coin comes up 'heads' and B if 'tails'. However we cannot simply repeat the procedure for the second unit because the probabilities for A and for B are no longer equal. We therefore abandon this approach.

The other approach of allocating units to each treatment can be achieved simply but tediously by numbering the units and writing the unit numbers on slips of paper. The slips are mixed in a container and are then drawn randomly in order. For the example with two treatments and 10 units we would have 10 slips of paper numbered 1 to 10. If the slips are drawn in the order 2, 4, 7, 8, 1, 6, 10, 3, 9, 5 then treatment A would be allocated to the first five units drawn, which are units 1, 2, 4, 7 and 8 and treatment B to the other five. The decision to allocate the first five to A and the last five to B is arbitrary and any predetermined order of allocation could be used provided the unit numbers are drawn in random order.

The procedure using slips of paper is tedious and an alternative procedure is to use tables of random numbers. These are sequence of the digits 0, 1, 2, 3, 4, 5, 6, 7, 8 and 9 constructed in such a way that at any point in the sequence the probability of the next digit taking any particular value is 1/10, independent of the previous digits in the sequence. Table A.7 gives such a sequence, the numbers being paired so that they can be regarded either as a sequence of single digits or as a sequence of two-digit numbers in the range 00 to 99. To allocate two treatments to 10 units we label the units 0, 1, 2, ... 8, 9. We then allocate the units corresponding to the first five, different, digits in the sequence to treatment A. The sequence in Table A.7 begins

10   27   53   96   23   71   50   54   36   23   54   51   50   14   28   02

The first five digits are all different so we allocate A to the units numbered 0, 1, 2, 5 and 7, with B allocated to the units 3, 4, 6, 8 and 9.

Suppose we have three treatments A, B and C and wish to allocate five units to each treatment. Then we need 15 unit identifiers and must use two-digit numbers 00, 01, ... 13, 14. The remaining two-digit numbers can be ignored. This is a laborious procedure because such a large proportion of the numbers in the sequence are ignored. After finding the number 10 at the beginning of Table A.7 we ignore the next 12 numbers before finding 14, then 02 and 12 and another long sequence of ignored numbers before completing the set of units for A with 06. To speed up the procedure we use a modified sequence of random numbers. For the design with 15 units and three treatments we have only 15 unit identifiers 00, 01, ... 13, 14. Instead of ignoring the remaining numbers we relabel them so that 15 becomes 00, 16 becomes 01, ... 29 becomes 14, 30 becomes 00, ... 44 becomes 14, 45 becomes 00 and so on up to 89 becomes 14. The numbers 90 to 99 are ignored. We can now rewrite the sequence from Table A.7 as

10   12   08 − 08   11   05   09   06   08   09   06   05   14   13   02

The first five different identified units (10, 12, 08, 11, 05) are allocated to treatment A; the next five (09, 06, 14, 13, 02) to B and the last five (00, 01, 03, 04, 07) to C.

This random allocation does not, of course, prevent the treatment allocated to the better units giving an overestimate of the treatment effect. What the method of allocation does do is to ensure that, in the long term, comparisons between treatments will be fair. This could not be assumed with any subjective method of allocation. In addition, randomization produces a statistical analysis which is mathematically valid. A systematic arrangement of treatments might be thought more advantageous because units chosen for the different treatments could then made as similar as possible. However, certain groups of plots then inevitably have the same treatment and the yields from these plots could not be considered individually. The plots no longer exist as individual units, being combined into larger units, and consequently we would only have one observation for each treatment. In addition the experimenter must assume that the particular systematic arrangement chosen does not coincide with any pattern in the uncontrolled variation. He is most unlikely to be able to assume this. We shall discuss in later chapters ways of using our knowledge of the experimental material to improve accuracy without invalidating the subsequent analysis of the experimental results.

4.2 ANALYSIS OF A COMPLETELY RANDOMIZED DESIGN WITH TWO TREATMENTS

Now suppose we have a very simple experiment with two treatments each applied to the same number of plots say r units each, giving $2r$ units in the whole experiment. For $r = 5$ random allocation of treatments to experimental plots might produce the design shown in Fig. 4.1.

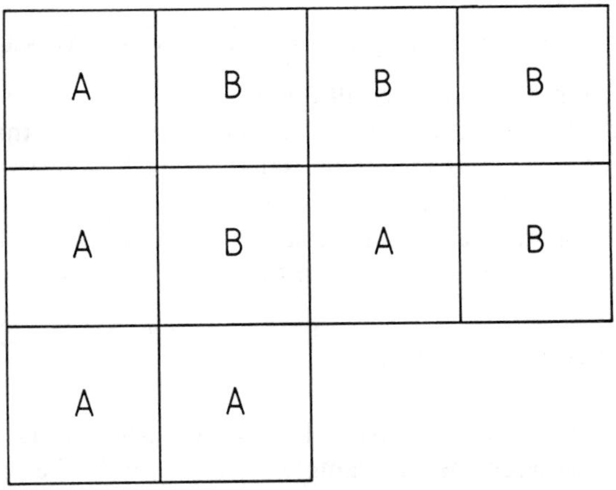

A	B	B	B
A	B	A	B
A	A		

Fig. 4.1 A completely randomized design for two treatments on five plots each.

We have already discussed the analysis of the yields from such a design since we can consider the yields for the two treatments as two samples and, as in the last chapter, we can obtain a pooled estimate of the common variance of the two populations from which the two samples are taken. Formally we calculate

$$s^2 = [\{\Sigma(x_1^2) - [\Sigma(x_1)]^2/r\} + \{\Sigma(x_2^2) - [\Sigma(x_2)]^2/r\}]/[(r-1) + (r-1)]$$

where $\Sigma(x_1)$ means the sum of all yields in the first sample, $\Sigma(x_2)$ the sum for the second sample, $\Sigma(x_1^2)$ the sum of squares of yields in the first sample and $\Sigma(x_2^2)$ the sum of squares for sample 2. We can rewrite this formula another way

$$s^2 = \left\{\Sigma(x^2) - \frac{[\Sigma(x_1)]^2 + [\Sigma(x_2)]^2}{r}\right\} \bigg/ [2(r-1)]$$

where now $\Sigma(x^2)$ means the sum of squares of all yields in both samples. Whichever way we write the formula we get the same value of s^2 and can use a t-test to compare the two sample means.

An alternative method of setting out the calculation for analysing the yields is to consider the total variation in the combined sample of yields and split it into different components in what is termed an **analysis of variance**. If we consider the yields as a single sample of $2r$ observations then the total variation $= \Sigma(x - \bar{x})^2$, the sum of squares of deviation from the overall mean, $\bar{x}$. We can use the same algebraic device as we used for the sample variance and write

$$\text{sum of squares} = \Sigma(x^2) - \frac{(\Sigma x)^2}{2r}$$

and we can also say that this sum of squares has $(2r - 1)$ degrees of freedom.

Now consider the variation between the two samples. This is simply represented as the sum of squared deviations of the treatment means from the overall mean and we can write

$$\text{treatment sum of squares} = \Sigma(\bar{x}_1 - \bar{x})^2 + \Sigma(\bar{x}_2 - \bar{x})^2$$

where the first summation is over all observations in the first sample, and the second is over all observations in the second sample (note that the quantity being summed, $(\bar{x}_1 - \bar{x})^2$, is in fact the same for each observation in the first sample). By using the same algebraic device of writing $(\bar{x}_1 - \bar{x})^2$ as $\bar{x}_1^2 - 2\bar{x}_1\bar{x} + \bar{x}^2$, summing each term separately and replacing $\bar{x}$ by $\Sigma x/n$ we can rewrite the treatment sum of squares in the form

$$\text{treatment sum of squares} = \frac{[\Sigma(x_1)]^2}{r} + \frac{[\Sigma(x_2)]^2}{r} - \frac{[\Sigma(x)]^2}{2r}$$

Now the treatment variation is part of the total variation and it measures all of the variation between the two samples. Consequently the remainder of the total variation measures the variation within the samples. In an analysis of variance this residual variation is termed **error variation**. In this case the error variation is:

error sum of squares (s.s.) = total sum of squares − treatment sum of squares

$$= \left\{ \Sigma(x^2) - \frac{[\Sigma(x)]^2}{2r} \right\}$$

$$- \left\{ \frac{[\Sigma(x_1)]^2}{r} + \frac{[\Sigma(x_2)]^2}{r} - \frac{[\Sigma(x)]^2}{2r} \right\}$$

$$= \Sigma(x^2) - \left\{ \frac{[\Sigma(x_1)]^2 + [\Sigma(x_2)]^2}{r} \right\}$$

In the same manner we can subdivide the total degrees of freedom, $2r - 1$, into the single degree of freedom for the comparison between the two samples and the remaining $(2r - 2)$ degrees of freedom for the error variation within the samples. Alternatively the $(2r - 2)$ degrees of freedom can be thought of as the sum of the two sets of $(r - 1)$ degrees of freedom within each sample. If we divide the error s.s. by the error degrees of freedom we shall obtain the pooled estimate of variance derived earlier in this section

$$s^2 = \left\{ \Sigma(x^2) - \frac{[\Sigma(x_1)]^2 + [\Sigma(x_2)]^2}{r} \right\} \Big/ [2(r - 1)]$$

To illustrate this form of calculation for numerical data we shall use a set of data corresponding to the design of Fig. 4.1.

Example 4.1

Suppose the design of Fig 4.1 was used for comparing yields for two varieties of wheat and that the yields (in cwt per acre) were as follows (variety given in brackets after the yield)

24(A)	28(B)	31(B)	26(B)
27(A)	23(B)	21(A)	29(B)
22(A)	20(A)		

Considering the samples for the two varieties separately we obtain the following results:

$$\Sigma(x_A) = 114 \quad \Sigma(x_A^2) = 2630 \quad s_A^2 = \frac{30.8}{4} = 7.7$$

$$\Sigma(x_B) = 137 \quad \Sigma(x_B^2) = 3791 \quad s_B^2 = \frac{37.1}{4} = 9.3$$

$$\text{Pooled estimate } s^2 = \frac{30.8 + 37.2}{8} = \frac{68.0}{8} = 8.5$$

Now using the analysis of variance technique,

$$\Sigma(x) = 251 \qquad \frac{[\Sigma(x)]^2}{10} = 6300.1$$

$$\Sigma(x^2) = 6421$$

$$\frac{[\Sigma(x_A)]^2}{5} = 2599.2 \qquad \frac{[\Sigma(x_B)]^2}{5} = 3753.8$$

$$\frac{[\Sigma(x_A)]^2}{5} + \frac{[\Sigma(x_B)]^2}{5} = 6353.0$$

Hence we can write out the analysis of variance:

total sum of squares	$6421 - 6300.1 = 120.9$ on 9 d.f.
treatment sum of squares	$6353 - 6300.1 = 52.9$ on 1 d.f.
error sum of squares = total − treatment	$= 68.0$ on 8 d.f.

and mean error variance $s^2 = 68.0/8 = 8.5$ based on 8 degrees of freedom. The standard error of the comparison between variety means is

$$\text{s.e. of difference} = \sqrt{\left(\frac{s^2}{5} + \frac{s^2}{5}\right)} = \sqrt{\left(\frac{2 \times 8.5}{5}\right)} = 1.84$$

The variety means are 22.8 for A and 27.4 for B and the value of the t-statistic for comparing means is

$$t = \frac{27.4 - 22.8}{1.84} = \frac{4.6}{1.84} = 2.5$$

which is greater than the 5% significance value of t for 8 d.f., 2.31. Hence we would conclude that there is a strong evidence of a difference in yield between varieties A and B, the estimate of this difference being 4.6 cwt per acre and a 95% confidence interval for the true value being $4.6 \pm 2.31 \times 1.84$ or $(0.35, 8.85)$.

4.3 A COMPLETELY RANDOMIZED DESIGN WITH SEVERAL TREATMENTS

The completely randomized design with two treatments is about the simplest experiment that can be analysed statistically and there are obviously a number of extensions that can be made. The most obvious which does not make any major alterations to the form of the experiment is to have more than two treatments. Suppose that we have a completely randomized design, i.e. one like that already described in which treatments are allocated completely at random to units, in which we are comparing t different treatments and in which each treatment is applied to r experimental units. The method of allocating treatments to this units has been discussed earlier in this chapter. Now consider the method of analysing the data, in particular how to obtain an estimate of the variance of the treated populations in order to make comparisons between treatments. We shall assume that the variance is the same for the different treatments, this is equivalent to assuming that the effect of each treatment is to increase or decrease the yield of the unit to which it is applied by a fixed amount.

For each treatment sample we can obtain an estimate of the common variance; for the units given treatment 1,

$$s_1^2 = \left\{ \Sigma(x_1^2) - \frac{[\Sigma(x_1)]^2}{r} \right\} \bigg/ (r-1)$$

and so on. Exactly as for the two-sample case we can pool these estimates to obtain a better estimate of σ^2.

$$s_1^2 = \left(\left\{ \Sigma(x_1^2) - \frac{[\Sigma(x_1)]^2}{r} \right\} + \left\{ \Sigma(x_2^2) - \frac{[\Sigma(x_2)]^2}{r} \right\} \right.$$

$$\left. + \cdots + \left\{ \Sigma(x_t^2) - \frac{[\Sigma(x_t)]^2}{r} \right\} \right) \bigg/ [(r-1) + (r-1) + \cdots + (r-1)]$$

and we can rewrite this

$$s^2 = \left\{ \Sigma(x^2) - \frac{[\Sigma(x_1)]^2 + [\Sigma(x_2)]^2 + \cdots + [\Sigma(x_t)]^2}{r} \right\} \bigg/ t(r-1)$$

Again, exactly as in the two-treatments design we can calculate an analysis of variance dividing the total variation into a component representing variation between treatments and the component representing variation within samples. The total variation, considering all the observations as a single sample, is

$$\text{total variation} = \Sigma(x^2) - \frac{[\Sigma(x)]^2}{tr}$$

and is based on $(tr-1)$ degrees of freedom. The treatment variation consists of comparisons between the different treatment means, $\bar{x}_1, \bar{x}_2, \ldots$ and $\bar{x}_t$, and to obtain the treatment variation we consider these treatment means as a sample. The sample variation would be $\Sigma(\bar{x}_i - \bar{x})^2$ where $\bar{x}_i$ represents the various sample means for $i = 1, 2, \ldots t$, $\bar{x}$ is the overall mean and the summation is over all samples. Each sample contains r observations, and to make the treatment variation comparable with the total variation we must sum the squared deviates of sample means from the overall mean over all experimental units. Hence the treatment variation is

$$\text{treatment variation} = \Sigma(\bar{x}_1 - \bar{x})^2 + \Sigma(\bar{x}_2 - \bar{x})^2 + \cdots + \Sigma(\bar{x}_t - \bar{x})^2$$

where each summation is over all units in the particular sample. The usual algebraic manipulation produces the more easily calculated form

$$\text{treatment variation} = \frac{[\Sigma(x_1)]^2 + [\Sigma(x_2)]^2 + \cdots + [\Sigma(x_t)]^2}{r} - \frac{[\Sigma(x)]^2}{tr}$$

This variation, which is essentially a comparison between t treatment means, has $(t-1)$ degrees of freedom.

The error variation and degrees of freedom are obtained by subtracting the treatment variation from the total variation, as in the two-treatment case, giving

$$\text{error variation} = \Sigma(x^2) - \frac{[\Sigma(x_1)]^2 + [\Sigma(x_2)]^2 + \cdots + [\Sigma(x_t)]^2}{r}$$

on $(tr-1) - (t-1) = t(r-1)$ degrees of freedom. The mean error variance = (error variation)/(error d.f.) gives the same estimate of σ^2 as would be obtained by pooling the t separate estimates of variation.

The standard method of presenting an analysis of variance is to give, for each source of variance (i.e. total, treatment or error), the sum of squares (s.s.) of deviations from the mean, for which the forms have been given above, the degrees of freedom and the mean square (m.s.), which is the sum of squares divided by the degrees of freedom. The mean square corresponding to the error variation, estimates the within-sample variance. The general form of an analysis of a completely randomized design for t treatments with r units per treatment is:

Source of variance	s.s.	d.f.	m.s.
Treatments	$\dfrac{[\Sigma(x_1)]^2 + [\Sigma(x_2)]^2 + \cdots + [\Sigma(x)]^2}{r} - \dfrac{[\Sigma(x_t)]^2}{tr}$	$t-1$	
Error	$\Sigma(x^2) - \dfrac{[\Sigma(x_1)]^2 + [\Sigma(x_2)]^2 + \cdots + [\Sigma(x_t)]^2}{r}$	$t(r-1)$	s^2
Total	$\Sigma(x^2) - \dfrac{[\Sigma(x)]^2}{tr}$	$tr-1$	

The quantity $[\Sigma(x)]^2/tr$, i.e. the overall total, squared, divided by the number of observations, used twice in this calculation, is referred to as the correction factor (c.f.) or sometimes as the adjustment for the mean. With the estimate s^2 of random variation between units receiving the same treatments we can make comparisons between the means for different treatments. The variance of a single treatment mean σ^2/r, is estimated by s^2/r, and the variance of the difference between two treatment means, $2\sigma^2/r$, is estimated by $2s^2/r$. The estimated standard error of a difference between any two treatment means is therefore $\sqrt{(2s^2/r)}$ and we can use this to test any particular difference or to obtain confidence intervals for the true difference between any two particular treatments. In making these tests and particularly in presenting the results of the analysis it is important to realize the limitations of the tests that can be made and of the corresponding confidence intervals. Because we have more than two treatments we have more possible comparisons between treatments than there are degrees of freedom for the treatment variation; in other words, not all our possible treatment comparisons are independent. For example, with five treatments we have 10 pairs of treatments and we could, quite validly, test each difference between the two treatments of a pair for significance. When we present the results, it is important to remember that statements about the importance of differences between treatments A and B, between B and C, and between A and C are necessarily related. Hence summaries in the form 'A was shown to be superior to B but not superior to C; there was no difference between B and C' are nonsensical, and even when a proper statement in terms of significance levels is used, care should be taken to avoid such absurdities. In many experiments the treatments have a structure and interpreting the treatment means in the context of this structure avoids some of these problems. We shall return to this point in Chapter 6.

4.4 TESTING OVERALL VARIATION BETWEEN THE TREATMENTS

In view of the potential difficulty of examining each pair of treatments, it seems desirable to test the overall variation between treatments to see whether the treatments, viewed as a group, are homogeneous. The obvious way to carry out this test is to compare the treatment mean square with the error mean square. If there are no differences between treatments then the average amount of variation between sample means should be simply related to the average amount of variation within samples. In fact, it can be shown that if the treatments are identical in effect then the treatment mean square, s^2 is an estimate of σ^2, the random variation between units. If the treatments are different then it can be shown that the expected value of the treatment mean square is

$$\sigma^2 + r \frac{\Sigma[(t_i - \bar{t})^2]}{(t - 1)}$$

where $t_1, t_2 \ldots$, are the true treatment effects and $\bar{t}$ is the average. As would be expected, the treatment mean square tends to increase as the true differences between treatments increase.

As usual when constructing a significance test we must state our null hypothesis. In this case it is that there are no differences between the effects of treatments. We must find the distribution of our test statistic,

$$\frac{\text{treatment mean square}}{\text{error mean square}}$$

under the assumption that these mean squares are both estimates of the random

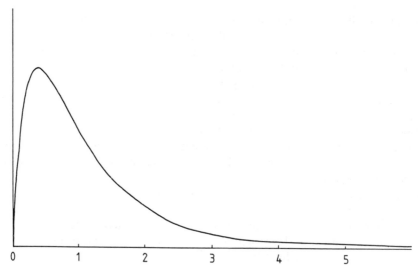

Fig. 4.2 The F-distribution on four and eight of freedom.

variation, σ^2. The distribution of such a ratio is the F-distribution and has been calculated mathematically; an example of such a distribution is shown in Fig. 4.2. The F-distribution, like the t-distribution, depends on the degrees of freedom of the variance estimates involved. For the F-distribution there are two variance estimates and consequently the F-distribution is tabulated according to the degrees of freedom of the upper mean square (treatments) and of the lower mean square (error). Because of the dependence of this F-distribution on two sets of degrees of freedom a tabulation of various percentage points similar to that for the t-distribution is impossible. Instead Tables A.3, A.4 and A.5 give the upper 5%, 1% and 0.1% points of the F-distribution for various degrees of freedom.

We need now to decide which part of the F-distribution we will use to reject the null hypothesis. As with the normal distribution and the t-tests we use the extreme values, but unlike these tests we use only the right-hand tail of the distribution, so that the 5% significance level of the F-distribution gives the point for which exactly 5% of the distribution takes greater values. The reason for using this different procedure stems from the different form of the alternative to the null hupotheses. We have seen that if there are treatment effects then the ratio of the mean squares will tend to increase. In other words if the null hypothesis is untrue then the distribution of values of the ratio will be more to the right than the F-distribution which holds under the null hypothesis. Consequently we are more likely to get large values of the ratio if the null hypothesis is untrue and we therefore use large values of the ratio to reject the null hypothesis. The logic of the test is illustrated in Fig. 4.3.

The F-test is an overall test of the significance of the differences that have been observed between the means of all the treatments in the experiment. If the

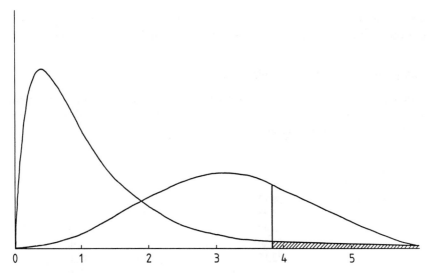

Fig. 4.3 The F-distribution on four and eight degrees of freedom with the 5% extreme values shaded and an example of the distribution of the ratio t.m.s./e.m.s. when treatment effects are present.

F-value is not significant then we must proceed with some caution in looking at individual differences between means. Tests of differences between pairs of means where the differences to be tested are chosen solely because they are large are clearly invalid. Often the structure of the treatments makes certain comparisons more meaningful and important than others and these comparisons can be made even if the F-test of all the differences is not significant: lack of difference among the other comparisons should not be allowed to conceal a difference observed in a comparison that is clearly important. Whether or not the F-test is significant, it is important to remember when making many t-tests that for each test there is a probability of rejecting the null hypothesis when it is true. Even if the comparisons are independent these probabilities will accumulate and the probability that at least one of the null hypotheses will be wrongly rejected will much exceed the significance level of each individual test.

With only two treatments there is no problem. The value of F will always equal the value of t^2; F and t are testing exactly the same difference and hence must be directly related. At the other end of the scale, the F-test may not be particularly relevant when the number of treatments is very large. A plant-breeder may include 100 varieties in a trial in the hope that one or two of them may be worth further testing or development. An F-test may say that the 100 varieties considered as a group are reasonably homogeneous, the superiority of the best one or two having been swamped by the lack of differences between all the other varieties. Here even the use of t-tests to compare the best varieties with the rest may not be relevant since the breeder is probably not interested in establishing the significance of the superiority of the best varieties but in maximizing the chances of finding a variety better than those already available, an entirely different concept which we cannot develop in a short elementary text.

Example 4.2 A completely randomized design

In an experiment to compare melon varieties, six plots of each of four varieties were grown, the plots being allocated to varieties in a completely random manner, and the results are given below:

Variety	A	B	C	D
Yields	25.12	40.25	18.30	28.05
	17.25	35.26	22.60	28.05
	26.42	31.98	25.90	33.20
	16.80	36.52	15.05	31.68
	22.15	43.32	11.42	30.32
	15.92	37.10	23.68	27.58
$\Sigma(x)$	122.94	224.42	116.95	179.38
$\Sigma(x)^2$	2629.23	8472.09	2434.12	5387.73
Mean $(\bar{x})$	20.49	37.40	19.49	29.90
Variance (s^2)	22.04	15.61	30.91	4.97

The pooled estimate of bariance is

$$s^2 = \frac{22.04 + 15.61 + 30.91 + 4.97}{4} = 18.38$$

Alternatively, we can obtain this estimate as follows:

Total of all observations = 643.69

Correction factor (c.f.) = $(643.69)^2/24 = 17\,264.03$

Total variation (of 24 plots) = $\Sigma(x^2) - $ c.f.

$= 2629.23 + 8472.09$
$+ 2434.12 + 5387.7$
$- 17\,264.03$
$= 1659.14$

Treatment variation

$$= \frac{122.94^2 + 224.42^2 + 116.95^2 + 179.38^2}{6} - \text{c.f.} = 1291.48$$

A simple experiment

Error variation = total variation − treatment variation = 367.66

Analysis of variance:

Sources of variation	s.s	d.f	m.s.
Between varieties	1291.48	3	430.49
Error	367.66	20	18.38
Total	1659.14	23	

The estimated standard error of a variety mean is $\sqrt{(s^2/6)} = 1.75$
The estimated standard error of a difference between two variety mean is
$\sqrt{(2s^2/6)} = 2.48$
We can therefore present the result as follows

Variety	Mean yield	
A	20.5	
B	37.4	s.e of difference between two
C	19.5	mean yields 2.48 (20 d.f)
D	29.9	

For an overall test of whether the varieties give different yields, we calculate the ratio

$$\frac{\text{treatment mean square}}{\text{error mean square}}$$

which, if the yield distribution has the same mean for all varieties, will be distributed according to the F-distribution with 3 and 20 d.f.:

$$F = \frac{430.49}{18.38} = 23.4$$

The 0.1% significance point for F with 3 and 20 d.f. is 8.10 and there is therefore strong evidence that there are real differences in yielding ability between the varieties.

Without further knowledge about the four varieties the making of particular comparisons between pairs of varieties is rather dangerous. However, it is clear that B is the best variety and that variety D is possibly really better than varieties A and C (the value of t for the difference between A and D is 9.4/2.48 = 3.79 as against the 0.1% significance point for t on 20 d.f. of 3.85).

One of the advantage of the simplicity of this design is that it is possible to have different numbers of units for different treatments. This will not, in general, be true for the various other designs we shall consider in the next two chapters. The desire to have more units for some treatments may stem either from practical reasons, e.g. the supply of seed for certain varieties is sufficient for only three field plots whereas other varieties have sufficient seed for six plots, or from theoretical arguments that a particular treatment, such as a control treatment, needs to be more accurately assessed than the other, untried, treatments so that the comparison between control and new treatments should be more precise than a comparison between two new treatments.

When the treatments have different numbers of units the analysis of variance is slightly changed. Suppose that treatment 1 has r_1 units, treatment 2, r_2 units, and so on, the total number of units being n. The total sum of squares is still

$$\Sigma(x^2) - \frac{[\Sigma(x)]^2}{n} \quad \text{on } (n-1) \text{ degrees of freedom}$$

The treatment sum of squares is now

$$\frac{[\Sigma(x_1)]^2}{r_1} + \frac{[\Sigma(x_2)]^2}{r_2} + \frac{[\Sigma(x_3)]^2}{r_3} + \dots - \frac{[\Sigma(x)]^2}{n} \quad \text{on } (t-1) \text{ degrees of freedom.}$$

The error sum of squares is obtained as usual by subtracting the treatment s.s from the total s.s. and is based on $(n-1)-(t-1)=(n-t)$ degrees of freedom. The estimate, s^2 of the variance, σ^2, between units with the same treatment is again taken from the error mean square and standard errors of differences between treatment means are calculated from s^2; e.g. for comparing the means for treatments 1 and 2 the standard error is

$$\sqrt{\left(\frac{s^2}{r_1} + \frac{s^2}{r_2}\right)}$$

EXERCISE 4.1 *A completely randomized design*

The percentage moisture content is determined for 10 samples for each of four different soils. Calculate the analysis of variance to obtain an estimate of the sampling variation within a soil and hence calculate the standard error of the mean moisture content of a soil and the standard error of a difference between two such means. Test the hypothesis that there is no variation of soil moisture content between different soils and summarize briefly the conclusions to be drawn from this data.

Solid A	Solid B	Solid C	Solid D
12.8	8.1	9.8	16.4
13.4	10.3	10.6	8.2
11.2	4.2	9.1	15.1
11.6	7.8	4.3	10.4
9.4	5.6	11.2	7.8
10.3	8.1	11.6	9.2
14.1	12.7	8.3	12.6
11.9	6.8	8.9	11.0
10.5	6.9	9.2	8.0
10.4	6.4	6.4	9.8

4.5 ANALYSIS USING A STATISTICAL PACKAGE

To analyse experimental data using a statistical package, the data must be presented to the package in a form which conveys the structure of the experiment. Most statistical computer packages perform analyses on columns of data. Throughout the text we will not be referring to specific ststistical packages but will be presenting the form of analysis that one should ideally expect from a good package. The reader is left to try our example analyses on his own system and to confirm that he obtains the same results.

To illustrate the form of analysis of the simplest experiment, we will look at the way the melon experiment presented in Example 4.2 would be analysed using a statistical package.

The information which we must give to the computer is the yield measurement for each plot and the variety sown on that plot. This is done by presenting the data in two columns which we shall call 'yield' and 'variety'. The rows of each column represent the yield and variety information for each plot. Data handling is usually facilitated by presenting the data in some systematic order. The data from this experiment could be expressed as:

'Yield'	'Variety'
25.12	1
17.25	1
26.42	1
16.08	1
22.15	1
15.92	1
40.25	2
35.25	2
.	.
.	.
.	.
30.32	4
27.58	4

The coding 1, 2, 3, 4 refers to varieties A, B, C, D. Some packages allow the varieties to be coded as characters A, B, C, D or even by variety names. We then need to identify that the variety column does not represent measurement data but is purely a coding column to identify which yield corresponds to which variety. This is achieved by setting 'variety' as a factor column with 4 levels.

Before computing an analysis of variance it is always advisable to look at the data and this can be done easily using the plotting facilities within a statistical package. For example a plot of yield against variety code is given in Fig. 4.4. This shows the variety differences and also shows that variety D has less variable

Yield

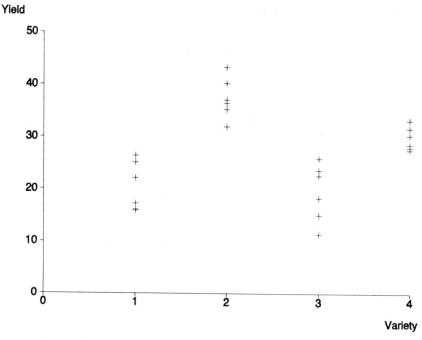

Fig. 4.4 Plot of yield against variety for the melon data, Example 4.2.

yields than the other varieties. This information is quantified in the variety means and standard deviations presented with the data.

We require an analysis of variance of yield by variety, although each package has its own commands to perform analysis of variance, the basic structure is the same. The variety column has to be declared as a treatment factor and the yield column as a response variable before the analysis of variance is calculated. Output from such analysis should be of the following form.

Analysis of variance

Variate: yield

Source	d.f.	S.S.	M.S.	*F*	Pr
Treatment	3	1291.48	430.49	23.42	<.001
Residual	20	367.65	18.38		
Total	23	1659.13			

Table of means

Variate: yield
Grand mean 26.82

Treatment	A	B	C	D	Standard Error
	20.49	37.40	19.49	29.90	Mean = 1.750

Standard errors of differences of means

Table	Treatment
rep.	6
s.e.d.	2.475

Individual packages will present the output in different ways but all should give an analysis of variance table, treatment means and at least one form of standard error.

5

Control of random variation by blocking

5.1 LOCAL CONTROL OF VARIATION

The lack of any restriction on the allocation of treatments to experimental units in the completely randomized design has obvious disadvantages. In a field experiment it would be possible, as a result of the randomization, for all the plots with a particular treatment to be grouped together in one corner of the experiment. With soil fertility trends possible this might lead to a most misleading result. Or, in an animal feeding trial, one diet might be allocated to all the heaviest animals, which again could give a very false result. Thus, while in the long run the completely randomized design gives a fair comparison of treatments a particular experiment might prove unfortunate. The logical way to overcome this possibility of an unfortunate design is to restrict the randomization in some way. This must be done in an objective manner: having obtained a design we cannot then decide whether to accept it or try again, or even worse, adjust it. The remedy lies in grouping the units in such a way that units in the same group are as similar as possible, and then allocate at random each treatment to one unit in each of the groups. Such a scheme requires that each group contains as many units as there are treatments.

Examples of such a grouping are:
(a) In a field experimental the natural groupings to choose would be to have plots in a small area constituting a group. Figure 5.1(a) shows a possible grouping for the two-treatment designs shown in Fig. 4.1, and Fig. 5.1(b) and (c) show two possible groupings in an experiment with 24 plots to compare six treatments.
(b) In animal trials the experimental units will usually be individual animals and these can be grouped either according to initial weight, or, particularly for young animals, according to litter.
(c) For botanical experiments with leaves, when only two treatments are being compared, it is common to use pairs of leaves from the same plant.
(d) In experiments with trees the past history of the trees can be used to group the trees into similar sets.

This form of grouping is referred to as **local control**, since the intention is (a) to control the allocation of treatments to units so that no treatment has a

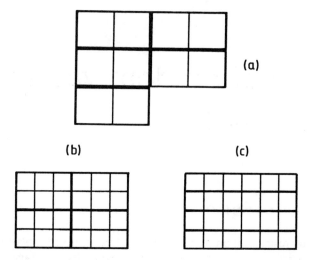

Fig. 5.1 Possible forms of grouping of plots for field trials; (a) ten plots for five treatments, (b) and (c) 24 plots for six treatments.

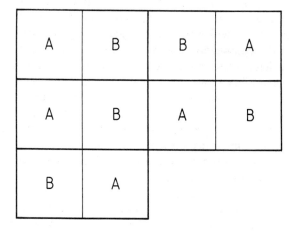

Fig. 5.2 Possible experimental design for comparing two treatments using ten plots in a randomized block design with five blocks.

monopoly of the extreme units, and (b) to reduce the amount of random variation by removing variation between blocks from the errors involved in comparing treatments. Within each group the treatments are allocated to units in a random manner using tables of random numbers as described for the completely randomized design. The groups of units are referred to as blocks and the resulting design is called a **randomized block design**. If a randomized block design were used for the example of Fig. 4.1, using the blocking, or grouping, of Fig. 5.1(a), the allocation of treatments to units might result in the design shown in Fig. 5.2. It is worth noting that any design using blocking could have occurred by chance if blocking were not used but the reverse is not true. In other words the number of designs possible for a randomized block experiment is considerably smaller than the number of designs possible for a completely randomized design using the same experimental units.

5.2 ANALYSIS OF A RANDOMIZED BLOCK DESIGN

The restriction on the allocation of treatments to units inevitably makes the analysis of yields from a randomized block experiment more complex but the objects of the analysis are the same, namely the estimation of the treatment effects and a measure of the random variation from which to obtain a standard error for comparing the treatment means. As previously, we assume that the effect of treatment is to increase (or decrease) the yield by a constant amount. This is equivalent to saying that we can express the yield of a particular experimental unit in the form

$$\begin{pmatrix} \text{observed} \\ \text{yield} \end{pmatrix} = \begin{pmatrix} \text{block} \\ \text{mean} \end{pmatrix} + \begin{pmatrix} \text{treatment} \\ \text{effect} \end{pmatrix} + \begin{pmatrix} \text{random unit} \\ \text{variation} \end{pmatrix}$$

where the random unit variation consists of variation between units within a block. Models of this form are called **additive** models.

Let us now consider an experiment with t treatments in b blocks of t units each, a total of $n = bt$ units. Working in the terms used in the above expression for yield, we can write the treatment total for treatment 1 as

$$\begin{pmatrix} \text{treatment 1} \\ \text{total} \end{pmatrix} = \Sigma \begin{pmatrix} \text{block} \\ \text{mean} \end{pmatrix} + b \begin{pmatrix} \text{treatment 1} \\ \text{effect} \end{pmatrix} + \Sigma_1 \begin{pmatrix} \text{random unit} \\ \text{variation} \end{pmatrix}$$

where Σ_1 implies the sum of random unit variations for units receiving treatment 1. This means that we can write the treatment mean as

$$\begin{pmatrix} \text{treatment 1} \\ \text{mean} \end{pmatrix} = \frac{1}{b} \Sigma \begin{pmatrix} \text{block} \\ \text{mean} \end{pmatrix} + \begin{pmatrix} \text{treatment 1} \\ \text{effect} \end{pmatrix} + \frac{1}{b} \Sigma_1 \begin{pmatrix} \text{random unit} \\ \text{variation} \end{pmatrix}$$

and the difference between the means for treatments 1 and 2 as

$$\begin{pmatrix} \text{treatment 1} \\ \text{mean} \end{pmatrix} - \begin{pmatrix} \text{treatment 2} \\ \text{mean} \end{pmatrix}$$

$$= \begin{pmatrix} \text{treatment 1} \\ \text{effect} \end{pmatrix} - \begin{pmatrix} \text{treatment 2} \\ \text{effect} \end{pmatrix}$$

$$+ \frac{1}{b} \left[\Sigma_1 \begin{pmatrix} \text{random unit} \\ \text{variation} \end{pmatrix} - \Sigma_2 \begin{pmatrix} \text{random unit} \\ \text{variation} \end{pmatrix} \right]$$

Hence, under the assumed model, the comparisons between treatments are unaffected by differences between blocks and the estimate of a difference between two treatments depends only on the true difference between the treatments and random variation between units within a block.

The model we have been discussing verbally is usually represented symbolically in the following manner:

$$Y_{ij} = b_i + t_j + e_{ij}$$

where

Y_{ij} = observed yield for treatment j in block i
b_1 = mean yield for block i
t_j = treatment effect on yield for treatment j, and
e_{ij} = represents the random unit variation within a block.

As is implied in the model for the observed yield, in the analysis of variance we divide the total variation into components for the variation between treatments, the variation between blocks, and the error variation remaining after these two sources of variation have been substracted from the total variation. As before,

$$\text{total variation} = \Sigma(x^2) - \frac{[\Sigma(x)]^2}{n}$$

based on $(n-1)$ d.f.

$$\text{treatment variation} = \frac{T_1^2 + T_2^2 + \ldots T_1^2}{b} - \frac{[\Sigma(x)]^2}{n}$$

based on $(t-1)$ d.f. since the treatment s.s. is the variation in a sample of t treatment means. T_1 is the sum of yields for all units receiving treatment 1, etc.

The variation between block means is calculated in just the same way as that between treatments

$$\text{block variation} = \frac{B_1^2 + B_2^2 + \ldots + B_b^2}{t} - \frac{[\Sigma(x)]^2}{n}$$

based on $(b-1)$ d.f., where B_1 is the sum of yields for all units in block 1. The degrees of freedom for the error s.s. are obtained by subtraction (like the error s.s. itself) and are $(n-1) - (b-1) - (t-1)$ which, since $n = bt$, is $(b-1)(t-1)$. Hence the formal analysis of variance for a randomized block experiment may be written

Source of variation	s.s.	d.f.	m.s.
Blocks	$\dfrac{B_1^2 + B_2^2 + \ldots + B_b^2}{t} - \dfrac{[\Sigma(x)]^2}{t}$	$b - 1$	
Treatments	$\dfrac{T_1^2 + T_2^2 + \ldots + T_t^2}{b} - \dfrac{[\Sigma(x)]^2}{n}$	$t - 1$	
Error	by subtraction	$(b-1)(t-1)$	s^2
Total	$\Sigma(x^2) - \dfrac{[\Sigma(x)]^2}{n}$	$n - 1$	

Control of the random variation

Standard errors of treatment means and of differences between treatment means are calculated from the error m.s., s^2, as follows. Each treatment mean is the mean of a sample of b observations, one from each block, and the standard error of this mean is $\sqrt{(s^2/b)}$, where s^2 is a measure of the variation between units within a block, because as explained before each treatment occurs once in each block and so differences between blocks do not contribute to the variability of the treatment mean. Similarly, for the difference between two treatment means, the standard error is $\sqrt{(2s^2/b)}$.

As in the completely randomized design, a test of the overall variation between treatments is provided by comparing the ratio of the treatment mean square to the error mean square with values from the appropriate F-distribution. In addition, there are three other quantities often calculated from an analysis of variance, which are used not for particular comparisons within the experiment, but as general measures of mean yield and precision for comparison with other, similar, experiments. These are

(a) The general mean of all observations, $\Sigma(x)/n$;
(b) The unit standard deviation, s.d. $= \sqrt{s^2}$; and
(c) The coefficient of variation, which is the unit standard deviation expressed as a percentage of the general mean.

The unit standard deviation is measured on the same scale as the yields themselves. The coefficient of variation, being the ratio of the standard deviation to the mean, is therefore a dimensionless quantity. The amount of variability in an experiment is often found to be related to the mean level of the yields in that experiment. For this reason, the coefficient of variation is sometimes found to be a more stable quantity than the unit standard deviation. The coefficient of variation will take the same value for a whole group of experiments if, in each experiment, the standard deviation is directly proportional to the mean yield in that experiment. The aim of an experiment is sometimes stated in terms of being able to detect, with a given probability, a stated percentage difference between treatment means. The coefficient of variation is then the relevant measure of variability to use in deciding how many replicates are needed. Unfortunately, the amount of variation is not always so simply related to the average level of yield and the coefficient of variation can often be a meaningless quantity. In particular, it should be used with considerable caution when measuring the amount of variation in quantities that can take both positive and negative values. The coefficient of variation is then liable to be very unstable because of its dependence on the exact balance in any particular experiment between the negative and positive values of the quantity being measured. The coefficient of variation only becomes a useful measure in situations where experience has shown that is does tend to take approximately the same value in different experiments. For example, in experiments with dairy cattle, the coefficient of variation of whole lactation milk yields has been found to be of the order of 25%. This means that we can calculate, with a certain amount of confidence, the number of cows that would be required to achieve a specified percentage degree of accuracy in the measurement of lactation milk yields. It is not possible to predict the standard deviation of the absolute yields with the same certainty. For other measurements, the standard deviation expressed in appropriate units may be just as stable as the coefficient of variation. In any particular case, the relevant research literature needs to be examined.

Example 5.1 Analysis of variance of a randomized block design

The experiment consisted of four randomized blocks of three treatments each, the treatments being

O   control (natural daylight only)
E   extended day (total day length 14 hours)
F   flash lighting (natural day + 2 × 20 second flashes per night)

The number of eggs laid by a pen of 6 pullets in the period 1 December 1950 to 22 February 1951 was recorded for each unit.

Blocks Treatments	1	2	3	4	Treatment totals
O	330	288	295	313	1225
E	372	340	343	341	1396
F	359	337	373	302	1371
Block totals	1061	965	1011	956	3993
Correction factor	$= \dfrac{(3993)^2}{12} = 1328671$				
Total s.s.	$= 330^2 + \ldots + 302^2 - \text{c.f.}$				$= 8864$

Blocks s.s. $= \dfrac{1061^2 + 965^2 + 1011^2 + 956^2}{3} - \text{c.f.}$ $= 2330$

Treatments s.s. $= \dfrac{1226^2 + 1396^2 + 1371^2}{4} - \text{c.f.}$ $= 4212$

Source of variation	s.s.	d.f.	m.s.	F
Blocks	2330	3		
Treatments	4212	2	2106.0	5.44*
Error	2322	6	$s^2 = 387.0$	
Total	8864	11		
c.f.	1328671			

s.d. $= \sqrt{s^2} = 19.7$	g.m. $= 332.8$	c.v. $= 5.9\%$
(standard deviation)	(general mean)	(coefficient of variation)

Treatment means:	Eggs per pen	Eggs per bird
O	306	51.1
E	349	58.2
F	343	57.1
s.e. for comparing any two treatment means		
(6 d.f.) $\sqrt{(2s^2/4)}$	$= 13.9$	$= 2.3$

Summary of results of analysis:

The F-ratio for treatments is significant at $5\%(*)$ and we therefore examine particular comparisons between treatments. Both the extended day and the flash lighting treatments resulted in a significantly ($P = 0.05$) higher number of eggs laid than the control treatment. The increase was 7.1 and 6.0 eggs per bird for the E and F treatments respectively. These increases have a standard error, based on six degrees of freedom, of 2.3. There was no significant difference between the E and F treatments in the number of eggs laid.

EXERCISE 5.1

In an experiment to compare the effects of four drugs, A, B, C and a placebo, or inactive substance, D on the lymphocyte counts in mice a randomized block design with four mice from each of five litters was used, the litters being regarded as blocks. The lymphocyte counts (thousands per mm^3 of blood) were:

Litters	1	2	3	4	5	Drug totals
Drugs						
A	7.1	6.1	6.9	5.6	6.4	
B	6.7	5.1	5.9	5.1	5.8	
C	7.1	5.8	6.2	5.0	6.2	
D	6.7	5.4	5.7	5.2	5.3	
Block	—	—	—	—	—	
Totals	—	—	—	—	—	

Complete the following analysis of variance table and summarize the results of the analysis. The uncorrected total s.s. $= \Sigma(x^2)$ is 720.51.

Var. due to	d.f.	s.s.	m.s.	F
Block				
Treatment				
Error				
Total				

c.f.

s.d. =   g.m. =   c. of v. =

Strain means:

A

B   s.e. for comparing any two

C

D   strain means (d.f.) =

5.3 MEANING OF THE ERROR MEAN SQUARE

It is rather more difficult in the randomized block design than in the completely randomized design to see just what the quantity, s^2, is estimating. To get a better idea, we shall consider two further ways of obtaining the error mean square. The first is in the analysis of a simple randomized block experiment in which there are only two treatments. Such an experiment is called a **paired comparison experiment**. In a test of a particular treatment supposed to induce growth, 20 plants were grouped into 10 pairs so that the two members of each pair were as alike as possible. One plant of each pair was chosen, randomly, and treated; the other was left as a control. The increases in height (in centimetres) of plants over a two-week period were as follows:

Pair (Block)	Treated plant	Control plant
1	7	4
2	10	6
3	9	10
4	8	8
5	7	5
6	6	3
7	8	10
8	9	8
9	12	8
10	13	10

To carry out the usual analysis of variance we calculate the block totals $(11, 16, 19, 16, 12, 9, 18, 17, 20,$ and $23)$, the treatment totals (89 for the treated plants, and 72 for the control), the overall total (161) and the sum of squares of the original observations (1415). Hence we obtain the following analysis of variance:

Source of variation	s.s.	d.f.	m.s.
Pairs	84.45	9	
Treatments	14.45	1	14
Error	20.05	9	$s^2 = 2.23$
Total	118.95	19	
c.f.	1296		

The standard error of the difference between the two treatment means is $\sqrt{(2s^2/10)} = 0.67$ cm on 9 d.f., and the difference between the treatment means is $8.9 - 7.2 = 1.7$ cm.

As alternative way of looking at these results is to consider, for each pair, the difference between the growth of the treated plant and the growth of the control plant. Since we are interested only in whether the treatment does increase growth, these differences are reasonable quantities to consider. In fact, we have a sample of ten differences. Note that by considering differences we have eliminated differences between blocks—in this case pairs of plants. If we consider this sample of differences, we can calculate the sample mean and sample variance and obtain, from the sample variance, the standard error of the sample mean. Since the methods of this approach and the analysis of variance approach are essentially the same, i.e. eliminate block differences and estimate variation within blocks, it would be surprising if the sample mean and its standard error were not the same as before.

The actual differences are $3, 4, -1, 0, 2, 3, -2, 1, 4$ and 3, giving a mean of 1.7 cm, a sample variance of 4.46 and the standard error of the mean $\sqrt{4.46/10} = 0.667$ cm as before. The sample variance is exactly double the previous error mean square because it is the variance of differences between two values (section 3.5). Thus for the particular case with two treatments in a randomized block experiment the random variation, s^2, can be regarded as a measure of the consistencies of the difference between the two treatments in the different blocks. When there are three or more treatments this interpretation of s^2 must be modified since with three treatments there are three possible pairs of treatments whose differences could be considered and these are not independent. As the number of treatments increases the problem becomes greater. However, it is still true that the error m.s., s^2 is a measure of the consistency of treatment effects from block to block.

Another way of understanding the meaning of the error s.s. is to return to the original model and to consider the effects of each component of the model in turn. The model can be written in the form

$$\begin{pmatrix} \text{observed} \\ \text{yield} \end{pmatrix} = \begin{pmatrix} \text{overall} \\ \text{mean} \end{pmatrix} + \begin{pmatrix} \text{block} \\ \text{effect} \end{pmatrix} + \begin{pmatrix} \text{treatment} \\ \text{effect} \end{pmatrix} + \begin{pmatrix} \text{residual} \\ \text{variation} \end{pmatrix}$$

where block effects and treatment effects are deviations of the particular block or treatment mean from the overall mean.

The method we are going to use is the 'sweeping' method in which we calculate the numerical values(s) for each component of the model and then subtract those values from the set of yields. We use the data of Example 5.1 and first calculate the overall mean.

	Block			
	1	2	3	4
Treatment				
O	330	288	295	313
E	372	340	343	341
F	359	337	373	302
	Overall mean = 332.75			

The overall mean is now subtracted from each observation. The sum of squares of the resulting values (residuals) is the total sum of squares ($= 8864.25$). Next the mean for each block is calculated from the residuals in that block.

	Block 1	2	3	4
Treatment				
O	−2.75	−44.75	−37.75	−19.75
E	+39.25	+7.25	+10.25	+8.25
F	+26.25	+4.25	+40.25	−30.75
Mean =	+20.92	−11.08	+4.25	−14.08

Note that the average of the means is zero. Now the block means are subtracted from the residual observations in each case using the block mean for the block in which the observation occurs. Next the mean for each treatment is calculated from the residuals for that treatment.

	Block 1	2	3	4	Mean
Treatment					
O	−23.67	−33.67	−42.0	−5.67	−26.25
E	+18.33	+18.33	+6.0	+22.33	+16.25
F	+5.33	+15.33	36.0	−16.67	+10.0

Note that the average of the means is again zero. The treatment means are subtracted from the residual observations producing the final residuals.

	Block 1	2	3	4
Treatment				
O	+2.58	−7.42	−15.75	+20.58
E	+2.08	+2.08	−10.25	+6.08
F	−4.67	+5.33	+26.00	−26.67

These residuals have had all differences between blocks and between treatments 'swept out'. The sum for each treatment row and for each block column is zero. The sum of squares of these residuals must therefore be the variation not explained by treatment and block differences. This sum of squares, which = 2321.70, is the residual, or error, sum of squares and we can see that, apart from rounding errors, it is the same as the error s.s. calculated previously.

The sums of squares for blocks and for treatments may also be calculated from the sweeping results. The sum of squares after removing the overall mean was 8864.25. The sum of squares after also removing the block effects may be calculated from the subsequent set of residuals and is 6534.00. The difference (the variation explained by the block differences is 2330.25, as previously calculated. The sum of squares after also removing the treatment effects is 2321.70. The difference (the variation explained by the treatment differences) is 4212.30.

The sweeping method is more laborious than the conventional method of calculating the analysis of variance and for manual calculations the conventional method would normally be used. However, the sweeping method does give further insight into the error s.s., is actually used in many computer packages, and does have other benefits to be seen later.

5.4 LATIN SQUARE DESIGNS

The randomized block design attempts to control one form of local variation. Often, though, there may be more than one type of local variation which it would obviously be desirable to control and we would like to use an experimental design which includes two forms of local control. Examples of such situations are (a) field experiments where either there are known to be two fertility trends running across each other, or where nothing at all is known about fertility trends and it is required to allow for a trend in either direction; (b) animal experiments where differences between litters and the variability of initial weights are both obvious sources of variation; or (c) glasshouse experiments where there may be variation across the house due to light differences and along the house due to temperature differences. A very clear, though non-biological, situation where a design with two criteria of blocking is required is in the testing of wear of different makes of car tyres. In any one trial the wear on the four different wheel position will vary and for any position the wear on different trials will vary. The problem of finding a design reduces to finding an allocation of the different makes, say A, B, C and D, so that each make appears once and once only for each trial and once and once only for each position. Consider a design such as that in Table 5.1. Conparisons between the amounts of wear for the different

Table 5.1 Design for comparing four makes of tyre using four positions on car, and four trials

	Trial 1	Trial 2	Trial 3	Trial 4
Front offside	A	B	C	D
Front near	C	A	D	B
Rear offside	D	C	B	A
Rear near	B	D	A	C

makes using such a design will clearly not be affected by differences between trials or between positions.

This form of design is called a **Latin square** design and it is possible to construct such a design for any number of treatments. Note that if we forget about one of the two blocking factors we are left with a design which might have been obtained as a randomized block design. As one might expect, the method of analysing a Latin square design is very similar to that for a randomized block design with one additional source of variation to be included. Instead of referring to blocks, in a Latin square design we refer to the two forms of control as 'rows' and 'columns' and hence we write our basic model in the form

$$\begin{matrix} \text{observed} \\ \text{yield} \end{matrix} = \begin{matrix} \text{row} \\ \text{effect} \end{matrix} + \begin{matrix} \text{column} \\ \text{effect} \end{matrix} + \begin{matrix} \text{treatment} \\ \text{effect} \end{matrix} + \begin{matrix} \text{random unit} \\ \text{variation} \end{matrix}$$

Again the model can be written in symbolic form:

$$Y_{ij} = r_i + c_i + t_{k(ij)} + \varepsilon_{ij}$$

Here the notation $t_{k(ij)}$ represents the way in which the treatment which occurs in row i and column j is defined by the design. As in the randomized block design it is easily verified that comparisons between treatments, as estimated by differences between yields, are independent of row or column differences. In the analysis of variance for a Latin square design we divide the total variation into row variation, column variation, treatment variation and error variation in exactly the same manner as the analysis of variance of the randomized block design. Thus, for a Latin square design with t treatments (and therefore t rows, t columns and t^2 total units) we have the following form of the analysis of variance

Source of variation	s.s.	d.f.	m.s.
Rows	$\dfrac{R_1^2 + R_2^2 + \ldots + R_t^2}{t} - \dfrac{[\Sigma(x)]^2}{t^2}$	$t - 1$	
Columns	$\dfrac{C_1^2 + C_2^2 + \ldots + C_t^2}{t} - \dfrac{[\Sigma(x)]^2}{t^2}$	$t - 1$	
Treatments	$\dfrac{T_1^2 + T_2^2 + \ldots + T_t^2}{t} - \dfrac{[\Sigma(x)]^2}{t^2}$	$t - 1$	
Error	by subtraction	$(t - 1)(t - 2)$	s^2
Total	$\Sigma(x^2) - \dfrac{[\Sigma(x)]^2}{t^2}$	$t^2 - 1$	

Standard errors of treatments are obtained from s^2, the error mean square, and since each treatment mean is based on a sample of t units the standard error of a treatment mean is $\sqrt{(s^2/t)}$, and the standard error of a difference between two treatment means is $\sqrt{(2s^2/t)}$. The overall variation between treatments is again tested by calculating

$$F = \frac{\text{treatment mean square}}{\text{error mean square}}$$

and comparing it with the F-distribution on $(t-1)$ and $(t-1)(t-2)$ degrees of freedom. Example 5.2 illustrates the analysis of a Latin square experiment in which the two blocking criteria were individual animals and stage of lactation.

Example 5.2 Analysis of variance of a Latin square design

An experiment to investigate the effect of winter feeding on milk production uses a Latin square design. Four diets, A, B, C and D, in order of increasing starch equivalent, were each fed for three weeks to each cow, and the total yield of milk in the third week of each period was recorded. The reason for using only the third week was that there might be some carry-over effects of the previous treatment during the first two weeks of a period. The yields (in pounds of milk) were

		Cow 1	Cow 2	Cow 3	Cow 4	Totals
	1	A192	B195	C292	D249	928
	2	B190	D203	A218	C210	821
Period	3	C214	A139	D245	B163	761
	4	D221	C152	B204	A134	711
Total		817	689	959	756	3221

Treatment totals: A, 683; B, 752; C, 868; D, 918

Var. due to	d.f.	s.s.	m.s.	F
Periods	3	6539	2180	
Cows	3	9929	3310	
Diets	3	8608	2869	21.2**
Error	6	811	135	
Total	15	25887		
c.f.	1	648428		

Mean milk yields (1lb): A, 171; B, 188; C, 217; D, 230

s.e. for comparing any two treatment means (6.d.f.) $= \sqrt{\left(\dfrac{2 \times 135}{4}\right)} = 8.2$

Summary of results of analysis:

Clearly there are statistically significant differences between the four treatment means. The yields increase with increasing starch equivalent in the diet. No other comment can be made without knowing more about the four diets.

We have not yet considered how we should choose a Latin square design. For each value of t, the size of the square, there are a large number of possible squares having the property that each letter appears in each row and each column once only. As with the randomized block design, in order to make the analysis of data from a design statistically valid, we must choose one design, in a random fashion, from a large set of possible Latin squares. Various methods of random selections of a Latin square are given in statistical textbooks all of which are complex, some unnecessarily so. A statistically acceptable method is the following:

1. Write down any Latin square of the required size.
2. Randomize the order of the rows.
3. Randomize the order of the columns.
4. Randomize the allocation of treatments to the letters of the square.

As an example of this procedure, let us consider the randomization of a Latin square of 4 treatments, T_1, T_2, T_3 and T_4. Suppose that we have the following sequence of random digits available:

$$6\ 4\ 1\ 7\ 1\ 8\ 4\ 3\ 9\ 7\ 3\ 7\ 6\ 6\ 5\ 5\ 8\ 6\ 0\ 8\ 7\ 4\ 5\ 0$$

We first write down our initial square, labelling the rows 1 to 4:

1(5)	A	B	C	D
2(6)	C	A	D	B
3(7)	D	C	B	A
4(8)	B	D	A	C

To randomize the order of the rows we use the random number sequence to select which of the four rows we place first in the randomization, then which second, third and finally fourth. To reduce the sequence of random numbers required we take the digits 1 and 5 to refer to row 1 in the original numbering, 2 and 6 for row 2, 3 and 7 for row 3 and 4 and 8 for row 4, ignoring 0 and 9. The first four numbers in the sequence then give the new order of the rows as (6) row 2, (4) row 4, (1) row 1, (7) row 3; hence we have the new square, in which we label the columns

row 2	C	A	D	B
row 4	B	D	A	C
row 1	A	B	C	D
row 3	D	C	B	A
	1(5)	2(6)	3(7)	4(8)

We now use the next numbers in the random number sequence to reorder the columns. The sequence after the last number used in reordering the rows reads 1, 8, 4, 3, 9, 7, 3, 7, 6, 6, 5, and so for our first column in the new square we take the old column 1, for the second we have the old column 4 (8), for the third the old column 3 (ignoring 4 as it has already been used) and finally the old column 2 (6, all other numbers, 9, 7, 3, 7 being ignored). Hence our new square is

Column 1	Column 4	Column 3	Column 2
C	B	D	A
B	C	A	D
A	D	C	B
D	A	B	C

Now we allocate the digits 1 and 5 to letter, A, 2 and 6 to B, 3 and 7 to C and 4 and 8 to D and, continuing with the random number sequence from where we left off, i.e. 6, 5, 5, 8, 6, 0, 8, 7, 4, we find which letter is to be replaced by T_1, then which by T_2, T_3 and T_4. This gives B(6) replaced by T_1, A(5) by T_2 D(8) by T_3 and C(7) by T_4 and hence our final square is

$$\begin{array}{cccc} T_4 & T_1 & T_3 & T_2 \\ T_1 & T_4 & T_2 & T_3 \\ T_2 & T_3 & T_4 & T_1 \\ T_3 & T_2 & T_1 & T_4 \end{array}$$

This procedure may seem unnecessarily complicated. It does, however, make the allocation of treatments to units entirely objective and, as has been stated previously, makes a valid statistical analysis possible. A complete scheme for randomization of Latin squares, together with an extensive though inevitably incomplete tabulation of the possible different Latin squares, is given in *Statistical Tables for Biological, Agricultural and Medical Research* by Fisher and Yates (1963).

5.5 ANALYSIS OF STRUCTURED EXPERIMENTAL DATA USING A COMPUTER PACKAGE

To analyse data from any experiment using a statistical package, columns must be set up containing codings for each experimental unit. To analyse a randomized block experiment columns must be set up containing codings for both treatment and block, which are then declared as experimental factors. For example the data from the egg experiment, Example 5.1, would be presented as three columns of data

Yield	Block	Treatment
330	1	1
288	2	1
295	3	1
313	4	1
372	1	2
340	2	2
343	3	2
341	4	2
359	1	3
337	2	3
373	3	3
302	4	3

The analysis of variance of yield is then calculated by specifying yield as the variable to be analysed and fitting the model terms:

Block + Treatment

The routine should also present treatment and block means with either standard errors of the means or standard errors of difference as presented in the analysis of example 5.1 earlier in this chapter.

To analyse an experiment arranged as a Latin square, coding columns must be set up to specify the codes for the treatment factor and the two blocking factors for each experimental unit in the design. For the Latin square design in Example 5.3 these columns would be:

Yield	Cow	Period	Diet
192	1	1	1
195	2	1	2
292	3	1	3
249	4	1	4
190	1	2	2
203	2	2	4
218	3	2	1
210	4	2	3
214	1	3	3
139	2	3	1
245	3	3	4
163	4	3	2
221	1	4	4
152	2	4	3
204	3	4	2
134	4	4	1

where the diet codes are given by $1 = A$, $2 = B$, $3 = C$ and $4 = D$. If the data is entered systematically, row by row or column by column, the block codes will be in regular patterns which can often be generated within the software thus saving typing the codes in but the treatment codes for a Latin square will not form this regular pattern and so must be entered directly into the computer. The model terms for this design are

$$Cow + Period + Diet$$

which will produce the analysis of variance table in the example.

5.6 MULTIPLE LATIN SQUARES DESIGNS

One disadvantages of the Latin square design is that by eliminating two sources of variation from the original units we inevitably reduce the degrees of freedom available for estimating the remaining random variation and this makes comparisons of treatments less precise. For instance, with three treatments there are only 2 d.f. for error, compared with 4 d.f. for the randomized block design with the same total number of units, and with four treatments there are 6 d.f. for error compared with 9 for the randomized block design. The 5% significance values of the t-distribution are 4.30 for 2 d.f. and 2.78 for 4 d.f.

To see the effect of this difference on the conclusions suppose we have an observed difference of two units between two means using a randomized block design with three treatments in three blocks of three plots each.

If the error m.s. is s_B^2 then the standard error of the difference between two means is $\sqrt{(2s_B^2/3)}$ and the difference of 2 will be detected as significant if

$$t = \frac{2}{\sqrt{(2s_B^2/3)}} \text{ is 2.78 or greater}$$

This means s_B^2 must be less than $(3/2(2/2.78)^2$ i.e. 0.78. Now suppose the same difference was obtained from a Latin square design with error m.s. $= s_L^2$. Now for the difference to be detected as significant we need

$$t = \frac{2}{\sqrt{(2s_L^2/3)}} \text{ to be at least 4.30}$$

Hence we need s_L^2 to be less than $(3/2)(2/4.)^2$ i.e. 0.32. Thus to detect a difference of a given size as significant at the same level for a 3×3 Latin square as for a randomized block of the same size the use of two blocking criteria in the Latin square must reduce the error m.s. to about 40% of the randomized block value.

One way of overcoming this difficulty, while retaining the advantage of two blocking factors, is to use more than one Latin square which will more than double the degrees of freedom available for error. There are two ways in which this can be done, illustrated in Fig. 5.3.

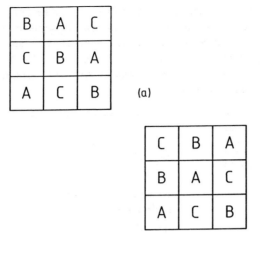

(a)

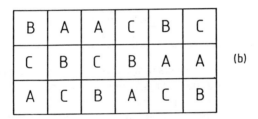

(b)

Fig. 5.3 Two forms of multiple Latin square design; (a) two independent Latin squares, (b) two Latin squares amalgamated to form a 3 × 6 rectangle.

In Fig. 5.3(a) the two (or more) squares are treated entirely independently, the randomizations being independent of each other, as is the case with randomization of treatments in different blocks of a randomized block design.

In Fig. 5.3(b) the two (or more) squares may be amalgamated to form a 'rectangle' in which each treatment appears once in each column and twice (or more) in each row. For this design two different squares are written down initially and the randomization of rows, columns and allocation of treatments proceeds as before, except that now we have six columns not two distinct sets of three columns each.

The essential difference between the two designs is that in (a) where we have two independent squares the row effect of a particular row in square 1 is independent of the row effect of the corresponding row in square 2. The same is true of column effects. In the 'rectangle' the row effects are assumed to be the same for both the original squares and thus for all columns. Column effects for the two original squares are still assumed to be unrelated.

This difference in the assumption inevitably leads to different forms of analysis. For (a) where we have two independent squares, the sums of squares for rows and columns are calculated for each square separately. A sum of squares for the overall difference between squares is calculated, and the treatment sum of squares is calculated using the treatment totals over both squares. The analysis of variance then takes the form:

Source of variation	s.s.	d.f.
Between squares	$\dfrac{G_1^2 + G_2^2}{9} - \dfrac{[\Sigma(x)]^2}{18}$	1
Rows	(row s.s.)$_1$ + (row s.s.)$_2$	4
Columns	(columns s.s.)$_1$ + (column s.s.)$_2$	4
Treatments	$\dfrac{T_1^2 + T_2^2 + T_3^2}{6} - \dfrac{[\Sigma(x)]^2}{18}$	2
Error	by subtraction	6
Total	$\Sigma(x^2) - \dfrac{[\Sigma(x)]^2}{18}$	17

(G_1, G_2 are the sums of the yields in squares one and two respectively; T_1, T_2, T_3 are treatment totals of 6 units each, 3 in each square).

For (b), the Latin rectangle, sums of squares are calculated for variation between row totals, column totals and treatment totals giving the folowing analysis of variance:

Source of variation	s.s.	d.f.
Rows	$\dfrac{R_1^2 + R_2^2 + R_3^2}{6} - \dfrac{[\Sigma(x)]^2}{18}$	2
Columns	$\dfrac{C_1^2 + C_2^2 + C_3^2 + C_4^2 + C_5^2 + C_6^2}{3} - \dfrac{[\Sigma(x)]^2}{18}$	2
Treatments	$\dfrac{T_1^2 + T_2^2 + T_3^2}{6} - \dfrac{[\Sigma(x)]^2}{18}$	2
Error	by subtraction	8
Total	$\Sigma(x^2) - \dfrac{[\Sigma(x)]^2}{18}$	17

The most important difference between the two analyses is the higher error degrees of freedom for design (b). This superiority of (b) is true for all sizes of

Latin square and also if more than two squares are used. It reflects the rather more severe assumption used in (b), that the differences between rows are unaffected by columns, whereas in (a) the row effects are allowed to be different in the two squares.

The following exercise consists of the analysis of two Latin squares in the form (b).

EXERCISE 5.2   *Analysis of variance of Latin squares*

An experiment to compare the effects of four light treatments, A, B, C and D, on the synthesis of mosaic virus in tobacco leaves was arranged in two Latin squares. Leaves from four positions for eight tobacco plants formed a 4×8 rectangle of units and two Latin squares were randomized together to produce a randomized design with each treatment appearing once for each plant and twice at each position. Sap from the 32 leaves was assayed on leaves of test plants and the square root of the number of lesions appearing taken as a measure of the treatment effect.

Plant	1	2	3	4	5	6	7	8
Leaf position								
1	45.4A	32.2D	34.6B	42.4C	38.1C	30.8A	58.4B	32.2D
2	33.4B	47.6B	44.0D	38.6D	27.2A	44.9C	24.8A	36.4C
3	45.6C	32.0A	42.4C	37.8A	40.8B	50.8D	46.8D	28.2B
4	42.7D	34.0C	39.0A	41.6B	35.8D	39.3B	45.8C	30.4A

Analyse the results of this experiment [the uncorrected s.s., $\Sigma(x^2)$ is 50027.40]. Write a summary of your conclusions.

5.7 THE BENEFIT OF BLOCKING AND THE USE OF NATURAL BLOCKS

In planning an experiment we cannot, of course, know exactly the likely variation between the results for different units if they had all received the same treatment but we can frequently recognize major potential differences between some groups of units. This often allows us to group the units in such a way that the units in a group are very similar. Consequently variation between units in a group should be small relative to variation between units in different groups.

The examples of groupings discussed earlier in this chapter can be divided into natural groupings where the number of units in a group is determined by the nature of that group and convenient groupings where the number of units is chosen first and the units are then sorted into groups of the chosen size. Examples of natural groupings are litters of animals, groups of leaves from the same plant or numbers of samples that can be processed in one batch. Groupings of field plots according to location and homogeneity of soil, or of animals according to birth weights are examples of groupings of convenience.

However the groups are defined, recognition of their existence enables us to improve the comparison between treatments to some extent. The benefit from blocking comes partly from ensuring that each treatment is assessed for a similar mixture of 'good' and 'bad' units. If blocking is ignored some treatments may be allocated, by chance, to predominantly 'good' or predominantly 'bad' units providing biased information about the differences between treatments.

Even if, in a completely randomized design, each treatment is allocated by chance, to a set of units representing the full range of 'good' and 'bad' units there is still a loss of precision compared with the blocked experiment. This is because the variation of yields from the units for a treatment is large simply because each treatment experiences the maximum variation between units. Treatments are effectively blocked but because this happened not by design but by chance the variation between units for a treatment is included in the Error variation in the Analysis of Variance (no block SS being included in the analysis).

Failing to group the plants into blocks will tend to produce less accurate estimates of treatment differences, or larger standard errors, or both!

So far in our discussion of blocking we have assumed that the number of units in each block is equal to the number of treatments. When this is easy to achieve then the resulting randomized complete block design is a particularly simple and useful design, and is quite properly used in, probably, a majority of experiments. However, it is important not to allow the simplicity of the randomized block design to be used as an argument to force units into inappropriate blocks of the same size as the number of treatments. The principle of dividing units into natural and sensible blocks so that units in the same block should be similar is more important than achieving a randomized block design.

Suppose that we are designing a taste testing trial in which several tasters are to taste and compare several different yogurts. Suppose also that it is well established that tasters cannot compare more than three yogurts in a single session. The number of yogurts for comparison will often be more than three. The best design that we can construct will be each taster (block) to taste three different yogurts. The combined information from the different tasters will enable us to assess, comparatively, the complete set of yogurts even though there may be consistent differences between tasters. Consider the situation where four yogurts (A, B, C, D) are to be tasted and four tasters (1, 2, 3, 4) are available. The design problem is to allocate three yogurts to each taster so that all yogurts are compared (by a taster) with all yogurts as equally as possible. The design shown is inevitable.

		Taster		
	1	2	3	4
Tastes	A	A	A	B
	B	B	C	C
	C	D	D	D

How do we sort out differences between yogurts from differences between tasters? Consider the data from the experiment (scores between 1 and 10):

		Taster		
	1	2	3	4
Yogurt				
A	8	5	9	
B	6	2		2
C	4		4	3
D		4	7	4

Differences between a particular pair of treatments where they occur together in a block are similar but not identical. This is exactly as we should expect in a randomized block design. It is obviously possible to estimate treatment differences roughly. Thus A is 2 or 3 better than B, 4 or 5 better than C and 1 or 2 better than D; B is about 2 better than D and very little different from C. A rough overall summary might be that A is 1.5 better than D, which is 1.5 better than B, which is 1 better than C.

We can use the sweeping technique from section 5.3 again here to estimate taster and yogurt differences more precisely, though because each taster tastes only 3 of the 4 yogurt, the calculations are a little longer

		Taster		
	1	2	3	4
Yogurt				
A	8	5	9	
B	6	2		2
C	4		4	3
D		4	7	4
Taster Means =	6.0	3.7	6.7	3.0

We first calculate taster means, and subtract each taster mean from the yogurt scores for that taster to give the first set of residuals.

Residuals (1) Yogurt	Taster 1	2	3	4	Yogurt means
A	+2.0	+1.3	+2.3		+1.9
B	0.0	−1.7		−1.0	−0.9
C	−2.0		−2.7	0.0	−1.6
D		+0.3	+0.3	+1.0	+0.5

We then calculate yogurt means and subtract each yogurt mean from the residual values for that yogurt

Residuals (2) Yogurt	Taster 1	2	3	4
A	+0.1	−0.6	+0.4	
B	+0.9	−0.8		−0.1
C	−0.4		−1.1	−1.6
D		−0.2	−0.2	+0.5
Taster Means (2)	+0.2	−0.5	−0.3	+0.7

Unlike the previous example in section 5.3 the residuals after subtracting the yogurt means do not sum to zero for each taster because our data does not have a complete set of taster × yogurt combinations. However, if we repeat the sweeping procedure, calculating and subtracting alternately taster and yogurt means the residuals gradually diminish and we can build up estimates of the taster means and yogurt differences allowing for the incompleteness of the data.

Residuals (3)				Yogurt Means (2)	Residuals (4)			
−0.1	−0.1	+0.7		+0.2	−0.3	−0.3	+0.5	
+0.7	−0.3		−0.8	−0.1	+0.8	−0.2		−0.7
−0.6		−0.8	+0.9	−0.2	−0.4		−0.6	+1.1
	+0.3	+0.1	−0.2	+0.1		+0.2	0.0	−0.3
				Taster means (3)	0.0	−0.1	0.0	0.0

The means of yogurt or taster residuals are almost reduced to zero and we perform one more sweep, using the most recent taster means.

Residuals (5)			Yogurt Means
−0.3	−0.2	+0.5	0.0
+0.8	−0.1		0.0
−0.4		−0.7	0.0
	−0.6	+1.1	0.0
+0.3	0.0	−0.3	0.0

We can build up the estimates of taster means and yogurt effects from the mean values at each successive step

	Taster			
	1	2	3	4
Initial means	6.0	3.7	6.7	3.0
First adjustment	+0.2	−0.5	−0.3	+0.7
Second adjustment	0.0	−0.1	0.0	0.0
Combined estimates	6.2	3.1	6.4	3.7

Yogurt	Initial means	Adjustment 1	Adjustment 2	Estimate
A	+1.9	+0.2	0.0	+2.1
B	−0.9	−0.1	0.0	−1.0
C	−1.6	−0.2	0.0	−1.8
D	+0.5	+0.1	0.0	+0.6

The estimates of yogurt differences are therefore:

A is 1.5 better than D, which is 1.6 better than B, which is 0.8 better than C.

We shall consider a general approach to the analysis of data from experiments where blocks do not include all treatments in Chapter 11. Having shown (we hope) that it is possible to extract estimates of treatment differences from such designs we now look at a few examples to demonstrate that the construction of incomplete block designs is quite straightforward. The reader may find it helpful to try to achieve a design for each example before reading the explanation of the construction.

Example 5.3

An experiment to compare six treatments A, B, C, D, E and F, has 30 experimental units. The natural blocking pattern of these units is in six blocks of five units each. How should treatments be arranged in blocks so that comparisons between treatments within blocks are as efficient as possible and are similarly precise for all treatment comparisons? In diagrammatic terms, in the six blocks of five units each, shown in Fig. 5.4(a) the treatments A, B, C, D, E and F must each appear five times, different treatments must appear as often as possible in the same block, and the joint occurrences in a block of a pair of treatments should be similar for all treatment pairs.

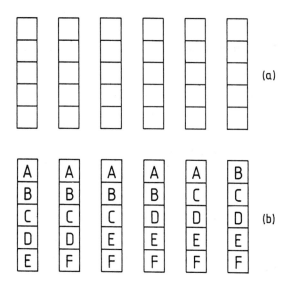

Fig. 5.4 (a) Natural blocking structure of 30 units in six blocks of five units each; (b) treatment allocation design optimizing the precision of all treatment comparisons equally.

In this example the solution is obvious because each block must have one treatment missing and because there are six blocks and six treatments, each treatment is omitted from precisely one block. The symmetry of the arrangement shown in Fig. 5.4(b) ensures that all treatments and all treatment pairs are equally represented: in fact each pair of treatments occurs together in precisely four blocks.

Example 5.4

Again there are six treatments and 30 experimental units but this time with a natural blocking pattern of ten blocks of three units each. How should

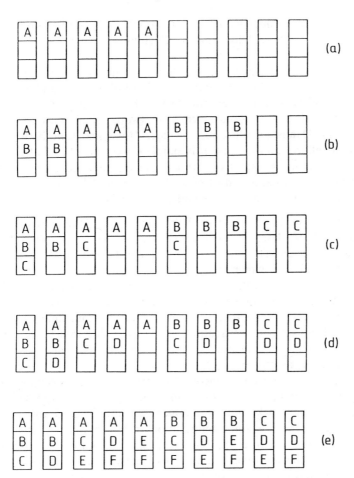

Fig. 5.5 Five stages in the construction of a design for six treatments in ten blocks of three units each.

the treatments be allocated to blocks, so that each treatment occurs five times and treatment pairs occur similarly often, if possible equally often?

The successive stages of the argument are shown in Fig. 5.5.

1. Treatment A must occur in five blocks, so we put it arbitrarily in the first five blocks (Fig. 5.5(a)).
2. In these five blocks with A there is a total of ten further units, enough to put each treatment of the remaining five twice. Pur B in the first two blocks, and then B must also occur in three blocks not including A (Fig. 5.5(b)).
3. The five occurrences of treatment C must occur twice with A, twice with B. We have a choice here and it seems sensible to put C with both A and B in one, but only one, block. The logic of the argument is that the treatment

must occur in the block with A and B, but with only 10 blocks no two blocks should be identical. This leads to C going into blocks 1, 3, 6, 9 and 10 (Fig. 5.5(c)).

4. In allocating D in we again have a choice. D must occur twice with A, twice with B and twice with C. If we put D in both of blocks 2 and 3 then blocks 4 and 5 will have to be identical (AEF). Therefore D goes in blocks 2 and 4. We have three more occurrences of D, two of which must be with C and one with B. D must therefore occur in block 7 (or 8) and 9 and 10 (Fig. 5.5(d)).

5. Now the allocation of E and F is simple if we remember that each should occur twice with A, then twice with B, but that the occurrence with C and D in blocks 3, 4, 6 and 7 must be such that each of E and F can still occur once with C and D in blocks 9 and 10 (Fig. 5.5(e)).

The designs in Fig. 5.4(b) and 5.5(e) are both in the class of balanced incomplete block designs discussed in Mead (1988) and other books on experimental design. The purpose of discussing them here is to demonstrate that the principles on which they are based are quite simple. The number of situations in which such designs exist is rather limited and often a design in which some pairs of treatments occur slightly more often than others have to be used. An example of six treatments in six blocks of four units each is shown in Fig. 5.6. Here six pairs of treatments occur together in a block thrice (AC, AD, BD, BE, CF, EF) and other pairs twice, giving as even a distribution of joint occurrences as possible.

When the blocks containing incomplete sets of treatments can be considered as a random sample of blocks from some, generally hypothetical, large population of such blocks and the treatment sets have been allocated at random to blocks, then there is some information to be obtained from an analysis of the differences between the block totals. In Example 5.3 (Fig. 5.4(b)), a comparison of the first and second block totals would be a comparison of treatments E and F since the other four treatments occur in both blocks. The estimates of the treatment differences from block totals generally have appreciably larger standard errors than those derived from the within blocks analysis. Also there are generally insufficient degrees of freedom to estimate the error

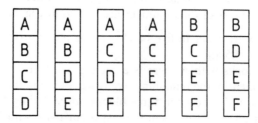

Fig. 5.6 Allocation of six treatments to units in six blocks of four units each, achieving the most even possible precision of treatment comparisons.

mean square appropriate to the between block totals analysis with the accuracy needed to decide how to weight the information about treatment effects available from the two analyses. For these reasons inter-block information is often ignored unless the number of blocks appreciably exceeds the number of treatments and the size of the blocks is not much greater than the number of treatments.

6
Particular questions about treatments

6.1 TREATMENT STRUCTURE

So far we have been concerned with the design and analysis of experiments from the point of view of reducing the random errors by arranging the units in groups. We have considered the treatments only as a heterogeneous group within which we can compare any pair of treament means using a t-test based on the error mean square from the analysis. In many experiments the treatments have some structure, so that we may wish to make particular comparisons between and within groups of treatments instead of the simple comparison of all possible pairs of treatments.

A simple example of treatment structure is an experiment in which a number of new strains of grass are being compared with a standard strain. In this situation there may be two basic hypotheses the experimenter wishes to test; first whether the new strains, taken as a group, show a difference in yield from the control, and second whether there is much variation between the new strains. To test these hypotheses we can subdivide the treatment s.s. into components representing (a) the difference between the control and the new strains, and (b) variation between the new strains. Suppose we have five new strains and a control in a randomized block experiment with four blocks, a total of 24 units. If we denote the treatment totals by T_1, T_2, T_3, T_4, T_5 and T_C the treatment s.s. will be

$$\frac{T_1^2 + T_2^2 + T_3^2 + T_4^2 + T_5^2 + T_C^2}{4} - \frac{G^2}{24} \tag{6.1}$$

where G is the overall total of yields. The comparison between new strains and control can be thought of as a comparison between a mean of 20 units (four for each of five new strains) and a mean of 4 units. The sum of squares corresponding to this difference is

$$\frac{T_N^2}{20} + \frac{T_C^2}{4} - \frac{G^2}{24} \tag{6.2}$$

where T_N is the sum of T_1, T_2, T_3, T_4 and T_5, i.e. the total of yields for all new strains. To measure the variation between new strains we calculate the sum of squares for variation between five treatments

$$\frac{T_1^2 + T_2^2 + T_3^2 + T_4^2 + T_5^2}{4} - \frac{T_N^2}{20} \qquad (6.3)$$

This is, of course, the treatment s.s. we would have calculated if the experiment had consisted of the five new strains only. These two sums of squares have 1 and 4 d.f. respectively (variation between two means, and between 5 means) and their sum $(6.2) + (6.3)$ is the complete treatment sum of squares (6.1). The two sums of squares correspond to different causes of variation and each mean square can be compared with the error mean square using the appropriate F-distribution to test the significance of the corresponding variation. The standard error of the difference between any two treatment means is $\sqrt{(2s^2/4)}$ as usual and the standard error of the difference between the control mean and the mean of the five new strains is $\sqrt{(s^2/4 + s^2/20)}$, since the two means are calculated from samples of size 4 and 20 units respectively (section 3.4).

If a control treatment is repeated several times in each block the same method can be applied. Suppose for example that we have five blocks each containing six new strains and four replicates of the control. Suppose further, that the four control treatments in each block can be identified and are labelled $7, 8, 9$ and 10. Our experiment could be viewed as having ten treatments, $1, 2, 3, 4, 5, 6, C_1, C_2, C_3$ and C_4; the full treatment s.s. could be calculated as

$$\frac{T_1^2 + T_2^2 + T_3^2 + T_4^2 + T_5^2 + T_6^2 + T_7^2 + T_8^2 + T_9^2 + T_{10}^2}{5} - \frac{G^2}{50}$$

This could be divided into three components:

1. Measuring variation between control and new

$$\frac{T_N^2}{30} + \frac{T_C^2}{20} - \frac{G^2}{50}$$

2. Measuring variation between new strains

$$\frac{T_1^2 + T_2^2 + T_3^2 + T_4^2 + T_5^2 + T_6^2}{5} - \frac{T_N^2}{30}$$

3. Measuring variation between controls

$$\frac{T_7^2 + T_8^2 + T_9^2 + T_{10}^2}{5} - \frac{T_C^2}{20}$$

Of course the mean square from this last sum of squares should be about the same as the error mean square since there are no differences between treatments 7, 8, 9 and 10. If this assumption that there are no differences between these treatments is made before the analysis, rather than tested in the analysis, then component (3) of the treatment s.s. can be combined with the error s.s. to give an improved estimate, s^2. The only reason, apart from sampling variation, for the mean square corresponding to (3) to be greater than the error mean square is that the repeatability of the effects of the control treatment is less that that of some of the other treatments. The mean square can be smaller than the error mean square because the effects of the control treatment are more repeatable than the effects of some of the other treatments. This last possibility is perhaps more likely than the first because the mean square corresponding to (3) is based on repeatability of effects within blocks whereas the error mean square is derived from the repeatability of the differences between all the treatments from block to block. The other two components, (1) and (2), are those previously calculated and each mean square can be compared with the error mean square using the F-distribution.

The comparison between controls and new treatments is the simplest form of treatment comparison after the basic comparison of two treatment means. When the treatments in an experiment have a more complicated structure it may be desirable to divide the treatment s.s. into a number of components each corresponding to comparisons between particular groups of treatments. This may be done in a similar manner to that used for the experiment with controls and new strains. For each comparisons between groups of treatments, the totals for the different groups are calculated, $T_1, T_2, \ldots$; the sum of these totals is $\Sigma(T) = T_1 + T_2 + \cdots$. The sum of squares is then calculated as

$$\frac{T_1^2}{r_1} + \frac{T_2^2}{r_2} + \cdots - \frac{[\Sigma(T)]^2}{\Sigma(r)}$$

where r_1 is the total number of units contributing to total T_1, r_2 to total $T_2, \ldots$ and $\Sigma(r) = r_1 + r_2 + \cdots$. The vital rule to remember is to divide each squared total by the number of units included in that total. The degrees of freedom for the sum of squares is the number of groups minus 1.

Example 6.1

As an example we will consider an experiment on spring cabbage in which there were eight treatments arranged in four randomized blocks of eight plots each. The experiment was to compare three sources of nitrogen, each at two levels, with a control treatment of no nitrogen, there being two replicates of the control in each block. The data were as follows.

| | Level | \multicolumn{4}{c}{Block} | |
		I	II	III	IV	Total
Nitro-chalk	1	70.3	72.5	79.0	86.2	308.0
	2	61.0	75.1	71.3	65.2	272.6
Sulphate of	1	75.5	63.0	65.4	67.7	271.6
ammonia	2	64.7	66.0	64.3	68.8	263.8
Nitrate	1	85.2	80.5	83.6	92.3	341.6
	2	65.9	78.7	71.1	76.8	292.5
Control	1	35.7	39.6	45.5	50.5	171.3
	2	38.5	36.3	42.1	61.4	178.3
		496.8	511.7	522.3	568.9	2099.7

The simple analysis of variance gives:

Source of variation	s.s.	d.f.	m.s.	F-ratio
Blocks	363.33	3		
Treatments	6 205.16	7	886.45	26.36***
Error	706.87	21	$33.65 = s^2$	
Total	7 275.06	31		
c.f.	137 773.12			

We can divide the treatment sum of squares into several useful components. For this we need several more totals:

Total for two control treatments = 349.6
Total for two nitro-chalk levels = 580.6
Total for two ammonia levels   = 535.4
Total for two nitrate levels   = 634.1
Total for six nitrogen treatments = 1750.1

The first component is for the comparison between controls and nitrogen treatments

$$\frac{(349.6)^2}{8} + \frac{(1750.1)^2}{24} - \text{c.f.} = 5123.15$$

Second, comparison between the three nitrogen sources

$$\frac{(580.6)^2 + (55.4)^2 + (634.1)^2}{8} - \frac{(1750.1)^2}{24} = 610.29$$

Third, the comparison between the two levels for each nitrogen source

(a) $$\frac{(308.0)^2 + (272.6)^2}{4} - \frac{(580.6)^2}{8} = 156.65$$

(b) $$\frac{(271.6)^2 + (263.8)^2}{4} - \frac{(535.4)^2}{8} = 7.61$$

(c) $$\frac{(341.6)^2 + (292.5)^2}{4} - \frac{(634.1)^2}{8} = 301.35$$

Finally we can calculate a component for the variation between the two control treatments (which should be similar to the error mean square since the control treatments were identical)

$$\frac{(171.3)^2 + (178.3)^2}{4} - \frac{(349.6)^2}{8} = 6.12$$

We can now summarize this subdivision of the treatment s.s. as follows:

Source of variation	s.s.	d.f.	m.s.	F-ratio
Control v. nitrogen	5123.15	1	5123.15	152.2***
Nitrogen sources	610.29	2	305.15	9.1
Levels of ammonia	7.61	1	7.61	0.2
Levels of nitrate	301.35	1	301.35	9.0***
Between controls	6.12	1	6.12	0.2
Total treatments	6204.16	7	886.45	26.4***
Error	706.57	21	33.65	

To present the results we calculate treatment means and various standard errors of differences between means:

	Treatment means		
	Level 1	Level 2	Mean
Nitro-chalk	77.0	68.2	72.6
Ammonia	67.9	66.0	67.0
Nitrate	85.4	73.1	79.9
Mean			72.9
Control			43.7

Standard errors:

(a) For comparing control treatment mean and mean of six nitrogen treatments

$$\sqrt{\left(\frac{s^2}{8}+\frac{s^2}{24}\right)}=\sqrt{\left[33.65\left(\frac{1}{8}+\frac{1}{24}\right)\right]}=2.37$$

(b) For comparing control with a source mean or comparing two source means

$$\sqrt{\left(\frac{s^2}{8}+\frac{s^2}{8}\right)}=\sqrt{(33.65/4)}=2.90$$

(c) For comparing control with any level of any source

$$\sqrt{\left(\frac{s^2}{8}+\frac{s^2}{24}\right)}=\sqrt{[33.65(3/8)]}=3.55$$

(d) For comparing any two means for particular levels of the same or different sources

$$\sqrt{(2s^2/4)}=\sqrt{(33.65/2)}=4.10$$

Summary of results: All nitrogen treatments increased yields substantially compared with the control treatment. The increase varied between 22 and 42 for the different sources, nitrate giving the largest average increase 35. Level 1 gave higher yields than level 2 for each source though the difference was significant only for nitro-chalk and nitrate.

6.2 TREATMENT CONTRASTS

The subdivision of the treatment sum of squares in this example is not unique. It was chosen because the individual components corresponded to comparison between treatments which were of interest from the experimenter's viewpoint. Many alternative subdivisions are possible according to the interest of the experimenter but some care must be taken in choosing a subdivision since writing down a set of comparisons and the corresponding s.s. will not necessarily give sums of squares which sum to the full treatment sum of squares. A set of comparisons whose corresponding s.s. must (by the form of definition of the comparisons) sum to the full treatment s.s. are said to be independent, or orthogonal, comparisons. As a simple example of orthogonal comparisons consider an experiment with three treatments, one of which is a control. Let the treatment totals for r replicates be T_1 (control), T_2 and T_3. The form of subdivision we have suggested is:

(a) Comparisons between treatments 2 and 3

$$\text{s.s.}=\frac{T_2^2+T_3^2}{r}-\frac{(T_2+T_3)^2}{2r}$$

(b) Comparison between control and rest

$$\text{s.s.} = \frac{T_1^2}{r} + \frac{(T_2 + T_3)^2}{2r} - \frac{(T_1 + T_2 + T_3)^2}{3r}$$

$$\text{sum of two components} = \frac{T_1^2 + T_2^2 + T_3^2}{r} - \frac{T_1 + T_2 + T_3}{3r} = \text{treatment s.s.}$$

At first sight a reasonable alternative subdivision is

(a) Comparison between 1 and 2

$$\text{s.s.} = \frac{T_1^2 + T_2^2}{r} - \frac{(T_1 + T_2)^2}{2r}$$

(b) Comparison between 1 and 3

$$\text{s.s.} = \frac{T_1^2 + T_3^2}{r} - \frac{(T_1 + T_2)^2}{2r}$$

But the sum of the two components:

$$= \frac{2T_1^2 + T_2^2 + T_3^2}{r} - \frac{(T_1 + T_2)^2 + (T_1 + T_3)^2}{2r}$$

This is in general not equal to the treatment s.s. and therefore this set of comparisons does not give a proper subdivision of the treatment s.s.

The mathematical theory of orthogonal comparisons is too complex to be discussed here but two rules may help to indicate when sets of comparisons are acceptable.

(a) Comparisons between groups of treatments and within groups form a satisfactory set.
(b) Individual comparisons (with single degrees of freedom) can be represented as linear combinations of treatment means, $\Sigma(l_i \bar{x}_i) = l_1 \bar{x}_1 + l_2 \bar{x}_2 + \dots$. Thus a comparison between treatments 1 and 2 can be written $(\bar{x}_1 - \bar{x}_2)$ and that between the mean of treatments 1, 2 and 3 and the mean of treatments 4 and 5 written.

$$\left(\frac{\bar{x}_1 + \bar{x}_2 + \bar{x}_3}{3} - \frac{\bar{x}_4 + \bar{x}_5}{2} \right)$$

If there are only five treatments we can represent the comparisons solely by the ls; the first comparison could be represented by $(1, -1, 0, 0, 0)$, the second by $(\frac{1}{3}, \frac{1}{3}, \frac{1}{3}, -\frac{1}{2}, -\frac{1}{2})$ or equivalently by $(2, 2, 2, -3, -3)$, Note that the sum of the ls will always be zero.

Two comparisons are orthogonal if the sum of the products of corresponding ls is zero. For this last example we have

$$\text{sum of products} = 1 \times 2 + (-1) \times 2 + 0 \times 2 + 0 \times -3 + 0 \times -3 = 0$$

and hence the comparisons are orthogonal. Returning to the previous example with three treatments, the first form of subdivision corresponds to the comparisons

$$(0, 1, -1) \qquad (2 \text{ compared with } 3)$$

and

$$(2, -1, -1) \qquad (1 \text{ compared with mean of } 2 \text{ and } 3)$$

so that

$$\text{sum of products} = 0 \times 2 + 1 \times (-1) - (-1) \times (-1) = 0$$

The comparisons for the second form of subdivision are

$$(1, -1, 0) \qquad (1 \text{ compared with } 2)$$
$$(1, 0, -1) \qquad (1 \text{ compared with } 3)$$

and it is easily seen that the sum of the products of corresponding ls is 1 so that the comparisons are not orthogonal, as we have already seen.

For interpretation, the advantages of orthogonal comparisons are that they can be interpreted separately because the errors involved in estimating any two orthogonal comparisons are independent of each other. Conversely, non-orthogonal comparisons need to be interpreted with care since the errors of estimation are not independent. Many statistical packages will estimate linear contrasts between treatments, specified in this way, either by partitioning the treatment sum of squares in the analysis of variance into components to test for the significance of the contrast or by calculating its estimate and standard error. The contrasts are specified by the appropriate coefficients of the treatment means.

EXERCISE 6.1 *Analysis of variance of a randomized block design with multiple controls.*

In an experiment on methods of seed production of cabbage, six methods were compared with the standard method (control treatment). A randomized block design was used with five blocks and four control plots in each block, the main aim of the experiment being comparison of the new methods with the control. Carry out the usual analysis of variance on the mean weights per plant (in kilograms) given below and subdivide the treatment s.s. into components for the comparisons (a) between control and new methods, (b) between controls and (c) between new methods. Summarize the conclusions.

Treatment	I	II	Blocks III	IV	I	Total
1 (a)	1.63	1.48	1.43	1.76	1.17	7.47
1 (b)	1.73	1.42	1.50	1.06	0.76	6.47
1 (c)	1.49	1.70	1.52	1.48	0.85	7.04
1 (d)	1.25	1.36	0.93	1.38	0.68	5.60
2	1.07	1.28	1.28	1.83	1.16	6.62
3	0.73	1.42	1.30	1.38	0.70	5.53
4	0.69	1.59	1.61	1.61	1.17	6.67
5	0.52	1.50	1.17	1.05	1.04	5.28
6	1.63	1.34	1.07	1.01	0.87	5.92
7	1.08	1.33	0.63	1.21	0.51	4.76
Block totals	11.82	14.42	12.44	13.77	8.91	61.36

(The uncorrected total sum of squares is 81.1078)

6.3 FACTORIAL TREATMENT STRUCTURE

In many experiments the proposed treatment structure will evolve only as the experimenter defines the purpose of his experiment and often there will be initially little apparent pattern in the choice of treatments and in their inter-relationships. There are, however, a number of advantages in using treatments with a strong structure pattern as may be simply demonstrated by considering the following set of yields of kale grown under four different fertilizer conditions.

Fertilizer	Yield (tons per acre)
(1) No P, no K	20.5
(2) No P, 1 cwt K_2O	19.8
(3) 0.6 cwt P_2O_5, no K	26.7
(4) 0.6 cwt P_2O_5, 1 cwt K_2O	30.5

The clear conclusion from these data, assuming that the standard errors of these yields are sufficiently small, is that the effect of the application of P on yield depends on the level of application of K and vice versa. Thus, if no K is applied, 0.6 cwt P_2O_5 increases yield by 6.2 tons per acre whereas when 1 cwt K_2O is applied, 0.6 cwt P_2O_5 increases yield by 10.7 tons per acre. Now consider the other sets of treatments which might have been used.

(a) Treatments (1) (2) and (3): these would have shown no apparent effect of K, and a reasonably large effect of P. They give no information on the effects of K (or P) at non-zero levels of P (or K).

(b) Treatments (5) 0.3 cwt P_2O_5, no K, and (6) 0.3 cwt P_2O_5, 1 cwt K_2O.

These might be expected to give yields of 23.6 and 25.2 (averaging the observed yields at the two levels of P) and would suggest a very small effect of K

(c) Treatments (7) 0.5 cwt K_2O, no P, and (8) 0.5 cwt K_2O, 0.6 cwt P_2O_5. These might be expected to give yields of 20.2 and 28.6 and would suggest a large effect of P.

The important point to observe is that none of these alternative experiments present a true picture of the situation shown by the original experiment. In each case the only possible deduction, which would have been incorrect, would have been that the effect of applying a particular fertilizer increases (or decreases) yields by an estimated amount.

In the original experiment the four treatments have a clearly defined structure being all possible combinations of one of two levels of P (0 or 0.6) with one of two levels of K (0 or 1). This is clearly shown by presenting the results in a two-way table:

	No P	0.6 cwt P_2O_5
No K	20.5	26.7
1 cwt K_2O	19.8	30.5

This form of treatment structure is called 'factorial' and is essential when the direct effects and combined cross-effects of a number of different types of treatments are to be examined.

At this stage it is desirable to introduce several definitions:

Factor: A type of treatment (e.g. variety, fertilizer, feed.)
Level: A particular treatment of a factor set (e.g. variety A, or 1 cwt K_2O, or 12% protein).
Experimental treatment: A combination of one level for each factor in the experiment.
Complete factorial experiment: An experiment containing all possible experimental treatments, possibly replicated a number of times.

To exemplify the use of these terms we shall consider an experiment to compare the effects of two levels of P and two levels of K on three varieties of kale. The experiment contains three factors, P, K and variety, and these factors have, respectively, two levels (absence and presence), two levels (absence and presence) and three levels (varieties V_1, V_2 and V_3). A typical experimental treatment would consist of plants of variety V_2, grown in soil with no P and 1 cwt K_2O added. There are $12 (= 2 \times 2 \times 3)$ different experimental treatments and the complete experiment might contain three randomized blocks of 12 plots each. Such an experiment is described as a $2 \times 2 \times 3$ factorial (or sometimes a $2^2 \times 3$ factorial).

When the experimental treatments in a factorial experiment have been decided, these treatments are allocated randomly to the units in a block, the treatment structure being ignored. A suitable way of doing this is to number the experi-

Block 1			Block 2	Block 3
1. No P	No K	V_1	5	6
7. 0·6 cwt P_2O_5	No K	V_1	4	2
4. No P	1 cwt K_2O	V_1	11	11
10. 0·6 cwt P_2O_5	1 cwt K_2O	V_1	1	7
12. 0·6 cwt P_2O_5	1 cwt K_2O	V_3	7	8
5. No P	1 cwt K_2O	V_2	12	1
8. 0·6 cwt P_2O_5	No K	V_2	2	12
3. No P	No K	V_3	3	9
6. No P	1 cwt K_2O	V_3	9	10
11. 0·6 cwt P_2O_5	1 cwt K_2O	V_2	6	4
9. 0·6 cwt P_2O_5	No K	V_3	10	5
2. No P	No K	V_2	8	3

Fig. 6.1 Possible layout of a 2 × 2 × 3 factorial experiment in three randomized blocks.

mental treatments consecutively from 1 upwards and then allocate the numbers randomly. For the 2 × 2 × 3 factorial considered earlier we might number the treatments as follows:

(1) No P, no K, variety V_1
(2) No P, no K, V_2
(3) No P, no K, V_3
(4) No P, 1 cwt K_2O, V_1
(5) No P, 1 cwt K_2O, V_2
(6) No P, 1 cwt K_2O, V_3
(7) 0.6 cwt P_2O_5, no K, V_1
(8) 0.6 cwt P_2O_5, no K, V_2
(9) 0.6 cwt P_2O_5, no K, V_3
(10) 0.6 cwt P_2O_5, 1 cwt K_2O, V_1
(11) 0.6 cwt P_2O_5, 1 cwt K_2O, V_2
(12) 0.6 cwt P_2O_5, 1 cwt K_2O, V_3

The randomized sequence of treatments might be: 1, 7, 4, 10, 12, 5, 8, 3, 6, 11, 9, 2 and the full, randomized experiment as shown in Fig. 6.1, the full treatment details being given for block 1.

6.4 MAIN EFFECTS AND INTERACTIONS

If we now consider the analysis of variance of a factorial experiment containing two factors we have to decide how to subdivide the treatment s.s. The obvious method of division, following on from methods used earlier in this chapter, is to take the first component s.s. to correspond to variation between the levels of one of the factors: other components will then represent the variation between the levels of the second factor for each level of the first factor. To take a simple, example, consider an experiment with three diets (0, 1 and 2) compared for male and female pigs (M, F). If the treatment totals are $T_{0M}, T_{0F}, T_{1M}, T_{1F}, T_{2M}, T_{2F}$, and the experiment consists or r blocks of six units each, then the full treatment s.s. would be

$$\frac{T_{0M}^2 + T_{0F}^2 + T_{1M}^2 + T_{1F}^2 + T_{2M}^2 + T_{2F}^2}{r} - \text{c.f.}$$

This could be subdivided into either:

(a) A component representing the overall difference between male and female pigs,

(i)
$$\frac{T_M^2 + T_F^2}{3r} - \text{c.f.}$$

where $T_M = T_{0M} + T_{1M} + T_{2M}$, and two components representing variation between diets for each sex,

(ii)
$$\frac{T_{0M}^2 + T_{1M}^2 + T_{2M}^2}{r} - \frac{T_M^2}{3r}, \text{ and}$$

(iii)
$$\frac{T_{0F}^2 + T_{1F}^2 + T_{2F}^2}{r} - \frac{T_F^2}{3r}, \text{ or}$$

(b) A component representing variation between the three diets,

(i)
$$\frac{T_0^2 + T_1^2 + T_2^2}{2r} - \text{c.f.}$$

and three components representing the difference between male and female for each diet,

(ii)
$$\frac{T_{0M}^2 + T_{0F}^2}{r} - \frac{T_0^2}{2r}$$

(iii)
$$\frac{T_{1M}^2 + T_{1F}^2}{r} - \frac{T_1^2}{2r}$$

(iv)
$$\frac{T_{2M}^2 + T_{2F}^2}{r} - \frac{T_2^2}{2r}$$

Either of these subdivisions might be appropriate for particular experiments. Neither, however, takes advantage of the symmetry of the factorial structure, both methods considering, first, major differences between levels of one factor and, second, for each levels of this factor, differences between levels of the second factor. The standard method of subdivision of a factorial treatment s.s.

treats each factor equally, the typical subdivision for the diets × sexes experiment being

(i)
$$\frac{T_M^2 + T_F^2}{3r} - \text{c.f.}$$

representing overall differences between sexes;

(ii)
$$\frac{T_0^2 + T_1^2 + T_2^2}{2r} - \text{c.f.}$$

representing overall differences between diets; and the remaining s.s.,

(iii)
$$\frac{T_{0M}^2 + T_{0F}^2 + T_{1M}^2 + T_{1F}^2 + T_{2M}^2 + T_{2F}^2}{r} - \text{c.f.} - \text{(i)} - \text{(ii)}$$

Since (i) and (ii) represent all the information about the average difference between the two sexes, and the average differences between the three diets, this last s.s. must correspond to differences between the responses of the two sexes to the different diets.

The two advantages of this method of subdividing the treatment s.s. are that it uses the symmetry of the factorial structure and that it is simply generalized to experiments with more than two factors. The three components s.s. are termed

(i) The s.s. for the main effect of sex,
(ii) The s.s. for the main effect of diets,
(iii) The s.s. for the interaction between sex and diets.

Formally we may define the main effect of a factor as differences between the mean yields for the different levels of that factor, averaging over all levels of all other factors, and the interaction between two factors as the variation of the differences between mean yields for different levels of one factor over different levels of the sother factor. The practical interpretation of these definitions is most easily explained diagrammatically. Consider the hypothetical mean weekly increases of weight for the pigs in the diets × sex experiment:

	Diet 0	Diet 1	Diet 2
Males	8.5 lb	10 lb	11.5 lb
Females	9 lb	12.5 lb	13 lb

These yields may be represented diagrammatically as shown in Fig. 6.2. Mean yields for diets (averaging over sexes) are 8.75, 11.25 and 12.25 for diets 0, 1 and 2 respectively and are shown by the circles in the three diet columns. The main effect comparisons of diets are the differences between these three values. Similarly mean yields for sexes (averaging over diets) are 10 and 11.5 for M and F, shown by the circles in the mean column in Fig. 6.2. The differences, F − M, for the diets are 0.5, 2.5 and 1.5 and the interaction between diets and

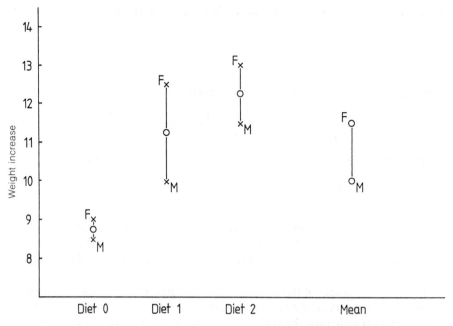

Fig. 6.2 Diagrammatic presentation of mean yields for six treatments, showing main effects and interactions.

sexes consists of the variation between these three values. Diagrammatically it is represented by the variation of the height of F above M over the three diets.

It is important to realize that interaction between two factors in symmetric. We have discussed the interaction between diets and sex in terms of the way the difference between male and female pigs changes according to which of the three diets is being considered. We could equally well have discussed how the differences between the three diets are different for male pigs from the corresponding differences for female pigs. For example, for male pigs the mean weekly weight increase for diet 1 was midway between the increases for diets 0 and 2, whereas for female pigs the increase for diet 1 was almost the same as for diet 2, and both were much larger than the increase for diet 0. More specifically, diet 1 gave an increase bigger by 1.5 lb than that for diet 0 for male pigs, but for female pigs the diet 1 increase was 3.5 lb bigger than that for diet 0. This comparison of differences of course gives exactly the same information as comparing the F − M difference of 0.5 for diet 0 with 2.5 for diet 1.

6.5 ANALYSIS OF VARIANCE FOR A TWO-FACTOR EXPERIMENT

The analysis of variance for a factorial experiment with two factors can be calculated as follows.

Suppose we have r blocks of t treatments, the treatments being all combinations of a levels of factor A and b levels of factor $B(ab=t)$. Let the total of r yields for level 1 of A and level 1 of B be T_{11}, the total for level 1 of A and level 2 of B be T_{12}, and so on, the total for level a of A and level b of B being T_{ab}. Also let T_{A1} be the total of the $r \times b$ yields for level 1 of factor A (summing over all levels of treatment B as well as over blocks) and T_{B1} the total of the $r \times a$ yields for level 1 of factor B. Then the first part of the analysis of variance, for which we treat the experimental treatments without reference to their structure, is the usual randomized block analysis.

Source of variation	s.s.	d.f.	m.s.
Blocks	$\dfrac{B_1^2 + B_2^2 + \cdots + B_r^2}{t} - \text{c.f.}$	$r-1$	
Treatments	$\dfrac{T_{11}^2 + T_{12}^2 + T_{13}^2 + \cdots T_{ab}^2}{r} - \text{c.f.}$	$t-1$	
Error	by subtraction	$(t-1)(r-1)$	
Total	$\Sigma(y^2) - \text{c.f.}$	$tr-1$	
c.f.	$[\Sigma(y)]^2/rt$		

The subdivision of the treatment s.s. gives:

Source of variation	s.s.	d.f.	m.s.
Main effect of A	$\dfrac{T_{A1}^2 + T_{A2}^2 + \cdots}{rb} - \text{c.f.}$	$(a-1)$	
Main effect of B	$\dfrac{T_{B1}^2 + T_{B2}^2 + \cdots}{ra} - \text{c.f.}$	$(b-1)$	
AB interaction	by subtraction	$(a-1)(b-1)$	
Total treatment variation	$\dfrac{T_{11}^2 + T_{12}^2 + \cdots T_{ab}^2}{r} - \text{c.f.}$	$(ab-1)$	

Note that, as usual, each squared total is divided by the number of units included in the total. Also note the symmetric form of the analysis, in particular the degrees of freedom for the interaction s.s. (obtained as $(ab-1)-(a-1)-(b-1)$). Each of the mean squares for the three components of the treatment s.s. can be compared with the error mean square, the ratio being compared with the corresponding F-distribution. For example, to test the main effect of A we calculate $F = [(A)$ mean square/error mean square] and compare it with the F-distribution for $(a-1)$ and $(t-1)(r-1)$ degrees of freedom. It is customary to combine the two sections of the analysis replacing the treatment s.s. in the first section by the three components from the second giving the analysis in the form:

Source of variation	s.s.	d.f.	m.s.	F
Blocks		$(r-1)$		
Main effect (A)		$(a-1)$		
Main effect (B)		$(b-1)$		
AB interaction		$(a-1)(b-1)$	s^2	
Total		$(abr-1)$		
c.f.				

If the only significant effects in a factorial experiment are main effects the results can be reduced to the mean yields for the different levels of each factor in turn, the mean yields having been averaged over the levels of the other factor. In other words, if the interactions between factors are not important then the mean yields for combinations of levels of different factors will not be more informative than the mean yields for each level of each factor separately. On the other hand, if the F-ratio for an interaction AB is significant then, when presenting the results, the mean yields for all combinations of levels of A and B should be included. When an interaction F-ratio is significant it may be much smaller than the corresponding main effect F-ratios, or it may be comparable in size. In the former case, which is the more usual, the interpretation should be that there is a major difference between the effects of the levels of, say, factor A, but that this difference is affected by the level of factor B. Consequently the presented results should include mean yields for each level of each factor, A and B, and also the two-way table of treatment means for all combinations of levels of A and B. If, however, the AB interaction F-ratio is similar to or much larger than the F-ratios for the main effects of A and B then the variation, over levels of A, of an effect of B is at least as large as the average effects and the main effect of each factor is clearly meaningless and only the two-way table of treatment means should be presented.

To illustrate these methods of presentation and interpretation of results, Table 6.1 shows treatment mean yields, main effect and interaction F-ratios and the form of results for presentation for three hypothetical experiments with four blocks of a 2×2 factorial experiment, the error mean square based on 9 d.f. being assumed to be 5.

Table 6.1 Results and form of presentation for three hypothetical 2×2 factorial experiments

Treatment means				F-ratios		Presented form		

(i)

	A_1	A_2	Main effect (A)	20.0**		A_1	19
B_1	16	20	Main effect (B)	39.2***		A_2	24
			Interaction	0.8		B_1	18
B_2	22	28				B_2	25

Standard error
for comparing
A_1 with A_2 or B_1
with $B_2 = 1.12$

(ii)

	A_1	A_2	Main effect (A)	20.0**
B_1	16	18	Main effect (B)	39.2***
B_2	20	28	Interaction	7.2*

	A_1	A_2	Mean
B_1	16	18	17
B_2	20	28	24
Mean	18	23	

s.e. for comparing any pair
of the four combination
means $= 1.58$ s.e. for
comparing main effect
means $= 1.12$

(iii)

	A_1	A_2	Main effect (A)	20.0**
B_1	13	24	Main effect (B)	12.8**
B_2	23	22	Interaction	28.8***

	A_1	A_2
B_1	13	24
B_2	23	22

s.e. for comparing any two
means $= 1.58$

Initial analysis of variance

(Ignoring factorial structure of treatments, i.e. treating experiment as three blocks $\times$ 18 treatments):

	s.s.	d.f.	m.s.	F
Block	0.01	2	0.005	
Treatments	85.64	17	5.038	124***
Error	1.38	34	0.041	
Total	87.03	53		
c.f.	1634.05			

Example 6.2 shows the layout of the analysis of variance of a typical factorial experiment with two factors, with one factor having four levels and the other three levels.

Example 6.2 Analysis of variance of a simple factorial experiment

An experiment was conducted to investigate the effects of three levels of sorbic acid and six levels of water activity (a_w) on survival of *Salmonella typhimurium*. A randomized block design was used, with three blocks and 18 treatment combinations. The data analysed are log (density/ml) measured seven days after the imposition of the treatments.

Sorbic acid	a_w	I	II	III	Total
0	0.9	8.19	8.37	8.33	24.89
	0.94	6.65	6.70	6.25	19.60
	0.90	5.87	5.98	6.14	17.99
	0.86	5.06	5.35	5.01	15.42
	0.82	4.85	4.31	4.52	13.68
	0.78	4.31	4.34	4.20	12.85
100 ppm	0.98	7.64	7.79	7.59	23.02
	0.94	6.52	6.19	6.51	19.22
	0.90	5.01	5.28	5.78	16.07
	0.86	4.85	4.95	4.29	14.09
	0.82	4.29	4.43	4.18	12.90
	0.78	4.13	4.39	4.18	12.70
200 ppm	0.98	7.14	6.92	7.19	21.25
	0.94	6.33	6.18	6.43	18.94
	0.90	5.20	5.10	5.43	15.73
	0.86	4.41	4.40	4.79	13.60
	0.82	4.26	4.27	4.37	12.90
	0.78	3.93	4.12	4.15	12.20
		98.64	99.07	99.34	297.05

Two-way interaction table for sorbic acid (s.a.) and water activity levels:

Sorbic acid	Water activity						Sorbic acid totals
	0.98	0.94	0.90	0.86	0.82	0.78	
0	24.89	19.60	17.99	15.42	13.68	12.85	104.43
100	23.02	19.22	16.07	14.09	12.90	12.70	98.00
200	21.25	18.94	15.73	13.60	12.90	12.20	94.62
a_w totals	69.16	57.76	49.79	43.11	39.48	37.75	297.05

s.s. main effect of water activity

$$= \frac{69.16^2 + 57.76^2 + \cdots + 37.75^2}{9} - \text{c.f.}$$

$$= 81.56$$

s.s. main effect of sorbic acid

$$= \frac{104.43^2 + 98.00^2 + 94.62^2}{18} - \text{c.f.}$$

$$= 2.75$$

interaction s.s.

$$= \frac{24.89^2 + 19.60^2 + \cdots + 12.20^2}{3} - \text{c.f.} - \text{s.s.}(a_w) - \text{s.s.(s.a.)} = 1.33$$

Notes:

S.S. main effect of sorbic acid is the same as would be obtained by ignoring a_w levels and treating the experiment as 12 replications of three s.a. levels.

The interaction s.s. measures the variation between the 18 treatment combinations which is not accounted for by the two main effect s.s.

Full analysis:

	s.s.	d.f.	m.s.	F
Blocks	0.01	2	0.005	
Water activity	81.56	5	16.312	401.9***
Sorbic acid	2.75	2	1.375	33.9***
a_w × s.a.	1.33	10	0.133	3.28**
Error	1.38	34	0.041	
Total	87.03	53		

As the interaction s.s. is significant the results are presented in a two-way table of treatment means. However, as the main effect mean squares are considerably larger than the interaction mean square the main effect yields are also presented.

Table of treatment means:

Sorbic acid	0.98	0.94	0.90	0.86	0.82	0.78	s.a. mean
			Water activity				
0	8.30	6.53	6.00	5.14	4.56	4.28	5.80
100	7.67	6.41	5.36	4.70	4.30	4.23	5.44
200	7.08	6.31	5.24	4.53	4.30	4.07	5.26
a_w mean	7.68	6.42	5.53	4.79	4.39	4.19	5.50

s.e. of difference between two values in body of table $= \sqrt{(2s^2/3)} = 0.165$
s.e. of difference between two water activity means $= \sqrt{(2s^2/9)} = 0.095$
s.e. of difference between two sorbic acid means $= \sqrt{(2s^2/18)} = 0.067$

Summary of results

For each sorbic acid level reducing water activity produced large and significant reductions in *Salmonella* density. Differences between the observed densities for different sorbic acid levels were largest at the highest water activity and were significant at a_w levels of 0.90 and 0.86 also. The overall pattern is of diminishing differences due to sorbic acid effects as the water activity is reduced.

EXERCISE 6.2 *Analysis of variance of a factorial experiment*

The data below are from an experiment with carrots to investigate the effect of sowing rate on yield for two stocks of seed. The experiment consisted of three randomized blocks of the eight treatment combinations. Calculate the analysis of variance, examining the effects of stock and sowing rate and the interaction between these two factors. Summarise the data in a table of means and report your conclusions.

Stock	Sowing rate (1 lb per acre)	I	II	III
			Block	
T	1.5 (A)	4.20	4.94	4.45
	2 (B)	4.36	3.50	4.17
	2.5 (C)	5.40	4.55	5.75
	3 (D)	5.15	4.40	3.90
H	A	2.82	3.14	3.80
	B	3.74	4.43	2.92
	C	4.82	3.90	4.50
	D	4.57	5.32	4.35

6.6 COMPUTER ANALYSIS OF FACTORIALS

If the treatments in an experiment have a factorial structure then codes must be set up for each experimental factor. For example if the 6 treatments in an experiment form a 3×2 factorial structure with factor A at three levels (A_1, A_2, A_3) and factor B at two levels (B_1, B_2), those would be specified by the following factor codes.

Treatment	Factor A	Factor B	Codes A	B
1	A_1	B_1	1	1
2	A_1	B_2	1	2
3	A_2	B_1	2	1
4	A_2	B_2	2	2
5	A_3	B_1	3	1
6	A_3	B_2	3	2

The treatment term in the analysis of variance model is then replaced by the terms:

$$A + B + A \times B$$

Ideally the statistical package should also present standard errors for both main effect and interaction means, however most simpler packages present the means but leave the user to calculate appropriate standard errors.

For the experiment of Example 6.2 the input to the computer would be

Log density	Block	a_w	Sorbic acid
8.91	1	1	1
8.37	2	1	1
8.33	3	1	1
6.65	1	2	1
6.70	2	2	1
6.25	3	2	1
5.87	1	3	1
.			
.			
.			
4.37	3	5	3
3.93	1	6	3
4.12	2	6	3
4.15	3	6	3

The model for the analysis of variance would be specified as

$$\text{block} + a_w + \text{s.a.} + a_w \times \text{s.a.}$$

and the computer package should produce the full analysis shown in Example 6.2.

7

More on factorial treatment structure

7.1 MORE THAN TWO FACTORS

The extension to more than two factors in a factorial experiment is fairly simple in the calculation of the analysis of variance; the difficulty is in the interpretation. Suppose that we have an experiment in r blocks with three factors A, B and C, having a, b and c levels respectively. The treatment s.s. for the total set of abc experimental treatments can be subdivided into

(i) S.S. due to the main effect of each factor, e.g. for factor A

$$\text{main effect s.s.} = \frac{T^2_{A_1} + T^2_{A_2} + \ldots}{rbc} - \text{c.f.}$$

where T_{A_1} is the total of all units at level 1 of factor A, summing over all levels of B and C.

(ii) Interaction s.s. for each pair of factors, AB, AC and BC; e.g. for factors A and B,

$$\text{interaction s.s} = \frac{T^2_{A_1B_1} + T^2_{A_1B_2} + T^2_{A_2B_2} + \ldots}{rc}$$

$$-\text{c.f.} - \text{main effect s.s. } (A) - \text{main effect s.s. } (B)$$

(iii) The remainder of the treatment s.s. which we shall designate the s.s. due to the interaction between A, B and C.

We have already discussed the meaning of a main effect of a factor and an interaction between two factors. When there are more than two factors the definition of an interaction between two factors must be extended: an interaction between two factors is the variation of the differences between mean yields for different levels of one factor over different levels of the second factor, each mean yield being averaged over all others factors. It is clear that the components included in (i) and (ii) include all the variation of pairs of factors but do not include the variation arising from the possibility that the pattern of variation over combinations of levels of two of the factors, say A and B, is different for different levels of C. This could validly be described as the interaction between the pair of factors AB and the factor C, or equally the pair AC and B, or, to use a symmetrical definition, as the three-factor interaction ABC. This must be the source of variation of the component (iii).

7.2 FACTORS WITH TWO LEVELS

A special set of factorial experiments consists of experiments in which all factors have two levels. Such experiments are referred to generally as 2^n factorials. Many factors naturally have two levels, e.g. presence and absence of the factor concerned, or sex of an animal. In addition, factors can often be usefully applied at two widely different levels to give the experiment a good chance of detecting interaction with other factors. A special feature of 2^n experiments is that all main effects and interactions can be defined in terms of simple differences between treatment means.

The main effect of a factor is the difference between the means for the two levels, e.g. for factor A, the main effect is

$$A = \frac{T_{A_1} - T_{A_2}}{(N/2)}$$

where T_{A_1}, T_{A_2} are the totals over all other factors for levels 1 and 2 of factor A, and N is the total number of units in the experiment ($N/2$ for level 1, $N/2$ for level 2 of A).

As mentioned in section 6.4, the interaction between two factors is the difference between the effects of one factor measured at the two levels of the other. For factors A and B the interaction is

$$\frac{(T_{A_1B_1} - T_{A_2B_1}) - (T_{A_1B_2} - T_{A_2B_2})}{(N/2)}$$

where, for exmple, $T_{A_1B_1}$ is the total over all other factors for the combination of level 1 of factor A and level 1 of factor B.

We have defined the interaction in terms of differences between levels of A compared at the two levels of factor B. As we saw earlier the interaction is symmetrical in A and B and this definition is identical with

$$\frac{(T_{A_1B_1} - T_{A_1B_2}) - (T_{A_2B_1} - T_{A_2B_2})}{(N/2)}$$

and also with

$$AB = \frac{T_{A_1B_1} + T_{A_2B_2} - T_{A_1B_2} - T_{A_2B_1}}{(N/2)}$$

For a 2^n factorial it rapidly becomes easier to define more complex interactions algebraically than in words. Thus the three-factor complex interaction for factors A, B and C is defined algebraically as

$$ABC = (AB)_{c_1} - (AB)_{c_2} = \left\{ \frac{\begin{array}{c} [(T_{A_1B_1C_1} - T_{A_2B_1C_1}) - (T_{A_1B_2C_1} - T_{A_2B_2C_1})] \\ - [(T_{A_1B_1C_2} - T_{A_2B_1C_2}) - (T_{A_1B_2C_2} - T_{A_2B_2C_2})] \end{array}}{(N/2)} \right\}$$

or, again using a symmetrical form

$$ABC = \frac{\begin{array}{c} T_{A_1B_1C_1} - T_{A_2B_1C_1} - T_{A_1B_2C_1} - T_{A_1B_1C_2} \\ + T_{A_2B_1C_2} + T_{A_1B_2C_2} + T_{A_2B_2C_1} - T_{A_2B_2C_2} \end{array}}{(N/2)}$$

It may be helpful in thinking about interaction in 2^n factorials to have a

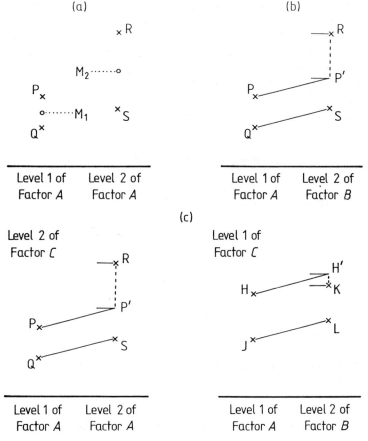

Fig. 7.1 Diagrammatic presentation of main effects, 2-factor interactions and 3-factor interactions.

diagrammatic respresentation of main effects and 2- and 3- factor interactions. This is given in Fig. 7.1. In Fig. 7.1(a) and (b) the points labelled P, Q, R and S represent yields, on a vertical scale, for the four combinations A_1B_1, A_1B_2, A_2B_1 and A_2B_2 respectively. In Fig. 7.1(a) the main effect of factor A is the difference $M_2 - M_1$ where M_2 is the mean of R and S, M_1 the mean of P and Q. In Fig. 7.1(b) the interaction between A and B is the difference $R - P'$ where PP' is drawn parallel to QS. In Fig. 7.1(c) the two diagrams show the AB interaction values separately for the two levels of C and the three-factor interaction ABC is the difference between R–P' and K–H', which in this case is equal to the sum of the two distances involved since K–H' is negative (i.e. the interaction AB has different signs for the two levels of C; in one case the combination of the second levels does 'better than expected' and in the other case worse).

In 2^n factorial experiments the main effect and interaction components of the full treatment s.s. can be calculated from treatment totals as in Chapter 6. There is also an alternative equivalent method of calculation based on the algebraic definitions of main effects and interactions given earlier. Each component s.s. is calculated as the square of the combination of totals given in the definition of the effect, divided by the total number of units in the experiment. Thus in a 2^3 factorial experiment with factors A, B and C and r blocks of 8 treatments the components may be calculated as follows

$$\text{s.s. main effect of } A = \frac{(T_{A_1} - T_{A_2})^2}{8r} \quad \text{or} \quad \frac{T^2_{A_1} + T^2_{A_2}}{4r} - \text{c.f.}$$

$$\text{s.s. main effect of } B = \frac{(T_{B_1} - T_{B_2})^2}{8r} \quad \text{or} \quad \frac{T^2_{B_1} + T^2_{B_2}}{4r} - \text{c.f.}$$

$$\text{s.s. } AB \text{ interaction} = \frac{(T_{A_1B_2} - T_{A_1B_2} - T_{A_2B_1} + T_{A_2B_2})^2}{8r}$$

$$\text{or} \quad \frac{T^2_{A_1B_1} + T^2_{A_1B_2} + T^2_{A_2B_1} + T^2_{A_2B_2}}{2r}$$

$$- \text{c.f.} - \text{s.s.}(A) - \text{s.s.}(B)$$

$$\text{s.s. } ABC \text{ interaction} = \frac{\begin{array}{c}(T_{A_1B_1C_1} - T_{A_2B_1C_1} - T_{A_1B_2C_1} - T_{A_1B_1C_2} \\ + T_{A_2B_2C_1} + T_{A_2B_1C_2} - T_{A_1B_2C_2} - T_{A_2B_2C_2})^2\end{array}}{8r}$$

or by subtracting the s.s. for all main effects and two-factor interactions from the total treatment s.s.

Example 7.1 Analysis of variance of a 2^3 factorial experiment

The data are taken from an experiment to investigate the water uptake of amphibia. Frogs and toads were kept in moist or dry conditions prior to the experiment. Half of the animals were injected with a mammalian water balance hormone. The percentage increases in weight after immersion in water for two hours were recorded for two animals in each of the eight treatment groups. No attempt was made to 'block' the animals.

Treatments	Replicates		Total
	1	2	
Toads wet control	+ 2.31	− 1.59	+ 0.72
Toads dry control	+ 17.68	+ 25.23	+ 42.91
Toads wet hormone	+ 28.37	+ 14.16	+ 42.53
Toads dry hormone	+ 28.39	+ 27.94	+ 56.33
Frogs wet control	+ 0.85	+ 2.90	+ 3.75
Frogs dry control	+ 2.47	+ 17.72	+ 20.19
Frogs wet hormone	+ 3.82	+ 2.86	+ 6.68
Frogs dry hormone	+ 13.71	+ 7.38	+ 21.09

Initial analysis of variance ignoring factorial nature of treatments:

	s.s.	d.f.	m.s.	F
Treatments	1510.28	7	215.75	6.25**
Error	276.05	8	34.51	
Total	1786.33	15		
c.f.	2357.10			

Two-way interaction tables:

	Wet	Dry	Total	Control	Hormone
Toads	43.25	99.24	142.49	43.63	98.86
Frogs	10.43	41.28	51.71	23.94	27.77
Total	53.68	140.52	194.20	67.57	126.63
Control	4.47	63.10			
Hormone	49.21	77.42			

$$\text{s.s. main effect of species} = \frac{142.49^2 + 51.71^2}{8} - \text{c.f.} = 515.06$$

$$\text{or} = \left[\frac{(142.49 - 51.71)^2}{16} \right]$$

$$\text{s.s. interaction species} \times \text{moisture} = \frac{43.25^2 + 99.24^2 + 10.43^2 + 41.28^2}{4}$$

$$- \text{c.f.} - \text{s.s. species} - \text{s.s. moisture}$$

$$= 39.50$$

$$\text{or} = \left\{ \frac{[(41.28 - 10.43) - (99.24 - 43.25)]^2}{16} \right\}$$

$$= \frac{(41.28 - 10.43 - 99.24 + 43.25)^2}{16}$$

Three-factor interaction by subtraction.

Analysis of variance of subdivision of treatment s.s.

	s.s.	d.f.	m.s.	F
Species	515.06	1	515.06	14.9**
Moisture	471.33	1	471.33	13.7**
Hormone	281.01	1	218.01	6.3*
S × M	39.50	1	39.50	1.1
S × H	165.12	1	165.12	4.8
M × H	57.83	1	57.83	1.7
S × M × H	43.43	1	43.43	1.3
Total	1510.28	7		
Error	276.05	8	34.51	

Main effect of species (= mean difference between toads and frogs)

$$= \frac{142.49 - 51.71}{8}$$

$$= 11.3\%$$

$$\text{Main effect of moisture} = 10.9\%$$

$$\text{Main effect of hormone} = 7.4\%$$

$$\text{s.e. of each of these differences} = \sqrt{(2s^2/8)} = 2.94\%$$

The only interaction that approaches significance, and therefore merits consideration, is that between hormone and species.

$$\text{Mean effect of hormone for toads} = \frac{98.86 - 43.63}{4} = 13.8\%$$

$$\text{Mean effect of hormone for frogs} = \frac{27.77 - 23.94}{4} = 1.0\%$$

$$\text{s.s. of each of these differences} = \sqrt{(2s^2/4)} = 4.15\%$$

Difference between effect of hormone for toads and frogs

$$(= \text{interaction between H and S})$$

$$= \frac{(98.86 - 43.63) - (27.77 - 23.94)}{4}$$

$$= 12.8\%$$

s.e. of this difference $= \sqrt{(4s^2/4)} = 5.87\%$

Summary of results

The individual treatment means (% weight increases) were as follows:

Toads wet control	0.36
Toads dry control	21.46
Toads wet hormone	21.26
Toads dry hormone	28.16
Frogs wet control	1.88
Frogs dry control	10.10
Frogs wet hormone	3.34
Frogs dry hormone	10.54

s.e. of difference of two means $= \sqrt{\dfrac{2s^2}{2}} = 5.87(8 \text{ d.f.})$

The only important interaction was that between species and hormone. Although this was not quite significant at the 5% level it is comparable in size to the main effect of hormone. The results are therefore presented as the main effect for (Dry-Wet) conditions and the four combinations for species × hormone (main effects being of little relevance when there is interaction).

Wet	Dry	Dry-Wet difference
6.7	17.6	10.9

s.s. of difference = 2.94

Dry condition before the experiment increased weight increases by 10.9%.

	Control	Hormone	Hormone-control
Toads	10.9	24.7	13.8
Frogs	6.0	6.9	0.9
Toad-frog difference	4.9	17.8	

s.e. of difference between two means in the body of the table = 4.15%. The hormone for toads showed 12.9% more increase in weight than for frogs. The difference has a s.e. of 5.87%.

EXERCISE 7.1 *Analysis of variance of a 2^3 factorial experiment*

The data are from a 2^3 factorial experiment in four blocks investing the effect of different amounts of zinc and copper added to basic diets of either maize of wheat for young chicks. The results are expressed as average weekly gains (g). The factors are two levels of zinc, two levels of copper and the two basic diets.

Carry out an analysis of variance ignoring the factorial nature of the eight treatments. Then split the seven degrees of freedom and s.s. for treatments into seven individual degrees of freedom and s.s. for main effects and interactions. Draw up tables of means with s.e.s for interesting comparisons. Summarise the conclusions to be drawn from the experiment.

Zinc	Copper	Basic diet	1	2	3	4	Treatment total
Z_0	Cu_0	M	21.6	22.6	25.7	23.3	93.0
		W	16.3	18.9	24.8	22.2	82.2
	Cu_0	M	23.7	22.0	26.7	25.0	97.4
		W	19.2	11.2	15.5	19.9	65.8
Z_1	Cu_0	M	21.2	26.3	28.1	23.0	98.6
		W	22.6	17.1	24.4	18.8	82.9
	Cu_1	M	20.9	22.3	26.7	21.5	91.4
		W	14.2	12.5	13.9	15.7	56.3
Block totals			159.7	152.7	185.8	169.4	667.6

(uncorrected total s.s. = 14534.94)

7.3 THE DOUBLE BENEFIT OF FACTORIAL STRUCTURE

The factorial structure of treatments is a major statistical contribution to the efficient design of experimental investigations. Factorial treatments offer two advantages by comparison with alternative sets of treatments, one of which we have already considered.

We have already seen that factorial structure and the corresponding analysis of the variation into components for main effects and interactions allow the possibility that effects of different factors are interdependent to be investigated. Without factorial structure it is difficult or impossible to investigate interactions and the interpretation of factor effects may be misleading. Conversely when inter-actions are small or non-existent the main effect estimates are of more general applicability because we then know that the effects of any one factor are the same whatever the levels of the other factors.

The other advantage of factorial experiments, which has not yet been discussed, and which is perhaps rather surprising, is that, when interactions are believed to be negligible, the factorial structure experiment is more economical than experiments looking at a single factor at a time. To demonstrate this we consider a situation where we have 24 experimental units available and we wish to look at the effects of three factors each of which has two levels.

If the three factors are denoted by A, B and C and their levels by a_0, a_1; b_0, b_1; c_0, c_1, then the simplest form of experimentation would be to do an experiment with each factor in turn with the two levels of the factor as two treatments, the other two factors being kept at their lower level. Since there are assumed to be no interactions the choice of levels for the other factors is actually irrelevant. With this form of experiment we have

(i) Exp 1: two treatments $(a_0 b_0 c_0)$ and $(a_1 b_0 c_0)$: four replications each
   Exp 2: two treatments $(a_0 b_0 c_0)$ and $(a_0 b_1 c_0)$: four replications each
   Exp 2: two treatments $(a_0 b_0 c_0)$ and $(a_0 b_0 c_1)$: four replications each
   The variance of the estimate of the difference between the two levels of A (or B or C) is $2\sigma^2/4$ where σ^2 is the variance per plot. Of course it would be more economical not to have $(a_0 b_0 c_0)$ in each of these experiments but instead to have a single experiment with each of the four distinct treatments from (i) represented equally. We now have a second form of experiment.

(ii) Four treatments $(a_0 b_0 c_0)$, $(a_1 b_0 c_0)$, $(a_0 b_1 c_0)$, $(a_0 b_0 c_1)$, with six replications each. The variance of the estimate of the difference between the two $2\sigma^2/6$. As expected, comparisons of the variance for (i) and (ii) shows that the second form of experiment is more economical since the variance for (ii) is 2/3 that for (i). However, the factorial form of experimentation is more economical still.

(iii) The eight factorial treatments $(a_0 b_0 c_0)$, $(a_0 b_0 c_1)$, $(a_0 b_1 c_0)$, $(a_0 b_1 c_1)$, $(a_1 b_0 c_0)$, $(a_1 b_0 c_1)$, $(a_1 b_1 c_0)$, $(a_1 b_1 c_1)$: three replications each. Now since the inter-actions are assumed to be zero, the difference $(a_1 - a_0)$ is estimated not only by $(a_1 b_0 c_0 - a_0 b_0 c_0)$ but also by $(a_1 b_1 c_0 - a_0 b_1 c_0)$, by $(a_1 b_0 c_1 - a_0 b_0 c_1)$ and by $(a_1 b_1 c_1 - a_0 b_1 c_1)$. Effectively the estimate of $(a_1 - a_0)$ is the mean of all observation for a_1 minus the mean of all observations for a_0. Since each of the two means is based on 12 observations the variance of the estimate of the difference between the two levels of a is $2\sigma^2/12$. Hence the factorial experiment (iii) is twice as precise as (ii) and three times as precise as (i).

The reason for this advantage is clear. In the factorial experiment all 24 units are involved in the $(a_1 - a_0)$ comparison, whereas in experiment (ii) only 12 units are used for this comparison, and in experiment (i) only 8 units. This benefit of factorial structure is referred to as 'hidden replication'. At first sight experiment (iii) provides only three-fold replication of the treatments. However, for the assessment of the main effect difference, $(a_1 - a_0)$ we have twelve-fold replication resulting from the apparent three-fold replication combined with the four-fold hidden replication from four (bc) combinations.

It should also be noted that the two advantages of factorial structure are retained, at a reduced level when, because some combinations are practically not feasible, incomplete factorial structures are used. Suppose, for example that the combination $(a_1 b_0 c_0)$ is known to be toxic for the experimental units. The other seven combinations:

$$a_0 b_0 c_0 \quad a_0 b_0 c_1 \quad a_0 b_1 c_0 \quad a_0 b_1 c_1$$
$$a_1 b_0 c_1 \quad a_1 b_1 c_0 \quad a_1 b_1 c_1$$

allow assessment of each two factor interaction (at one level of the third factor) and provide three-fold hidden replication for each main effect.

We can also note that the benefits of hidden replication apply more widely than the situation where there are no interactions at all. Consider the experiment in Example 7.1 again. The eight treatments are each replicated twice. The analysis of variance suggests that the three main effects and the species × hormone interaction are sizeable. We therefore summarized the results in a two-way table of means for the species and hormone combinations and a one-way table of means for the two moisture levels.

None of the interactions with moisture appear to be important. Therefore the comparison of wet-dry has eight-fold replication (two direct replicates × four hidden replicates). The four species-hormone means have four-fold replication (two direct × two hidden). Even if the two factor interactions involving moisture had been important there would still be hidden replication for each two-way table of means, because of the third factor, provided that the three-factor interaction was negligible.

7.4 MANY FACTORS AND SMALL BLOCKS

Often the number of factors which the experimenter would like to include, once he has been convinced of the virtues of factorial experimentation, is quite large and the question then becomes 'Is there a limit to how many factors can be included? There has been a lot of work in the development of complex designs for large factorials and the experimenter should discuss the situation with a statistician or consult a more advanced textbook such as Mead (1988) before deciding to limit the scope of the experiment. Although we cannot go into details of advanced designs in an introductory text we can outline some of the principles on which such experiments are based.

The most obvious problem arises because of the need to use blocking to obtain more precise comparison of treatment means. Usually the number of units per block should not exceed about 12 or the advantage of blocking to reduce variation between units within blocks will be lost. Often the block size will be smaller than 12. But to include 5 or more factors, even with only two levels each, produces at least 32 treatment combinations, too many to allocate one to each unit in a block of at most 12 units. The device used to overcome this problem is called **confounding** and we shall illustrate its use with two examples.

Example 7.2

This is a small artificial example to demonstrate the basic principles. Assume that the natural block size is four units and that we wish to examine the effects of three factors each at two levels. For this 2^3 experiment we have eight treatment combinations which we can write as $a_0b_0c_0$, $a_0b_0c_1$, $a_0b_1c_0$, $a_0b_1c_1$, $a_1b_0c_0$, $a_1b_0c_1$, $a_1b_1c_0$, $a_1b_1c_1$. The idea of confounding is to allocate treatments to units in different blocks in such a way that for all the more important effects or comparisons the differences between blocks 'cancel out' in the same way that they do for comparisons between treatments in a randomized block design. For example the main effect of factor A is the difference between the means for treatments with the a_1 level and the a_0 level. In symbols,

$$\text{effect } A = \frac{(a_1b_0c_0 + a_1b_0c_1 + a_1b_1c_0 + a_1b_1c_1)}{4}$$

$$- \frac{(a_0b_0c_0 + a_0b_0c_1 + a_0b_1c_0 + a_0b_1c_1)}{4}$$

This will be independent of differences between blocks if each of the two blocks contains two of the positive (a_1) terms in the definition of the effect and two of the negative (a_0) terms. By considering the various effects of primary importance, the three main effects and then the three two-factor interactions we can find an appropriate allocation of treatment combinations to blocks. The AB interaction effect is

$$\left[\frac{(a_1b_1c_0 + a_1b_1c_1)}{2} - \frac{(a_1b_0c_0 + a_1b_0c_1)}{2} \right]$$

$$- \left[\frac{(a_0b_1c_0 + a_0b_1c_1)}{2} - \frac{(a_0b_0c_0 + a_0b_0c_1)}{2} \right]$$

$$\left[\frac{a_1b_1c_0 + a_1b_1c_1 + a_0b_0c_0 + a_0b_0c_1}{2} \right]$$

$$- \left[\frac{a_1b_0c_0 + a_1b_0c_1 + a_0b_1c_0 + a_0b_1c_1}{2} \right]$$

And again we need two positive (a_1b_1 or a_0b_0) and two negative (a_1b_0 or a_0b_1) terms of the effect in each block.

We can construct the required two groups of four treatment combinations by ensuring that each group includes a_1 twice and a_0 twice, b_1 and b_0 twice each, a_1 and a_0 twice each, each (ab) combination once, each (ac) combination once and each (bc) combination once. The following division of treatment combinations is inevitable.

Block 1	Block 2
$a_0b_0c_0$	$a_0b_0c_1$
$a_0b_1c_1$	$a_0b_1c_0$
$a_1b_0c_1$	$a_1b_0c_0$
$a_1b_1c_0$	$a_1b_1c_1$

Here each level of each factor and each pair of levels for each of two factors appears equally often in each block. The only effect which is confounded with blocks in the sense that the apparent size of the effect will be affected by differences between blocks is the three-factor interaction. This is written in terms of the treatment combinations as

$$[(a_1b_1c_0 - a_1b_1c_1) - (a_1b_0c_0 - a_1b_0c_1)]$$
$$- [(a_0b_1c_0 - a_0b_1c_1) - (a_0b_0c_0 - a_0b_0c_1)]$$

i.e.

$$(a_1b_1c_0 + a_1b_0c_1 + a_0b_1c_1 + a_0b_0c_0)$$
$$- (a_1b_1c_1 + a_1b_0c_0 + a_0b_1c_0 - a_0b_0c_1)$$

in which all the positive terms are in block 1 and all the negative terms in block 2. In fact one way to ensure that all the important effects are unconfounded with blocks is, deliberately, the confound an effect of little interest with blocks.

In the design defined above we cannot estimate the three-factor interaction effect because it depends on the difference between the two blocks. If, for instance, we accidentally added 10 to each yield in block 2 the three-factor interaction would be apparently 40 less. But since we do not know what the difference between blocks is we cannot estimate the three-factor interaction. With the exception that we have no information, and no sum of squares, for the three-factor interaction the analysis of variance for a confounded experiment is exactly as for the corresponding unconfounded experiment.

Example 7.3

The previous example was on a small scale. We now consider a real design problem for an experiment to investigate the intercropping of maize and cowpea in India.

Block I	Block II	Block III
$C_1M_1D_1$	$C_1M_1D_1$	$C_1M_1D_1$
$C_1M_1D_2$	$C_1M_1D_2$	$C_1M_1D_2$
$C_1M_1D_3$	$C_1M_1D_3$	$C_1M_1D_3$
$C_1M_2D_1$	$C_1M_2D_1$	$C_1M_2D_1$
$C_1M_2D_2$	$C_1M_2D_2$	$C_1M_2D_2$
$C_1M_2D_3$	$C_1M_2D_3$	$C_1M_2D_3$
$C_2M_1D_1$	$C_2M_1D_1$	$C_2M_1D_1$
$C_2M_1D_2$	$C_2M_1D_2$	$C_2M_1D_2$
$C_2M_1D_3$	$C_2M_1D_3$	$C_2M_1D_3$
$C_2M_2D_1$	$C_2M_2D_1$	$C_2M_2D_1$
$C_2M_2D_2$	$C_2M_2D_2$	$C_2M_2D_2$
$C_2M_2D_3$	$C_2M_2D_3$	$C_2M_2D_3$

Fig. 7.2 Design structure (unrandomized) of $2 \times 2 \times 3$ factorial in three randomized blocks.

The experimenter wished to investigate the effects of four treatment factors. Threse were:

Cowpea variety (C)—two levels (c_1, c_2)
Maize variety (M)—two levels (m_1, m_2)
Cowpea density (D)—three levels (d_1, d_2, d_3)
Nitrogen fertilization (N)—three levels (n_0, n_1, n_2)

He realized that this was a rather ambitious set of objectives, particularly as he had available space for only 36 plots. When he approached the statistician for advice his provisional experimental plan was to include only the first three factors (C, M and D) in a randomized complete block design with three blocks of 12 plots per block. He believed that blocks of this size would be acceptablv homogeneous. His question was simply whether there was any way that the different nitrogen levels could be included.

The provisional design (unrandomized) is shown in Fig. 7.2 and the analysis of variance structure would be

Source of variation	d.f.
Blocks	2
Cowpea varieties (C)	1
Maize varieties (M)	1
Cowpea density (D)	2
CM	1
CD	2
MD	2
CMD	2
Error	22

We can add the nitrogen levels to the experiment in such a way that all the information in the provisional experiment would be retained with additional information on the nitrogen main effect and two-factor interactions provided that in each block of the final design:

(1) Each nitrogen level occurs four times
(2) Each combination of C and N levels occurs twice
(3) Each combination of M and N levels occurs twice
(4) Each combination of D and N levels occurs at least once

A further requirement is that in the complete experiment each of the 36 possible combinations occurs exactly once, and the construction of the design is easier if we also impose the condition that in each block each combination of C, M and N levels occurs once.

The reader may find it helpful to try to construct an appropriate design before looking at the solution in Fig. 7.3. Notice that we could not require the equal occurrence in each block of the nine D and N combinations because, obviously nine combinations cannot occur equally in blocks of 12 plots. However the N levels have been allocated so that each D N combination occurs at least once in each block, thus achieving the best possible spread of the nine combinations between the three blocks.

The modified design in Fig. 7.3 provides full information on the main effect of N and on the two-factor interactions of N with C and with M. In addition, because the $D \times N$ combinations are allocated to blocks as nearly equally as is possible, it would be possible to extract information about the $D \times N$ two-factor interaction, adjusting for block differences as discussed in section 5.7 and discussed in more detail in Chapter 11.

There are many extensions to the idea of confounding to deal with larger numbers of factors. Some of the concepts involved are:

(i) If we have so many factors that we connot afford to have more than one observation per treatment combination (as in Example 7.3) and if we can

Block I	Block II	Block III
$C_1M_1D_1N_0$	$C_1M_1D_1N_1$	$C_1M_1D_1N_2$
$C_1M_1D_2N_1$	$C_1M_1D_2N_2$	$C_1M_1D_2N_0$
$C_1M_1D_3N_2$	$C_1M_1D_3N_0$	$C_1M_1D_3N_1$
$C_1M_2D_1N_1$	$C_1M_2D_1N_2$	$C_1M_2D_1N_0$
$C_1M_2D_2N_2$	$C_1M_2D_2N_0$	$C_1M_2D_2N_1$
$C_1M_2D_3N_0$	$C_1M_2D_3N_1$	$C_1M_2D_3N_2$
$C_2M_1D_1N_2$	$C_2M_1D_1N_0$	$C_2M_1D_1N_1$
$C_2M_1D_2N_0$	$C_2M_1D_2N_1$	$C_2M_1D_2N_2$
$C_2M_1D_3N_1$	$C_2M_1D_3N_2$	$C_2M_1D_3N_0$
$C_2M_2D_1N_0$	$C_2M_2D_1N_1$	$C_2M_2D_1N_2$
$C_2M_2D_2N_1$	$C_2M_2D_2N_2$	$C_2M_2D_2N_0$
$C_2M_2D_3N_2$	$C_2M_2D_3N_0$	$C_2M_2D_3N_1$

Fig. 7.3 Design structure (unrandomized) of $2 \times 2 \times 3 \times 3$ factorial in three incomplete blocks.

assume that high level interactions are negligible then we can estimate the random variation using the sums of squares for the high level interactions in place of the 'error' sum of squares.

(ii) If we have yet more factors then again if high level interactions can be ignored all the important effects can be estimated even though not all the treatment combinations are used. This is known as **fractional replication.**

Thus the statistician has techniques for constructing experiments much more complicated than are discussed fully in this book and the experimenter who finds it difficult to reconcile the experiment he wants to do with the material available should seek advice. There are of course limits to the number of factors in addition to those imposed by practical considerations but it is important that an experimenter should keep an open mind about the possibility of his ideal experiment until he has seen a statistician.

7.5 THE ANALYSIS OF CONFOUNDED EXPERIMENTS

There is really only one point of difference between the analysis of an experiment with a complete factorial treatment structure in complete blocks and a con-founded design in which each block contains an incomplete set of treatment combinations. In the confounded design we calculate sums of squares only for those main effects and interactions which have been arranged to be not confounded. The actual calculation of the S.S. for many effect or interaction is exactly the same as for the same effect in a complete block design.

Thus for the design in Example 7.3 the sum of squares for the density main effect would be

$$\text{D main effect S.S.} = \frac{T_{D_1}^2 + T_{D_2}^2 + T_{D_3}^2}{12} - \text{c.f.}$$

where T_{D_1} is the total of the yields from the 12 plots with D_1. The sum of squares for the two-factor interactions for C and M could be calculated, as in section 7.1 as either

$$\text{S.S. CM interaction} = [T_{C_1M_1} - T_{C_1M_2} - T_{C_2M_1} + T_{C_2M_2}]^2/36$$

$$\text{or as } \frac{T_{C_1M_1}^2 + T_{C_1M_2}^2 + T_{C_2M_1}^2 + T_{C_2M_2}^2}{9} - \text{c.f.} \qquad - \text{S.S. (C)}$$

$$- \text{S.S. (M)}$$

Because the analysis is almost the same as that for a factorial treatment structure in complete blocks the information required for the analysis by statistical computer package is also in the usual format with columns for the yield, for blocks and for each treatment factor

For the four-factor confounded design of Fig. 7.3 the yields of maize (kg/ha) and the block and treatment coding are as follows

Block	C	M	D	N	Yield
1	1	1	1	0	2421
1	1	1	2	1	2955
1	1	1	3	2	3722
1	1	2	1	1	4289
1	1	2	2	2	4542
1	1	2	3	0	2054
1	2	1	1	2	4155
1	2	1	2	0	2435
1	2	1	3	1	2475
1	2	2	1	0	4509
1	2	2	2	1	4689
1	2	2	3	2	4883

Block	C	M	D	N	Yield
2	1	1	1	1	3562
2	1	1	2	2	3855
2	1	1	3	0	2475
2	1	2	1	2	4435
2	1	2	2	0	3528
2	1	2	3	1	3789
2	2	1	1	0	2835
2	2	1	2	1	3302
2	2	1	3	2	3615
2	2	2	1	1	5236
2	2	2	2	2	4396
2	2	2	3	0	3789

Block	C	M	D	N	Yield
3	1	1	1	2	4395
3	1	1	2	0	3062
3	1	1	3	1	3549
3	1	2	1	0	4369
3	1	2	2	1	4329
3	1	2	3	2	4442
3	2	1	1	1	3422
3	2	1	2	2	3435
3	2	1	3	0	2088
3	2	2	1	2	4002
2	2	2	2	0	2221
3	2	2	3	1	3142

The set of terms to be included in the analysis of variance includes all effects which are both important and known to be not confounded (entangled) with blocks. In this case a reasonable set of terms would be:

$$\text{Blocks} + C + M + D + N + C \times M + C \times D + C \times N + M \times D + M \times N$$

All sums of squares for the terms are calculated as usual. For manual calculation we would need various totals as follows

Blocks	1	2	3	Total
	43129	44817	42456	130402

	D_1	D_2	D_3	Total	N_0	N_1	N_2
C_1	23471	22271	20031	65773	17909	22473	25791
C_2	24159	20478	19992	64629	17877	22266	24486
M_1	20790	19044	17924	57758	15316	19265	23177
M_2	26840	23705	22099	72644	20470	25474	26700
Total	47630	42749	40023		35786	44739	49877

	C_1	C_2
M_1	29996	27762
M_2	35777	36867

The analysis of variance is

	S.S.	d.f.	m.c.	F
Blocks	246572	2	123286	
C	36354	1	36354	0.09
M	6155361	1	6155361	15.51
D	2475602	2	1237801	3.12
N	8475320	2	4237660	10.68
CM	306916	1	306916	0.77
CD	271122	2	135561	0.34
CN	35554	2	17777	0.04
DM	157810	2	78905	0.20
CN	305216	2	152608	0.38
Error	7142832	18	396824	
Total	25608659	71		

The results are very simply interpreted. There are large differences between maize varieties, between nitrogen levels and between densities. There is no convincing evidence of two-factor interactions (nor are the three-factor interactions CMD, CMN at all significant). Indeed apart from the three significant main effects all the other F-ratios are so small that there must be some concern whether the

Error M.S. is surprisingly large. The presentation of the relevant means is as follows

	M_1	M_2	Standard Error (18 d.f.)
	3209	4036	± 210
D_1	D_2	D_3	
3969	3562	3335	± 257
N_0	N_1	N_2	
2982	3728	4156	± 257

7.6 SPLIT PLOT EXPERIMENTS

Sometimes in factorial experiments the smallest units to which different levels of one factor can be applied may be rather large, certainly larger than would be suitable for application of the different levels of other factors. One example of this situation is found in cultivation experiments where each cultivation method must be applied to sizeable areas of ground whereas the other factors such as variety or plant spacing can be applied to much smaller units. Another example from experiments on milk yields might be a factorial experiment with different pasturages and milking methods as the two factors. The ideal unit for the different milking methods would be a single cow, but the smallest reasonable unit for different pasturages would be considerably larger than would normally be grazed by a single cow. To use the larger unit will tend to produce larger standard errors and may mean that the replication had to be reduced to a level at which the number of degrees of freedom for error is small and consequently estimation is not so precise.

What we would like to do, of course, is to fix the randomization of the cultivation × variety experiment so that all the plots with the same cultivation treatment occur together as shown in Fig. 7.4. Obviously it is not legitimate to do this within the normal randomization procedure, but it is perfectly reasonable to define a new randomization procedure so that if we have four cultivation treatments and three varieties we first divide each block into four large plots

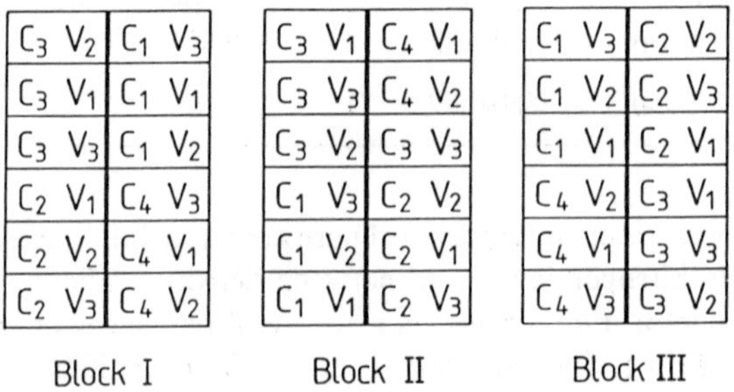

Block I　　　Block II　　　Block III

Fig. 7.4 Factorial experiment with four cultivation methods (C_1, C_2, C_3, C_4) and three varieties (V_1, V_2, V_3), the cultivation levels being grouped together in each block.

and allocate the four cultivation treatments randomly to these four plots; then we subdivide each large plot into three small plots and allocate the three varieties randomly to the three small plots, the randomization procedure would in fact give the same form of design as that shown in Fig. 7.4, the difference from the crude method of grouping similar plots being that we determined the form of the design before the allocation of the treatments to units within the framework of the design.

This form of design is called a split plot design, the large plots being termed **main plots** and the smaller plots, within main plots, **split plots**. We have only considered the use of split plot experiments for factorial experiments with two factors but the extension to more factors does not cause any complications. The factors are grouped into those to be applied the main plots and those for split plots. The randomization proceeds in two parts as before, the experimental treatments consisting of all combinations of levels of factors in the main plot group being allocated randomly to the main plots in each block and then all combinations of the split plot factors are allocated randomly to the split plots in each main plot.

As an example of this procedure we shall consider an experiment on celery in which three methods of seedling propagation and two levels of nutrient are compared at four harvest dates, the propagation and nutrient factors being applied to main plots and the harvests taken as split plot treatments. For each block we have six main plot treatments and the order of these is randomized, e.g. $5, 3, 4, 1, 6, 2$. Each main plot is divided into four split plots for the four different harvest times, H_1, H_2, H_3, H_4, and the times are allocated randomly to the split

P_3	P_3	P_3	P_3	P_2	P_2	P_2	P_2
N_0	N_0	N_0	N_0	N_0	N_0	N_0	N_0
H_1	H_4	H_3	H_2	H_3	H_2	H_1	H_4
P_2	P_2	P_2	P_2	P_1	P_1	P_1	P_1
N_1	N_1	N_1	N_1	N_0	N_0	N_0	N_0
H_3	H_1	H_4	H_2	H_1	H_2	H_3	H_4
P_3	P_3	P_3	P_3	P_1	P_1	P_1	P_1
N_1	N_1	N_1	N_1	N_1	N_1	N_1	N_1
H_3	H_2	H_4	H_1	H_2	H_4	H_3	H_1

(1) P_1N_0
(2) P_1N_1
(3) P_2N_0
(4) P_2N_1
(5) P_3N_0
(6) P_3N_1

Fig. 7.5 Layout of the first block in a split plot experiment with two main plot factors (P, N) and one split plot factor (H).

plots for each main plot, e.g. 1st main plot $1, 4, 3, 2$; 2nd main plot $3, 2, 1, 4$; 3rd main plot $3, 1, 4, 2$. The two-stage randomization procedure is used independently for each block. Figure 7.5 shows the full randomization for the first block.

7.7 ANALYSIS OF A SPLIT PLOT EXPERIMENT

Since the form and randomization procedure for a split plot experiment differ from those of a randomized block experiment we should expect that the form of the analysis of variance will also be different. In fact the analysis of variance is divided into two parts, one concerned with variation between main plots and the other with variation between split plots within main plots. If we consider the main plots only and calculate for each main plot the total of the yields for the split plots contained in that main plot, we have, effectively, a randomized block experiment and can calculate an analysis of variance with s.s. for blocks, main plot treatment and error from the main plot totals. Since this error s.s. is calculated in terms of main plot variation only, it is called the main plot error s.s. The variation between split plots within main plots can also be divided, in an analysis of variance, into variation between split plot treatments, the variation due to the interaction between split plot treatments and main plot treatments and a second error variation, the split plot error s.s. As it is obviously desirable that the two analyses should be comparable, both analyses are calculated in terms of the variation of the smallest unit, the split plot. This means that in the main plot analysis of variance, although all the calculations are in terms of main plot totals, when deciding the divisor for sums of squared totals, we use the number of split plots contained in the totals. The full analysis for a split plot experiment with r blocks, a main plot factor A with a different level, and a split plot factor B with b different levels, is as follows

(i) *Main plot analysis*

Source of variation	s.s.	d.f.	m.s.
Blocks	$\dfrac{B_1^2 + B_2^2 + \cdots}{ab} - \text{c.f.}$	$r - 1$	
Main effect, A	$\dfrac{T_{A_1}^2 + T_{A_2}^2 + \cdots}{rb} - \text{c.f.}$	$a - 1$	
Main plot error (a)	by subtraction	$(a-1)(r-1)$	s_a^2
Main plot totals	$\dfrac{Y_1^2 + Y_2^2 + \cdots}{b} - \text{c.f.}$	$ra - 1$	

where $Y_1, Y_2, \ldots$ are the totals of yields in the various main plots and the correction factor is the squared grand total of all yields divided by the total number of split plots in the experiment, rab.

(ii) *Split plot analysis*

Main effect, B	$\dfrac{T_{B_1}^2 + T_{B_2}^2 + \cdots}{ra} - \text{c.f.}$	$b - 1$	
Interaction AB	$\dfrac{T_{11}^2 + T_{12}^2 + T_{21}^2 + \cdots}{r}$ $- \text{c.f.} - \text{s.s.}\,(A) - \text{s.s.}\,(B)$	$(a-1)(b-1)$	
Split plot error (b)	by subtraction	$(r-1)a(b-1)$	s_b^2
Total	$\Sigma(y^2) - \text{c.f.}$	$rab - 1$	

Note that when calculating the sum of squares for split plot error (b), by subtraction, all s.s. calculated previously must be subtracted from the total s.s. including those in the main plot analysis: effectively we subtract from the total s.s., the main plot total s.s., the B main effect s.s. and the interaction s.s. Similarly the error d.f. are most easily obtained by subtraction.

The results of a split plot experiment will usually be presented as a two-way table of means (assuming there is some evidence of interaction) as in Table 7.1.

Now since some comparisons are entirely in terms of main plots, e.g. the comparison between the means of A_1 and A_2, averaging over the levels of factor B, and some are entirely in terms of split plots, e.g. the comparison between B_1 and B_2 for level A_1 of factor A, it is necessary to provide several different standard errors for making different comparisons in different parts of the two-way table. In fact five different standard errors are required if all possible comparisons are required.

Table 7.1 Presentation of treatment means for a split plot experiment

	Factor A			
	A_1	A_2	$A_3 \ldots A_a$	Mean
B_1	(5)		(7)	(3)
B_2				
B_3	(6)		(8)	(4)
$\vdots$				
B_b				
Mean	(1)		(2)	

(a) For the comparison between (1) and (2) (see Table 7.1), i.e. between two main plot treatment means which involves only main plot variation and therefore only the main plot standard error, the standard error of the difference between the two means is $\sqrt{[2s_a^2/(rb)]}$.

(b) For the comparison between (3) and (4) which average over main plots (consider main plots as blocks) and is therefore only concerned with the split plot error the standard error of the difference between the two means is $\sqrt{(2s_b^2/ra)}$.

(c) For the comparison between (5) and (6), again a strictly split plot comparison, the standard error of the difference between the two means is $\sqrt{(2s_b^2/r)}$.

(d) For the comparison of the difference $(5) - (6)$ with the difference $(7) - (8)$, an interaction comparison, the standard error of the difference of the differences between pairs of means is $\sqrt{(4s_b^2/r)}$.

(e) For either of the comparison (5) and (7) or (5) and (8) both variation between main plots and variation between split plots within main plots are involved. Consequently the standard error of the difference between the two means involves both s_a^2 and s_b^2 and is

$$\sqrt{\{2[(b-1)s_b^2 + s_a^2]/(rb)\}}$$

As we found in section 3.4 when a standard error involves two estimates of variation it is not possible to assign an exact number of degrees of freedom to the combined standard error for making t-tests or calculating confidence intervals. In this situation it is best to use the conservative lower values of degrees of freedom which will be that for the main plot error.

In general we have assumed that larger units were more variable than smaller units (this is the justification for blocking in field, trials) so that we would expect that the main plot error mean square, s_a^2, would usually be larger than the split plot error mean square, s_b^2. This means that comparisons between split plot treatments and interactions with main plot treatments will usually be more precise than comparisons between main plot treatments and also more precise than if the usual randomized block design had been used. This provides another reason for using split plot designs in certain investigations where the main effect of one factor is of little interest either because the effect has already been thoroughly investigated or because it is intrinsically of little importance. In such a case it may be thought desirable to improve the precision of comparisons between split plot treatments and also of interaction comparisons at the cost of less precise information on the comparisons between main plot treatments and this is what a split design may be expected to do. It is, however, only rarely that a split plot experiment will give a large improvement in precision of the split plot comparisons and split plot designs should be chosen for this advantage only in special cases and after some thought. In our view, the only convincing reason for using a split plot design is the practical requirement of large plots for some treatment factors.

The analysis of split plot experiments using statistical software is inevitably more complicated than the analysis of a simple randomization block factorial design. Software written specifically for analysis of experiments may have facilities for specifying the structure of the design so that the error terms for main plots and split plots are separated and the appropriate error terms used as divisor in F-tests. The best software will also calculate standard errors for comparisons between main effect and interaction means.

To use a simpler package to analyse a split plot or more complicated experiment the user must understand the model to be fitted and specify the error terms to be used as divisions in the variance ratio tests, in many packages the standard errors for comparisons at different levels must be calculated manually. This is illustrated by analysis of the following example:

Example 7.4 Analysis of a split plot experiment

The response of six varieties of lettuce, grown in frames, to various uncovering dates was investigated in a split plot experiment with four blocks. The main plot treatments were three uncovering dates and each main plot was split into six split plots for the six varieties.

Uncovering date	Variety	I	Block II	III	IV
x	A	11.8	7.5	9.7	6.4
	B	8.3	8.4	11.8	8.5
	C	9.2	10.6	11.4	7.2
	D	15.6	10.8	10.3	14.7
	E	16.2	11.2	14.0	11.5
	F	9.9	10.8	4.8	9.8
y	A	9.7	8.8	12.5	9.4
	B	5.4	12.9	11.2	7.8
	C	12.1	15.7	7.6	9.4
	D	13.2	11.3	11.0	10.7
	E	16.5	11.1	10.8	8.5
	F	12.5	14.3	15.9	7.5
z	A	7.0	9.1	7.1	6.3
	B	5.7	8.4	6.1	8.8
	C	3.3	6.9	1.0	2.6
	D	12.6	15.4	14.2	11.3
	E	12.6	12.3	14.4	14.1
	F	10.2	11.6	10.4	12.2

The data is set up by specifying codings for each experimental factor:

Unit	Yield	Block	Date	Variety
1	11.8	1	1	1
2	7.8	2	1	1
3	9.7	3	1	1
4	6.4	4	1	1
5	8.3	1	1	2
6	8.4	2	1	2
7	11.8	3	1	2
8	8.5	4	1	2
9	9.2	1	1	3
10	10.6	2	1	3
11	11.4	3	1	3
12	7.2	4	1	3
13	15.6	1	1	4
14	10.8	2	1	4
15	10.3	3	1	4
16	14.7	4	1	4

and so on for all 72 units in the experiments using the uncovering date codes $x = 1$, $y = 2$ and $z = 3$ and variety codes $A = 1$, $B = 2$, $C = 3$, $D = 4$, $E = 5$ and $F = 6$.

When using computer software to analyse split plot experiments it is advisable to write down the structure of the analysis of variance with degrees of freedom before attempting to fit the model.

For this experiment the analysis of variance required is of the form:

	Source	d.f.
Main Plots	Block	3
	Date	2
	Main Plot Error	6
Split Plots	Variety	5
	Date × Variety	10
	Split Plot Error	45
	Total	71

Since the main plot design is a randomized block design with 12 units, the main plot error is the block × date interaction. The model fitted is then:

Block + Date + Block × Date + Variety + Variety × Date

leaving the remaining residual variation as the split plot error. Simpler software packages will automatically use this split plot error as the divisor in F-tests in the analysis of variance table and so the printed tests for significance of main effects will be incorrect and must be calculated by the user. The complete analysis of variance is:

Source	d.f.	s.s	m.s.	F	P
Block	3	29.343	9.781	1.35	0.656
Date	2	38.003	19.002	2.62	0.845
Block × Date (Main Plot Error)	6	43.566	$7.261 = s_a^2$		
Variety	5	260.508	52.102	10.32	0.000
Variety × Date	10	163.698	16.370	3.24	0.003
Split Plot Error	45	227.277	$5.051 = s_b^2$		
Total	71	762.395			

Table of means:

Variety	Uncovering date			Mean
	x	y	z	
A	8.8	10.1	7.4	8.8
B	9.2	9.3	7.2	8.6
C	9.6	11.2	3.4	8.1
D	12.8	11.6	13.4	12.6
E	13.2	11.7	13.4	12.8
F	8.8	12.6	11.1	10.8
Mean	10.4	11.1	9.8	10.3

s.e. for comparing two date means $\quad = \sqrt{(2s_a^2/24)} = 0.78$ (6 d.f.)

s.e. for comparing two variety means $= \sqrt{(2s_b^2/12)} = 0.91$ (45 d.f.)

s.e. for comparing two varieties at a
single date $\qquad\qquad\qquad\qquad = \sqrt{(2s_b^2/4)} = 1.59$ (45 d.f.)

s.e. for comparing two differences
between two varieties for two dates $= \sqrt{(4s_b^2/4)} = 2.25$ (45 d.f.)

s.e. for comparing two dates, either $= \sqrt{[2(s_b^2 + s_a^2)/4 \times 6]} = 1.65$
for the same variety or for different $\qquad\qquad\qquad$ (no exact d.f.)
varieties

Summary of results: Differences between varieties varied with uncovering date. For uncovering date × varieties D and E outyielded all other varieties significantly; for y differences between varieties were small; for z varieties, D, E and

F gave significantly higher yields than varieties A, B and C. Although the mean yield declined between y and z for four of the six varieties the decline was significant for variety C only. (All significance statements refer to the 5% significance level.)

EXERCISE 7.2 *Analysis of variance of a split plot experiment*

Four strains of perennial ryegrass were grown as swards at each of two fertilizer levels. The four strains were, S23, New Zealand, Kent and X (a 'hypothetical' strain introduced to illustrate some points of statistical interest). The fertilizer levels were denoted by H, heavy and A, average. The experiment was laid out as four blocks of four whole plots for the varieties each split in two for the application of fertilizer.

The midsummer dry matter yields, in units of 10 1b/care, were as follows:

	Manuring	Block 1	2	3	4	Manuring total	Strain total
S23	H	299	318	284	279	1180	1983
	A	247	202	171	183	803	
New Zealand	H	315	247	289	307	1158	1952
	A	257	175	188	174	794	
X	H	403	439	355	324	1521	2281
	A	222	170	192	176	760	
Kent	H	382	353	383	310	1428	2220
	A	233	216	200	143	792	
Total		2358	2120	2062	1896		8436

Check a few of the 'manuring', 'strain' and 'replicates' totals. Carry out the analysis of variance. The uncorrected total s.s. is 2420734. Construct a table of means with appropriate standard errors and write a short report on the conclusions to be drawn from the experiment.

8

The assumptions behind the analysis

8.1 OUR ASSUMPTIONS

We have become accustomed, in the last four chapters, to calculating an analysis of variance. From the analysis of variance we obtain an estimate, s^2, of the random variance between plots within blocks which we then use to calculate standard errors of differences between the mean values for different treatments combinations. This is very different from our original use of, for example, the t-test to compare the difference between the means of two distinct samples with a hypothetical zero difference. Consequently some of the assumptions which appeared intuitively acceptable, even obvious, for the original simple situation need restating at this stage of the book, and need to be considered each time an analysis of variance is carried out.

The first assumption we made when introducing the t-test was that in the population from which each sample was drawn the observations were normally distributed. The second assumption was that the variability, as measured by the variance, was the same for both populations being sampled. This second assumption was necessary in order to 'pool' the variance estimates so that a t-test and confidence intervals could be calculated. The third assumption implicit in the analysis of variance came in later when we introduced the concept of blocks. It was then necessary to assume that the difference expected between the effects of the two treatments was the same for all the units irrespective of the block to which they belong. These assumptions are known as

(1) The assumption of normality,
(2) The assumption of homogeneity of variance, and
(3) The assumption of additivity.

We shall look more closely at each assumption in turn so as to be able to recognise those situations in which they are likely to break down. We shall look at empirical methods for detecting when the assumptions do not hold for a given set of data, and we shall discuss how we can modify our analysis when the assumptions are plainly untrue.

One, even more basic assumption that is often forgotten is that the data are, in some sense, correct. Unfortunately, recording and copying errors do occur, particularly when data have to be collected under difficult field conditions. Despite careful checking, errors can also occur when data are input to a computer. Obviously, such recording and transcribing errors are unlikely to result in observations satisfying the assumptions on which the analyses are to be based. Also, 'rogue' values or outliers can occur. These are observations that are outside the expected range of values. In a field experiment, one plot may be excessively fertile, perhaps because the treatment was wrongly applied or because of the past history of the plot. In an animal experiment, an animal may be diseased and because of this give an anomalous measurement.

What should be done about values that are so extreme as to raise doubts about their correctness or relevance? The first rule is that detailed records should be kept about the plots or animals in the experiment so that reasons can be sought if anomalous values are found later. A second rule is to decide, before the data are available to bias our decisions, what values of the observations should arouse suspicions that some of them are incorrect in relation to be the assumed model.

In deciding whether or not to reject observations as incorrect, the two possible consequences need to be considered. If the observation is 'correct' but is rejected, then valuable information is wasted and a bias introduced into the comparisons of the treatments. In particular, if occasional extreme values are a feature of the particular treatment then omitting them clearly introduced serious biases into the estimation of the effects of that treatment. The other danger—of accepting observations when they are in fact incorrect—has more obvious consequences. The decision to reject or not should always be based on what is known or has been recorded about the experimental units and the measurements made. If errors or recording are involved, the nature of the error, e.g. a missing decimal point, can sometimes be so obvious that a correction can be made. Otherwise a judgement to reject or not has to be made or analyses have to be conducted with and without the suspect observations and the results compared. Only if the conclusions are different from the two analyses does the decision to reject or not have to be made with any great care. Sometimes both analyses, with and without the suspect values, should be presented so that readers can make their own judgement on which analysis is most likely to be correct. The methods to be used in analysing incomplete data have already been discussed in Chapter 5.

8.2 NORMALITY

The assumption of normality means that it must be reasonable to suppose that the distribution of the population of observations from which we have obtained our sample of observations is approximately normal, looking something like Fig. 8.1. The idea of the distribution of a population of observations is another concept that has been neglected in our discussion of block experiments. To

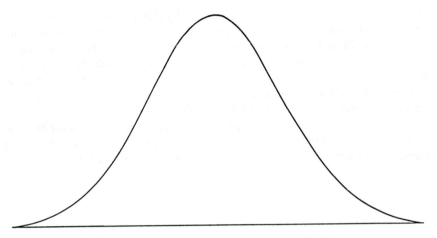

Fig. 8.1 Assumed form of population distribution.

recreate the idea, consider an experiment on pig growth where we are using litters as blocks. This means we believe that there would be substantial differences between the 'yields' we would get for animals from different litters even if the animals were to receive identical treatments; the differences between yields for pigs from the same litter receiving identical treatments would be expected to be smaller. Suppose we treated all pigs from all litters in the same way; then we would expect to get a cluster of yields for each litter, the clusters being more widely separated though with some degree of overlap. A possible set of results of four litters is shown in Fig. 8.2(a). If we now correct the yields for each litter to allow for the average differences between litters then we would get the distribution shown in Fig. 8.2(b). If we now think of many more litters of pigs and the yields under this identical treatment, correlated to allow for the average of each

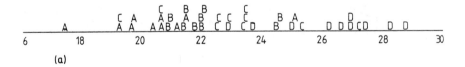

(a)

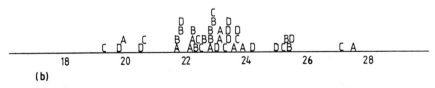

(b)

Fig. 8.2 (a) Hypothetical yields for four litters (A, B, C, D) of pigs. (b) Yields corrected to a common overall mean.

litter, then we get the population distribution of yields analogous to the population distribution when we were considering only a single sample of observations.
Thus, the distribution we are assuming to be approximately normal is that of
the yields of units treated identically, allowances being made for block differences. Note that we do not assuming that the original set of yields recorded in
an experiment should show a normal distribution. These yields will include
treatment differences as well as the block differences just discussed.

Now the nature of the normal distribution means that we can never have an
exactly normal distribution, particularly since the tails of the normal distribution
extend forever with increasingly small proportions of observations. In practice

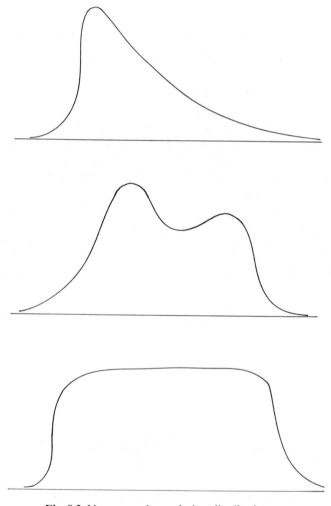

Fig. 8.3 Non-normal population distributions.

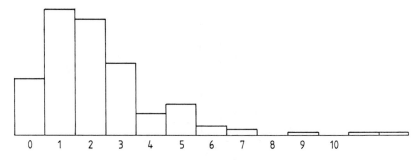

Fig. 8.4 Typical frequency distribution of counts.

as described above, we can always define upper and lower limits outside which we could reject an observation as 'impossible' or a 'freak' irrelevant to the main body of our investigation. In practical terms the assumptions of normality implies that the distribution shall be roughly symmetrical about a single peak and a shape something like that of Fig. 8.1 as opposed to the distribution curves of Fig. 8.3.

In most situations it is impossible to decide by examining the data whether the assumption of normality is reasonable and one has to rely on common sense in arguing whether the assumption is biologically likely. Fortunately the *t*-test is robust—that is, even when the underlying distribution is not approximately normal the significance points of the *t*–distribution approximately cut of the correct percentages of the distribution. The two common forms of measurement for which the assumption of normality is markedly untrue are when the measurement being analysed is a count, examples being weeds in a plot or insects on plants or animals, or a score from some simple scoring system, examples being tasting or colour assessments. Ecological investigations of the distribution of plants in randomly placed quadrats and of insects on randomly selected units show aggregation fairly consistently. That is, there is a relatively large number of quadrats or units with fairly few individuals and a smaller number of quadrats with large numbers of individuals; Fig. 8.4 shows a typical distribution of observations. This difficulty with analysing counts often disappears when the numbers involved are large. We shall be discussing methods of analysing counts in sections 8.5 and 8.6 and in Chapters 13 and 14.

Now consider the analysis of scores. The usual difficulty with scores is that the range of possible values is severely limited. On, say a five-point scoring system it is unlikely that a particular treatment will lead to observations for all five possible scores. Hence the population distribution will look like one of those in Fig. 8.5, none of which can seriously be argued to look like a normal distribution. It is of course true that what we ultimately compare after an analysis are the mean scores for two treatments and the distribution of a mean of scores will be rather closer to normality. However, a sound rule in using scoring to produce measurements to be analysed is not to use scales with small

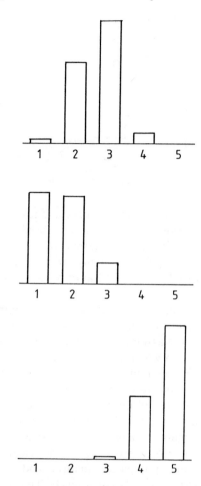

Fig. 8.5 Typical frequency distributions for three treatments when a five point scoring system is used.

ranges but use the mean of several scores for each experimental unit by getting several people to score the sample independently or by scoring several independent samples. This is effectively another way of extending the scoring scale but is often much easier than trying to assess a single sample in terms of a 20-point scale.

8.3 VARIANCE HOMOGENEITY

The assumption of homogeneity of variance requires that the different treatments applied to the units do not change the variability of the yields, but only the

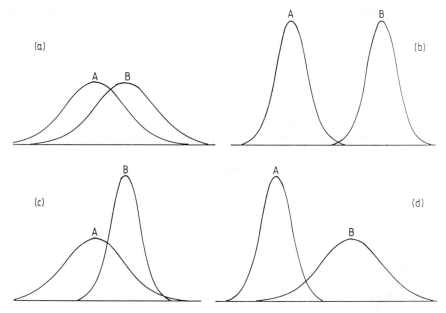

Fig. 8.6 Illustrations of variance homogeneity (a, b) and non-homogeneity (c, d) of treatment population distributions.

average yields. We must be able to assume that if we have two treatments A and B the two population distribution curves may look like those shown in Fig. 8.6(a) or (b) but not like those in Fig. 8.6(c) or (d). For investigations where the changes due to the treatments are small compared with the average value of yield this assumption is probably near enough to the truth for the analysis to proceed and since many agricultural experiments are of this nature this explains why many analyses of agricultural experiments never consider the assumption. Ignoring the possible failure of this assumption can, however, lead to completely nonsensical results. For example the following results were found in a paper on insecticide effects, submitted for publication:

	Mean egg counts per cubic metre
Control (no insecticide)	150
Insecticide A	85
Insecticide B	8
Insecticide C	1.2

Standard error of a mean = 6.4 (12 d.f.).

(If the full horror does not strike you at once, think back briefly to the meaning of a standard error.) What has happened is that the variances for the four treatments have been pooled during the analysis of variance without the experimenter looking at them to see whether it was realistic to assume that the population variances were the same. Obviously in this instance the variance for insecticide C cannot have been much more than about 3 and that for insecticide B not more than about 40. But the pooled estimate of variance, σ^2, must have been about 200 to give the standard error quoted and therefore the variance for the other treatments must have been of the order of 200–500, which is rather different from 3 or even 40.

Failures of assumption such as those of the insecticide data should have been detected before any analysis started. Two situations which should always induce doubt about the validity of the homogeneity assumption are (i) the analysis of counts and, more generally, (ii) the analysis of any variable for which the treatment means are considerably different. A factor of two for the difference of two treatment means or two block means is probably a limit which should not be exceeded without a formal check on the homogeneity of variances. To illustrate again the dangers of ignoring the assumption, consider the analysis of the following data on poppy counts.

Treatment	Block 1	Block 2	Block 3	Block 4	Total
A	538	422	377	315	1652
B	438	442	319	380	1579
C	77	61	157	52	347
D	115	57	100	45	317
E	17	31	87	16	151
F	18	26	77	20	141
Total	1203	1039	1117	838	4187

Analysis of variance

	s.s.	d.f.	m.s.
Blocks	12 877	3	
Treatments	6 41 024	5	
Error	39 799	15	2653
Total	6 93 700		

s^2, the estimate of the variance of the distribution of yields from different plots

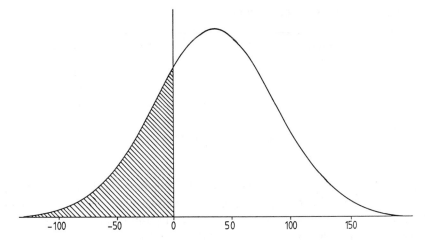

Fig. 8.7 Treatment F population distribution assumed in analysis of variance of poppy data.

with a common treatment, is 2653, giving s, the standard deviation per plot, of 51.5. Consider now treatment F which has a mean yield of 35 and is now supposed to have a standard deviation of yield ± 51.5. If we draw a normal distribution, as in Fig. 8.7, of yields with mean 35 and standard deviation 51.5 we find 25% of yields should be less than zero. For poppy counts! Clearly at least one assumption is wrong, probably in this case both the normality and the homogeneity of variance assumptions.

8.4 ADDITIVITY

The assumption that the difference between yields for two different treatments should on average be the same in each block is probably tenable, like the homogeneity assumption, for experiments in which treatment difference are not large. In addition, it is almost certainly acceptable when differences between blocks are small. However, when the blocks represent different situations and the treatments have sizeable effects then it seems to be biologically more reasonable to assume that treatment effects are proportional rather than additive. In other words it seems more likely that one treatment will give results about 50% higher than a second treatment regardless of the absolute level of the results than that one treatment will give values about 20 higher than the other regardless of the absolute level. An example of the sort of results that occur in practice is the following data on insect counts:

Type of trap	Site 1	Site 2	Site 3	Total
A	11	47	105	163
B	13	62	170	245
C	15	87	165	267
D	7	43	86	136
Total	46	239	526	811

Despite the strong pattern in the values for different traps at each site the F-ratio for testing differences between traps is nowhere near significant simply because the effects of traps are not additive but approximately proportional (look at the ratio of C to D, which is almost exactly 2).

An interesting comparisons can be drawn with some forms of interaction in factorial experiments. Suppose that the above figures were means over several blocks for different insecticides (instead of sites) on different varieties of some species of plant (instead of trap types) so that the results looked like:

Variety	Insecticide 1	2	3	Mean
A	11	47	105	54
B	13	62	170	82
C	15	87	165	89
D	7	43	86	45
Mean	12	60	132	

Then, ignoring the probability that because of the large differences between the 12 treatment combinations the assumption of homogeneity of variance is invalid, the conclusions liable to be drawn from the analysis of variance are that there is a large main effect of insecticide, a relatively small main effect of variety and a large degree of interaction between the two factors. Obviously in view of the proportional nature of the difference between the variates this is an unnecessarily complex summary of the results and an analysis which emphasized the simple proportional effects of the combinations of factor levels would be much to be preferred. In fact it is common to find that when a modification of the analysis is made to allow for the failure of the additivity assumption then the pattern and the interpretation of results becomes simpler by virtue of the reduction in interaction effects.

8.5 TRANSFORMATIONS OF DATA FOR THEORETICAL REASONS

If the analysis of variance cannot be validly applied to the original data what can we do? Often the appropriate solution is to change the scale on which the measurements were recorded. The measurements we take are often on a some-what arbitrary scale. For example, consider an experiment on the effect of growth substances on plants. The experimenter may be interested in the effect of different substances on leaf size. There is a choice of several measurements of size: length, breadth, area or weight. If each is tried in turn then there will be differences in the analyses and in the conclusions to be drawn from the analyses. It is difficult to say *a priori* that one measurement is right and the other wrong; they are just, inevitably, different.

From the statistical point of view it is virtually certain that the assumptions will not be valid for all analyses. Usually leaves maintain their shape so that over time breadth will increase roughly proportionally to length, and area,

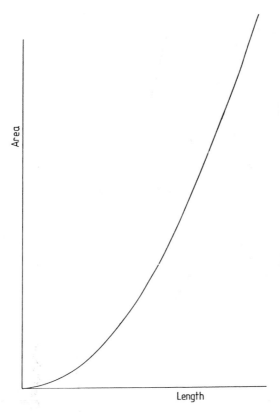

Fig. 8.8 Hypothetical area–length relationship.

being proportional to breadth × length, will increase approximately as the square of the length, the relationship being roughly that shown in Fig. 8.8. Suppose now that the assumptions for the analysis of leaf length are valid. Then it is unlikely that any of the assumptions will be valid for the leaf area analysis. Take the normality assumption. If the population distribution of leaf lengths looks like Fig. 8.9(a), the population of leaf areas will look like Fig. 8.9(b). Obviously larger leaves have the bigger areas but a difference of 1 cm in length for two small leaves will correspond to a smaller difference in area between the leaves than will the same difference of 1 cm in length for two large leaves. Consequently in the area distribution, the smallest values will get 'bunched up' compared with the larger values as shown in Fig. 8.9(b). The same disruption happens to the homogeneity assumption. If we look again at Fig. 8.8 we see that variation of 1 cm in leaf length between 3 and 4 cm corresponds to variation from 5 to 8 cm^2 in leaf area but the same variation of 1 cm in leaf lengths between 5 and 6 cm corresponds to variation between 12 and 18 cm^2 in area. Thus equal amounts of variation on the length scale correspond to unequal

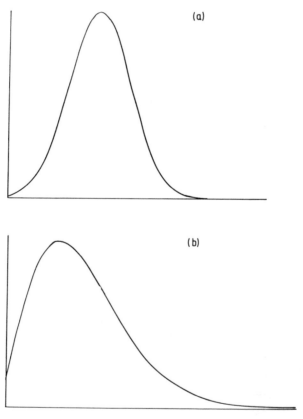

Fig. 8.9 Population distributions (a) for length and (b) for area of leaves.

amounts of variation on the area scale. The experimenter might have ignored all measurements except area and analysed area only. If we believe from biological considerations that the leaf lengths satisfy the assumptions for the analysis then we should not analyse measurements of area but instead analyse the square root of area as this should be approximately on the same scale as length.

The important point is that the natural scale is not always the simplest for analysis and interpretation. Let us return to the insect counts for which the treatment effects were proportional rather than additive and suppose that we have a treatment which doubles the mean of the distribution of counts for each situation in which it is applied. Thus if the count in the control situation was 2, the treatment would be expected to increase it to 4; if 10 to 20; if 100 to 200. Put like this the natural scale of measurement seems simple as well as natural. But the analysis of variance works in terms of differences and the corresponding differences are 2, 10 and 100 which are certainly not equal. Clearly the assumption of additivity has failed. If we change to log (count), however, we would expect the differences to be equal because the log scale is the scale on which multiplicative changes become additive.

	Natural scale			Log scale		
	Control	Treated	Difference	Control	Treated	Difference
Set 1	2	4	2	0.3010	0.6020	0.3010
Set 2	10	20	10	1	1.3010	0.3010
Set 3	100	200	100	2	2.3010	0.3010

Another situation where the log scale produces a simplification is the growth of young seedlings. In the early stages of growth the rate of growth appears to be proportional to the current size of the plant. That is if we have two plants, one twice as big as the other, then, in these early stages of growth the larger one will grow at twice the rate so that it will still be twice as big. The actual weight increase will also be twice as big. However, if we consider a log transformation, the increase of log weight will be the same for both plants. If we are comparing the effects of different treatments during the early stages of growth these treatments may have different initial effects on growth and subsequently the plants may grow at the same rate or at different rates. To examine the possibility of different rates of subsequent growth the experimenter takes measurements of plant weight at different times. If he simply compares the changes in weight, he will have to make allowances for the fact that plants which are larger at the beginning of the period will naturally grow faster than the smaller plants. If he uses log weight and compares the differences in log weight then, under the null hypothesis that the initial treatments do not affect the pattern of growth during this subsequent period, the differences in log weight should be the same for the different treatments.

Hoping that the ideas of working on a log scale or a square root scale are now germinating in the reader's mind, if not actually blossoming and producing fruit, we now suggest that, when the assumptions in the analysis of variance break down, transformation of the raw data to a different scale may yield an analysis for which the assumptions are valid. From our earlier discussion of how assumptions fail, we can expect particular transformations to be successful in certain situations. Obviously when treatment or block effects are likely to be proportional rather than additive then use of the log scale will often lead to a more valid analysis than the natural scale. When the measurement of interest is some form of count or an area, then the square root transformation is often effective. A theoretical argument in favour of considering square roots of counts is that area has the dimensions of $(length)^2$ and therefore transformation of counts to a square root scale produces a form of 'linear count'. This rather vague philosophical argument is paralleled by an algebraic argument too complex for this book. To illustrate the effect of square root and log transformations on, respectively, the poppy count and insect count data, we give the original and transformed data together:

Poppy counts

| Treatment | Untransformed | | | | Square roots of counts | | | |
	Block 1	Block 2	Block 3	Block 4	Block 1	Block 2	Block 3	Block 4
A	538	422	377	315	23.2	20.5	19.4	17.7
B	438	442	319	380	20.9	21.0	17.9	19.5
C	77	61	157	52	8.8	7.8	12.5	7.2
D	115	57	100	45	10.7	7.6	10.0	6.7
E	17	31	87	16	4.1	5.6	9.3	4.0
F	18	26	71	20	4.2	5.1	8.4	4.5

Insect counts

| Trap type | Untransformed | | | Log counts | | |
	Site 1	Site 2	Site 3	Site 1	Site 2	Site 3
A	11	47	105	1.04	1.67	2.02
B	13	62	170	1.11	1.79	2.23
C	15	87	165	1.18	1.94	2.22
D	7	43	86	0.85	1.63	1.93

Just glancing over the transformed and untransformed data there seems to be a much more consistent pattern about the transformed data though there is, of course, still plenty of random variation.

If we consider again the particular assumptions whose apparent failure led us to consider a transformation then we see that the transformed data seem to agree with the assumption more nearly, but also the other assumptions also seem to be more realistic for the transformed data. Thus for the poppy counts for which the homogeneity assumption seemed invalid, the transformed values on the square root scale seem credibly homogeneous, the within-treatment differences being very similar with ranges about 4 or 5. However, the differences between treatments seem very similar in the different blocks on the transformed scale whereas on the original scale the differences in blocks 3 and 4 seem clearly smaller than in blocks 1 and 2; the transformation also appears to improve the additivity assumption. Similarly for the insect count data the transformation achieves the desired improvement in the additivity assumption but also appears to make the variation more homogeneous for the four trap types. This double gain should not be surprising. Suppose there is a biological justification for believing that there is a scale of measurement on which all the assumptions are satisfied. Then if we select a transformation to that scale to make one assumption valid this implies that assumption and the other assumptions would not be valid on the original scale and that the same transformation might improve the other assumptions also.

8.6 A MORE GENERAL FORM OF ANALYSIS

The use of transformed data as the basis of the statistical analysis is sensible when there are reasons to believe that there might be a transformed scale, which is biologically meaningful, for which it is credible that all the assumptions required for the analysis should be correct. The logarithmic transformation with its natural interpretation of percentage effects and proportional variation is the most obviously sensible example; the square root transformation is less obvious but still sensible in some situations. However, there are also other situations where our understanding of the biological processes suggest strongly that the statistical assumption would be incorrect, and where no simple transformation would 'correct' the assumptions. Most examples of such difficult situations involve frequency data and we shall discuss some methods for analysing such data later in Chapters 13 and 14.

There are now available new forms of statistical analysis, analogous to the analysis of variance, which allow us to analyse data making much more general assumptions than those for the analyses of variance which we have discussed so far. Consider the three basic assumptions separately again. The assumption of additivity concerns the construction of the model for the mean values of our measurement. We assume that it should be expected that

$$\begin{array}{llll}
\text{Observed} & = \text{Overall} & + \text{block} & + \text{treatment} \\
\text{measurement} & \text{Mean} & \text{effect} & \text{effect}
\end{array}$$

The other two assumptions relate to the random variation, or error term. We assume that the error terms are homogeneous and normally distributed. The assumptions about the mean values (additivity) and error terms (homogeneity and normality) do not have to be considered simultaneously in the new, more general forms of analysis.

The benefit of separate modelling of the mean and the error term distribution may be perceived by considering again the use of the logarithmic transformation. If we decide to analyse $\log(y)$ then we assume that for the original measurement, y, the treatment effects are to be interpreted as producing the same percentage change in each block, and the error terms are proportional to the treatment-block mean. An analysis on the original measurement scale would require assumptions that treatment effects produce the same absolute change in each block and the error terms are independent of the treatment–block means. With separate modelling of mean and error term we could assume that the treatments have constant percentage effects while the error terms are independent of the treatment–block means. Or it could be assumed that treatments produce the same absolute change in each block but that error terms are proportional to the treatment–block mean.

As we said earlier these more general models are most beneficial for frequency data and examples of these models are discussed in Chapter 14. A general description of generalized linear models is given by McCullagh and Nelder (1990).

8.7 EMPIRICAL DETECTION OF THE FAILURE OF ASSUMPTIONS AND SELECTION OF APPROPRIATE TRANSFORMATIONS

As we have said many failure of assumptions should be anticipated on theoretical grounds. However, when there are apparently no theoretical reasons for choosing a particular transformation it is still necessary to examine the data and consider empirically whether the assumptions are credible. The assumption whose failure is most easily detected is that of homogeneity of variance and much of this section will be concerned with detecting homogeneity. It is important simultaneously to consider evidence for non-additivity though correcting for failure of the homogeneity assumption also frequently improves the additivity assumption.

There are two approaches that are useful for detecting failures of assumptions.

The first is to look at the residuals after the analysis of variance to see whether there is any form of pattern in them. The residuals, which we discussed briefly in Chapter 5, are the deviations of the actual observed yields from the expected yields based on block and treatment average yields. To use the residuals the analysis of variance has to have been calculated, possibly invalidly, on the original scale of measurement. Even when the residuals have been calculated there are no clear rules of how to infer from a particular pattern in the residuals that a particular failure of the assumptions has occurred. Nevertheless, patterns are frequently obvious to the eye. In the poppy count and insect count data if the analysis had been calculated before thinking about assumptions the following residuals should bring the experimenter back to his assumptions:

| | Poppies | | | | | Insects | | |
	1	2	3	4		1	2	3
A	+100	+11	−47	−61	A	+13	+1	−13
B	+18	+49	−87	+22	B	−11	+10	+21
C	−35	−24	+59	+2	C	−20	+4	+15
D	+11	−20	+10	+3	D	+18	+6	−23
E	−46	−5	+36	+15				
F	−42	−7	+31	+22				

For the poppies the pattern of + and − is very nearly the same or completely reversed in each column; also the larger residuals are in the earlier rows, for which the treatment means are higher, suggesting heterogeneity. For the insects the larger residuals are found for the more extreme sites (1 and 3) and for each row the residuals in these columns are of opposite signs, suggesting non-addditivity.

The other method for empirical detection of failure of assumptions is cruder and is applied before the analysis; it is concerned primarily with detecting heterogeneity patterns, with non-additivity of second importance. For each treatment, and possibly also for each block, the mean yield and the range of the yields are calculated. If the ranges are reasonably constant and show no relationship with the means then no transformation is indicated. If the ranges are not constant then we calculate range/mean for each treatment (or block) and also range/$\sqrt{}$mean. If range/mean is constant then a log transformation is appropriate; if range/$\sqrt{}$mean is constant, a square root transformation. In each case a further check is provided by applying the proposed transformation and looking at the consistency of the ranges of the transformed values. It is usually found that the ranges of the transformed values are more constant than the values of range mean for the untransformed data. To examine the method let us resurrect, for the penultimate time, the poppy and insect count.

Poppy counts

Treatment	Mean	Range	Range/mean	Range/$\sqrt{\text{mean}}$
A	413	223	0.54	11.0
B	395	123	0.31	6.2
C	87	102	1.17	10.9
D	79	70	0.89	7.9
E	38	71	1.87	11.5
F	35	59	1.69	10.0

Block	Mean	Range	Range/mean	Range/$\sqrt{\text{mean}}$
1	200	521	2.60	37.0
2	173	416	2.40	31.6
3	186	300	1.61	22.0
4	140	364	2.60	31.0

Looking first at the calculations for treatment means and ranges it is clear that the range of values for a treatment is related to the mean; the ranges are not constant, there being almost a fourfold difference, so that some transformation is needed to reduce the variation for treatments with larger mean yields. The values of range/mean show even less constancy (a sixfold difference) but the values of range/$\sqrt{\text{mean}}$ are much more nearly constant; there is no trend with the treatment mean and the biggest difference is 6.2 to 11.5, less than a factor of two. The ranges of the square root transformed values (given in section 8.5) are 5.5, 3.1, 5.3, 4.0, 5.3 and 4.2 respectively. In this case the block ranges and means offer little information on possible informations. This is because differences between blocks in the original data are small.

Insect counts

Trap type	Mean	Range	Range/mean	Range/$\sqrt{\text{mean}}$
A	54	94	1.74	12.8
B	82	157	1.91	17.3
C	89	150	1.69	15.9
D	45	79	1.76	11.8
Site				
1	12	8	0.67	2.31
2	60	44	0.73	5.68
3	132	84	0.64	7.31

This time the clearest evidence comes from the site ranges and means though the calculations for trap type ranges and mean show the same pattern. The ranges vary very closely with the means; range/mean is remarkably constant, range/$\sqrt{\text{mean}}$ rather less so, and the obvious transformation is to the log scale.

Looking at the transformed values in section 8.5 we find the ranges for trap types to be 0.98, 1.12 1.04, 1.08, and for sites 0.33, 0.31 and 0.30.

The advice to an experimenter about to analyse experimental data can be summed up in three rules:

(a) **Before** you start to analyse the data think about the biology of the measurements and whether you would expect the assumptions to be satisfied and if not, how not? Think about the reasonableness of each of the three assumptions for your measurements. Be prepared to ask for statistical advice about generalized linear models.

(b) **When doing** the analysis (or presenting the data to a computer for analysis) look at the figures to see whether the assumptions appear credible and if in any doubt calculate ranges, and investigate the relationship between treatment means and ranges. If there is a pattern then a transformation is probably needed.

(c) **After** the analysis, if you have not thought about the assumptions before (and even if you have it may be a useful check) calculate the residuals and look at them to see whether there is any pattern.

Example 8.1 Transformation of data in a Latin square design

An experiment on the repellent effect of lime sulphur on the honeybee consisted of eight treatments (A, B, ... H) replicated spatially as a Latin square within an experimental chamber. The eight treatments were different concentrations of the lime sulphur. The data, taken from an early edition of Finney's *Probit Analysis*, are values of uptake of sucrose in mg/cell. An analysis of the original and the table of treatment means are given below.

Original data

										Row total
D	57	C	84	F	87	H	130	E	43	A 12	B 8	G 80	501
E	95	B	6	H	72	A	4	D	28	C 29	G 72	F 114	420
B	8	H	127	A	5	E	114	G	60	F 44	C 13	D 39	410
H	69	D	36	E	39	C	9	A	5	G 77	F 57	B 14	306
G	92	E	51	D	22	F	20	C	17	B 4	A 4	H 86	296
F	90	A	2	C	16	G	24	B	7	D 27	H 81	E 55	302
C	15	F	69	G	72	B	10	H	81	E 47	D 20	A 3	317
A	2	G	71	B	4	D	51	F	71	H 76	E 61	C 19	355

Column totals

| 428 | 446 | 317 | 362 | 312 | 316 | 316 | 410 | 2907 |

Treatment totals

A	B	C	D	E	F	G	H
37	61	202	280	505	552	548	722

Analysis of variance

	s.s.	d.f.	m.s.	F-ratio
Rows	4768	7	681.1	
Columns	2808	7	401.1	
Treatments	56160	7	8022.9	21.1
Error	15994	42	380.8	
Total	79730	63		

Table of treatment means

A	B	C	D	E	F	G	H
4.6	7.6	25.2	35.0	63.1	69.0	68.5	90.2

Standard error of a treatment mean $= \sqrt{(380.8/8)} = 6.90$

As mentioned by Finney, examination of the data arouses suspicion that the assumptions underlying the analysis are not satisfied. The analysis of variance estimates the standard deviation of individual observations in the experiment as $\sqrt{380.8} = 19.5$. This standard deviation obviously cannot apply to the lower values in the data because although low, they clearly have to be positive. The large range of the treatment means also suggests that the assumptions of the analysis of variance may not be satisfied. A treatment mean of 4.6 (A) should be expected to have a smaller standard error than a mean of 90.2 (H). Thus the variation in the data may well be different for different treatments. To explore this we examine the relationship between the ranges of treatment values and the corresponding treatment means. In practice, and certainly with this particular set of data, these calculations would probably have been done before the analysis of variance is attempted.

Treatment mean	Range	Range/mean
4.6	10	2.17
7.6	10	1.32
25.2	75	2.98
35.0	37	1.06
63.1	75	1.19
69.0	94	1.36
68.5	68	0.99
90.2	61	0.68

Apart from the large range for the third treatment (due to one extreme value) the ratio of range:mean stays reasonably constant, indicating that the variation is proportional to the mean, i.e. that the variation of log (value) should be fairly uniform. Accordingly we transform the data to the log scale.

Log_{10} data

								Row total
G 1.90	D 1.76	C 1.92	F 1.94	H 2.11	E 1.63	A 1.08	B 0.90	13.24
F 2.06	E 1.98	B 0.78	H 1.86	A 0.60	D 1.45	C 1.46	G 1.86	12.05
D 1.59	B 0.90	H 2.10	A 0.70	E 2.06	G 1.78	F 1.64	C 1.11	11.88
B 1.15	H 1.84	D 1.56	E 1.59	C 0.95	A 0.70	G 1.89	F 1.76	11.44
H 1.93	G 1.96	E 1.71	D 1.34	F 1.30	C 1.23	B 0.60	A 0.60	10.67
E 1.74	F 1.95	A 0.30	C 1.20	G 1.38	B 0.85	D 1.43	H 1.91	10.76
A 0.48	C 1.18	F 1.84	G 1.86	B 1.00	H 1.91	E 1.67	D 1.30	11.24
C 1.28	A 0.30	G 1.85	B 0.60	D 1.71	F 1.85	H 1.88	E 1.79	11.26

Ccolumn total

12.13	11.87	12.06	11.09	11.11	11.40	11.65	11.23	92.54

Treatment totals

	A	B	C	D	E	F	G	H
	4.76	6.78	10.33	12.14	14.17	14.34	14.48	15.54
Range	0.78	0.55	0.97	0.46	0.47	0.76	0.58	0.27

Analysis of variance on transformed scale

	s.s.	d.f.	m.s.	F-ratio
Rows	0.06004	7	0.0858	
Columns	0.1545	7	0.0221	
Treatments	13.7305	7	1.9615	50.9
Error	1.6201	42	0.0386	
Total	16.1055	63		

Table of treatment means

A	B	C	D	E	F	G	H
0.60	0.85	1.29	1.52	1.77	1.79	1.81	1.94

Standard error of a treatment mean $= \sqrt{(0.0386/8)} = 0.069$

Notes

The results of using a transformation so that the assumptions of the analysis are satisfied are interesting. The *F*-ratio for treatments increases considerably. Differences between mean values of A and B and of C and D are now significant though the difference between G and H is no longer significant.

However, the use of a transformation is justified not by the improved (or otherwise) results, but by the fact that inferences deduced from the results are valid. Often the use of a correct analysis will also give more clear-cut results, as in this case.

8.8 PRACTICE AND PRESENTATION

There is an understandable reluctance to use transformations, arising mainly from the difficulties of presenting results on the transformed scale. It must be admitted that for many readers the presentation of results for different treatments as mean log weights instead of mean weights erects a barrier to understanding out of all proportion to the difference in the pattern of results. But the transformed results must be used if valid statistical comparisons are to be made. A reasonable solution is to present two sets of results—first, the means calculated on the untransformed scale with no standard errors for comparisons of the means, and alongside these means the corresponding set of means on the transformed scale with standard errors. The advantage of this system is that the reader can look first at the untransformed means and can also draw statistical conclusions about particular differences by looking at the corresponding differences on the transformed scale. A further advantage of presenting both is that an explanation of why the data were transformed for analysis is necessary, enabling the less numerate reader to begin to understand the principles of the analysis behind the results!

To illustrate the proposed form of presentation of results we present the results for Example 8.1 in this form:

Treatment	Untransformed weights (mg/cell)	Log weights
A	4.6	0.58
B	7.6	0.58
C	25.2	1.29
D	35.0	1.52
E	63.1	1.77
F	69.0	1.79
G	68.5	1.81
H	90.2	1.94
Standard error of difference between any two treatment means:		0.069 (42 d.f.)

Notice that the transformation has reversed the order of two of the treatments, F and G. This generally only occurs when the treatment means are very close together on both scales.

EXERCISE 8.1 *Transformation of data in a randomized block design*

In an experiment on weed control in red beet, emergence counts were taken for an area of 3 square feet in each plot. The data and analysis are given below.

It is often recommended, in situations of this kind, that the data should be transformed to square roots before analysis. Examine the data for indications that any transformation is needed in the case; transform the data appropriately and carry out the analysis of the transformed data.

Compare the analyses and also the tables of treatment means with their standard errors. Summarize the results and present your conclusions.

Treatments	I	II	Block III	IV	Total
1	2	7	12	18	39
2	4	12	23	22	61
3	29	61	56	64	210
4	44	61	85	94	284
5	30	62	71	93	256
Totals	109	203	247	291	850

Analysis

	s.s.	d.f.	m.s.	F-ratio
Blocks	3 631	3	1 210.3	
Treatments	12 758	4	3 189.5	34.00***
Error	1126	12	93.8	
Total	17 515	19		
c.f.	36 125			

9
Studying linear relationships

9.1 LINEAR REGRESSION

We turn now from the analysis of observations of a single variable to consider simultaneous observations on several variables. Instead of estimating differences caused by the application of different treatments we shall now be interested in the relationship between two or more variables. There are various possible reasons for this interest but the two principal ones are: (a) to investigate and test hypothetical mathematical models for biological systems, and (b) to predict the values of one variable from another. Often a particular investigation may stem from both of these motives. Examples of relationships that may need to be investigated are:

(a) The response of a crop to varying amounts of several fertilizers; the objects may be to establish the form of the relationship, or to predict the optimum combination of fertilizers;
(b) The relationship between various meteorological measurements and crop yield; the most obvious objective would be to try to understand possible causative mechanisms of meteorological effects on crop growth;
(c) The lactation curves of cows and the effect on these of varying the diet of the cows; here the objective could again be to understand the pattern of lactation yield. Another possible objective could be to establish the general form of lactation curves with the intention of examining subsequently how the fitted lactation curves are changed by different treatments; essentially the fitted curves are being used as a criterion for evaluating treatment effects; or
(d) The relationship between leaf area and leaf weight for several varieties of the plant and various ages of leaves for each plant. In this situation we would be almost exclusively interested in predicting leaf area, an important variable in photosynthesis which is difficult to measure, using leaf weight, which is relatively simple to measure.

The simplest form of relationship between two variables is a straight line and we shall consider in some detail methods of estimating a linear relationship and predicting from the estimated line. We shall go on to discuss the use of these methods in the analysis of experimental data. In the next two chapters we shall consider, briefly, more complex relationships.

In the simplest situation we have a series of pairs of observations of y and x, where y, the **dependent variable**, is assumed to depend on x, the **independent variable**. As in experiments we must assume that the units on which observations are made are variable, so that if we plot the values of y against the corresponding values of x we do not get a series of points exactly on a straight line but a scatter of points on an apparently straight line. An illustration of the kind of data we should expect to find is given in Fig. 9.1 in which weight gains of cattle (y) are plotted against level of vitamin added to diet (x), the values of y and x being

y	6	9	13	12	14	18
x	1	2	3	4	5	6	(artificial data)

We now have to choose a line which in some sense represents the set of pairs of y and x values most closely. There are obviously a large number of lines we could choose. We could choose, for example:

1. The line joining the two extreme points;
2. The line joining the first two points;
3. The line joining the two middle points; or
4. The line through the mean values $\bar{y}$ and $\bar{x}$ with some estimate of the slope, say 2.

Some of these lines, if superimposed on the plot of Fig. 9.1, will be felt intuitively to fit the points well, other obviously not.

In order to choose a best line we must define a criterion for judging the adequacy of the fit of any particular line. The standard criterion used is that of minimizing the sum of squared vertical deviations of the observed points about the line. This criterion is known as the principle of least squares. The

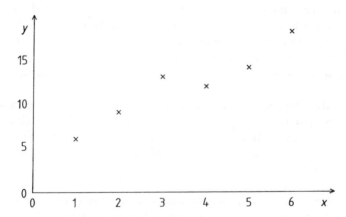

Fig. 9.1 Weight gains of cattle (y) plotted against the level of added vitamin (x).

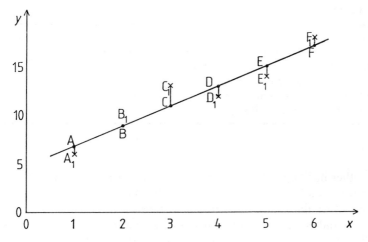

Fig. 9.2 An eye-line fitted to the data of Fig. 9.1.

use of squared deviations should not be surprising after the earlier chapters of this book. The criterion is illustrated in Fig. 9.2 for an arbitrary line fitted by eye to the data of Fig. 9.1. The measure of the adequacy of the line A–F in fitting the points $A_1, B_1, \ldots F_1$, is given by the sum

$$S_r = (A_1 - A)^2 + (B_1 - B)^2 + (C_1 - C)^2 + (D_1 - D)^2 + (E_1 - E)^2 + (F_1 - F)^2$$

where S_r is referred to as the residual sum of squares, since it measures the residual variation of the y observations about the line, i.e. that part of the original variation between the y observations which is not attributable to the estimated linear dependence of y on x. The line which gives the smallest residual sum of squares, S_r is defined to be the best fitting line. We can calculate this criterion for any given line by calculating for each value of x, for which there is an observation, the value of y on our given line, obtaining the difference between the observed and line values of y and summing the squares of such differences. Thus, for line (i), the line joining the two extreme points, we have the values $y = 6$ when $x = 1$ and $y = 18$ when $x = 6$, so that the slope of the line is $(18 - 6)/(6 - 1) = 2.4$. Hence the values of y on the line at $x = 1, 2, 3, 4, 5, 6$ are $6, 8.4, 10.8, 13.2, 15.6$ and 18. We can present the calculations as follows:

Value of x	1	2	3	4	5	6
Observed value of y	6	9	13	12	14	18
Value of y on line	6	8.4	10.8	13.2	15.6	18
Difference	0	0.6	2.2	−1.2	−1.6	0
(Difference)^2	0	0.36	4.84	1.44	2.56	0
Sum of squared deviations = 9.2						

Table 9.1

Line	Sum of squared deviations
(i) Joining the two extreme points	9.2
(ii) Joining the first two points	45.0
(iii) Joining the two middle points	170.0
(iv) Through $(\bar{y}, \bar{x})$ with slope 2	8.0

For the other lines suggested earlier the calculations are similar and the sums of squared deviations are summarized in Table 9.1.

Various other lines may be tried and it seems reasonable to suppose that such a trial-and-error method will bring us somewhere near the best line, the fourth line tried being the best so far. However, it is clearly advisable to have a more direct approach. Returning to the least squares principle of choosing the line to minimize the sum of squared deviations.

$$\Sigma(y - a - bx)^2$$

it can be shown mathematically that the best fitting line is

$$y = a + bx$$

where

$$b = \frac{\Sigma[(x - \bar{x})(y - \bar{y})]}{\Sigma[(x - \bar{x})^2]}$$

and

$$a = \frac{\Sigma(y) - b\Sigma(x)}{n} = \bar{y} - b\bar{x}$$

n being the number of observations of y. This line is the **regression line** of y on x. The terms $\Sigma[(x - \bar{x})(y - \bar{y})]$ and $\Sigma[(x - \bar{x})^2]$ are referred to as the corrected sum of product of x and y and the corrected sum of squares of x, and are written S_{xy} and S_{xx}. As in the calculation of the sample variance, these sums can be written in alternative form for ease of calculation:

$$S_{xy} = \Sigma(xy) - \frac{\Sigma(x)\Sigma(y)}{n}$$

$$S_{xx} = \Sigma(x^2) - \frac{[\Sigma(x)]^2}{n}$$

We can calculate the regression line for the data on weight gains of cattle illustrated in Fig. 9.1.

$$n = 6 \quad \Sigma(x) = 21 \quad \Sigma(x^2) = 91 \quad S_{xx} = 91 - \frac{21^2}{6} = 17.5$$

$$\Sigma(y) = 71 \quad \Sigma(xy) = 289 \quad S_{xy} = 289 - \frac{72 \times 21}{6} = 37$$

$$b = \frac{37}{17.5} = 2.114$$

$$a = \frac{72 - 21\,(2.114)}{6} = 4.601$$

Therefore the fitted line is

$$y = 4.601 + 2.114x$$

predicting an additional gain in weight of 2.114 for every additional unit of vitamin.

By setting $x = 1, 2, 3, 4, 5$ and 6 in this equation we can obtain the values of y on the regression line corresponding to the observed yields and can then calculate the sum of squared residuals. Thus the expected weight gain when two units of vitamin is applied is $4.601 + 2(2.114) = 8.829$. The sum of squared residuals is obtained as follows:

Value of y on line	6.715	8.829	10:943	13.057	15.171	17.285
Difference	−0.715	+0.171	+2.057	−1.057	−1.171	+0.715

Sum of squared deviations = 7.76

As is inevitable from the method of fitting, this sum of squared residuals is smaller than the values for the four lines tried previously, and is smaller than for any other line which can be drawn.

9.2 ASSESSING THE REGRESSION LINE

The first question we would expect to ask about the regression line we have just fitted is whether it represents a real relationship. In other words, is the slope of the assumed linear relationship genuinely different from zero? There are several ways of answering that question, all numerically equivalent. Initially we shall consider the analysis of variance for fitting a regression line. We divide the total variation in the sample of y values into the residual variation about the fitted line and the variation between the fitted y values along the line. This latter variation can be thought of as the variation explained by, or attributable to, the regression of y on x, and is referred to as the regression sum of squares.

In practice the analysis of variance is obtained by calculating the total s.s.,

$$S_{yy} = \Sigma(y^2) - [\Sigma(y)]^2/n$$

and the regression s.s.,

$$\text{regression s.s.} = \frac{(S_{xy})^2}{S_{xx}}$$

where S_{yy}, S_{xy}, S_{xx} are, as previously, the corrected sum of squares of y, the corrected sum of products and the corrected sum of squares of x. The residual s.s. which is the minimum value of the sum of squared deviations is then obtained by

$$\text{residual s.s.} = \text{total s.s.} - \text{regression s.s.} = S_{yy} - S_{xy}^2/S_{xx}.$$

The division of the $(n-1)$ degrees of freedom for the total variation is 1 for the regression s.s. and $(n-2)$ for the residual s.s. To understand why the degrees of freedom for the residual sum of squares are $(n-2)$, consider again the original introduction of the concept of degrees of freedom for a sample variance. The divisor, and therefore the degrees of freedom, for a sample variance is $(n-1)$ because we are considering n deviations from a single fitted value, $\bar{y}$. For a regression line we are considering n deviations from a set of values on a line, the line being defined by two values, a and b, which are estimated from the data. As in the analysis of variance for experiments the residual mean square is denoted by s^2 and provides an estimate of the variance of the observations having corrected for the effect of differences in the x-values, **assuming that the relationship between y and x is linear**. For the data on weight gains the analysis of variance is constructed as follows:

$$\Sigma(y^2) = 950 \quad S_{yy} = 950 - (72)^2/6 = 86$$
$$\text{regression s.s.} = (S_{xy})^2/S_{xx} = 37^2/17.5 = 78.23$$

Analysis of variance

Source of variation	s.s.	d.f.	m.s.	F
Regression	78.23	1	78.23	40.3
Residual	7.77	4	$1.94 = s^2$	
Total	86			

Note that the residual s.s. agrees, apart from rounding-off errors, with the sum of squared residuals calculated earlier.

The significance of the regression is tested by comparing the ratio of the regression and residual mean squares with the F-distribution having 1 and $(n-2)$ d.f. The hypothesis being tested is that there is no dependence of y on

x, or, in other words, that the slope of the linear relationship is zero. In this case the F-ratio of $78.2/1.94 = 40.3$ is very significant so that there is clearly a trend of y increasing with x.

9.3 INFERENCES ABOUT THE SLOPE OF A LINE

The value we have calculated for b is our estimate of the rate at which y increases for a unit increase in the value of x. In some situations we may be directly interested in this rate. For example if we are feeding different levels of vitamin we could assess whether the inclusion of extra vitamin in the diet was economically justified. We clearly need to know the precision of the estimate of the rate of increase, or slope. Since b is a linear combination of the y observations we can show that the variance of b is

$$\text{Var}(b) = \frac{\sigma^2}{S_{xx}}$$

where σ^2 is the variance of observations about the linear relationship between y and x. We estimate σ^2 by the residual mean square, s^2, and hence have a standard error for b

$$\text{s.e. } (b) = \sqrt{(s^2/S_{xx})}$$

The general form of this standard error is sensible. Smaller standard errors are obtained when S_{xx} is larger, or in other words, when the spread of x values is large. The greater the spread of the x values the better the line is defined.

Given our estimate of the slope, b, and its standard error, we can calculate a confidence interval for the slope of the population relationship between y and x in exactly the same way that we earlier calculated a confidence interval for a population mean, μ, from a sample meam, $\bar{x}$, and its standard error. Thus the 95% confidence interval for the slope of the true relationship of y on x is

$$b \pm t\sqrt{(s^2/S_{xx})}$$

where the t value is the 5% point for the t-distribution with $(n-2)$ degrees of freedom.

For our example of weight gains and vitamin levels in the diet, the standard error of b is

$$\begin{aligned}
\text{s.e.}(b) &= \sqrt{(s^2/S_{xx})} \\
&= \sqrt{(1.94/17.5)} \\
&= \sqrt{0.1108} \\
&= 0.33
\end{aligned}$$

The 95% confidence interval for the slope is $2.114 \pm 2.78 \times 0.333$ which is $(1.19, 3.04)$.

We can also use the standard error of b to test the strength of the relationship between y and x. To test the hypothesis that there is no dependence of y on x we calculate

$$t = \frac{b - 0}{\text{s.e.}(b)}$$

and compare this with the t-distribution on $(n-2)$ d.f. This t-test is exactly equivalent to the F-test discussed in the previous section, the value of t ($= 6.34$) being the square root of the previously calculated F-statistic.

9.4 PREDICTION USING A REGRESSION LINE

Consideration of the standard error of b leads us to the idea of predicting the value of y for a given value of x and the precision of such prediction. The predicted value of y for a specific x, say x_0, is obviously $y_0 = a + bx_0$. Thus the predicted weight gain for a vitamin level of 3 from the preceding data is $4.601 + 3(2.114) = 10.943$; for a level of $4\frac{1}{2}$ the predicted value would be 14.114; for a level of 12 units the predicted value would be 29.969, though since this vitamin level is so far outside the range of values on which the prediction is based we may expect the precision of this last predicted weight gain to be poor.

Any error in the predicted value, $a + bx_0$, arises entirely from the errors in the estimates of the constant term, a, and the slope of the line, b. The standard error of the predicted value can be obtained, mathematically, as

$$\text{s.e.}(a + bx_0) = \sqrt{\left\{ s^2 \left[\frac{1}{n} + \frac{(x_0 - \bar{x})^2}{S_{xx}} \right] \right\}}$$

where $\bar{x}$ is the mean of the x values used in the estimation of a and b. We must be careful to differentiate between this standard error and the standard error of the predicted weight gain of a single animal receiving x_0 units of vitamin. The expected weight gain of a single animal is still $a + bx_0$ but, in addition to the variation arising from the fact that this predicted value is only an estimate, we must also take into account the variation of individual cattle in the population about the true, or average, weight gain for the particular value of x_0. The standard error for the predicted value of a single animal is

$$\text{s.e.}(\text{individual value}) = \sqrt{\left\{ s^2 \left[1 + \frac{1}{n} + \frac{(x_0 - \bar{x})^2}{S_{xx}} \right] \right\}}$$

Thus, in predicting from the data on the relationship between weight gain and vitamin level, the predicted mean weight gain for a large population of cattle being given a vitamin level of 3 units is 10.943 and the standard error of this predicted mean value is

$$\sqrt{\left\{ 1.94 \left[\frac{1}{6} + \frac{(3 - 3.5)^2}{17.5} \right] \right\}} = 0.591$$

However, the predicted weight gain for a single animal receiving a vitamin level of 3 units is also 10.943 but the standard error is now

$$\sqrt{1.94\left[1+\frac{1}{6}+\frac{(3-3.5)^2}{17.5}\right]}=1.513$$

The standard error of a predicted value depends on the value of x and is least when $x_0 = \bar{x}$ the mean of the x values on which the prediction is based; as x_0 moves towards the extremes of this set of x values the standard error of a predicted value increases. To illustrate this consider the predicted values, for $x = 4.5$ and 12, calculated previously. The standard errors of these predicted mean values are

$$\text{for } x_0 = 4.5, \quad \text{s.e.} = \sqrt{\left\{1.94\left[\frac{1}{6}+\frac{(4.5-3.5)^2}{17.5}\right]\right\}}=0.659$$

$$\text{for } x_0 = 12, \quad \text{s.e.} = \sqrt{\left\{1.94\left[\frac{1}{6}+\frac{(12-3.5)^2}{17.5}\right]\right\}}=2.89$$

Since for any value x_0 of x we can calculate the estimate of an individual weight gain and also its standard error, we can obtain a 90% confidence interval for the true weight gain

$$a + bx_0 \pm t \times \text{s.e. (individual predicted value)}$$

For the single animal prediction for 3 units this gives an interval of

$$10.943 \pm 2.78 \, (1.513)$$
$$\text{i.e. } 6.74 \text{ to } 15.15$$

All the errors and standard errors of prediction mentioned so far arise from difficulties in estimating the values of a and b in the equation of the line, $y = a + bx$, and from the variation of individual values about the line. We have assumed that the relationship is linear. Methods for testing the linearity and for fitting more complex relationships will be discussed in the next chapter. At this point we would emphasize the importance of checking the assumptions of the analysis, particularly the linearity, by drawing a graph.

The dangers of extrapolating a relationship beyond its known range of validity are well known. We have already seen how the standard errors of prediction increase as the value of x moves away from the mean value of x in the data. Doubts about the validity of the relationship at the extremes of the range of the data and beyond are additional reasons for caution in predicting values of y in these areas. Predictions for other populations will, of course, only be valid if the same values of a and b apply to these populations. This generally needs justification or, better still, verification.

In regression calculations, the independent variable x is generally assumed to be measured without observational error and, in some sense, to influence the values taken by the y variable. The values of x may be chosen by the experimenter or may, like the y-values, be a random sample. As an example of the latter situation, the pairs of x and y values might be the weights of the roots and the above ground components of a number of plants. If the values of x and of y can both be treated as random samples representative of some larger population of values, then questions can arise about which, if either, should be treated as the dependent variable and which as the independent variable. The lines representing the regression of y on x and the regression of x on y are not the same; they minimize the sums of squares of the deviations about the line in two different directions, the y direction and the x direction respectively.

If, when the x and y values are both random samples, the purpose of the investigation is to predict value of y from values of x then y should be the dependent value and x the independent variable. More advanced texts, such as Sprent (1969), must be consulted if the x or y variable, or both, are measures, subject to observational errors, of some more fundamental variables and it is the relationship between these more fundamental variables that is of interest.

Example 9.1 Linear regression

Air with varying concentrations of CO_2 is passed over wheat leaves at a temperature of 35 °C and the uptake of CO_2 by the leaves is measured. Uptake values (y) for seventeen leaves at different concentrations (x) are obtained and are as follows:

Concentration	Uptake ($cm^2/dm^2/hour$)
75	0.00
100	0.65
100	0.50
100	0.40
120	1.00
130	0.95
130	1.30
160	1.80
160	1.80
160	2.10
190	2.80
200	2.50
200	2.90
200	2.45
200	3.05
240	4.30
250	4.50

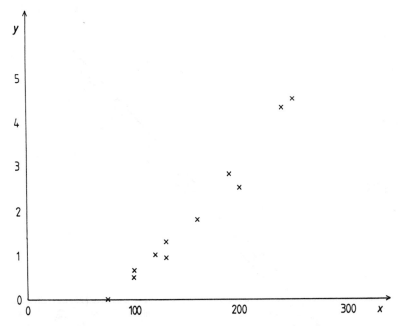

Fig. 9.3 Data on the relationship of CO_2 uptake (y) in $cm^3/dm^3/hour$, against CO_2 concentration in air (x).

The graph of y against x is shown in Fig. 9.3.

$$n = 17$$
$$\Sigma(x) = 2715 \qquad \bar{x} = 159.7$$
$$\Sigma(y) = 33.00 \qquad \bar{y} = 1.941$$
$$\Sigma(y^2) = 91.8600 \qquad \Sigma(xy) = 6348.50 \qquad \Sigma(x^2) = 476825$$

The corrected sums of squares and products are

$$S_{yy} = 91.8600 - \frac{(33.00)^2}{17} = 27.80$$

$$S_{xy} = 6348.50 - \frac{(33.00)(2715)}{17} = 1078.21$$

$$S_{xx} = 476825 - \frac{(2715)^2}{17} = 43224$$

Regression coefficient

$$b = S_{xy}/S_{xx} = 1078.21/43224 = 0.0249$$

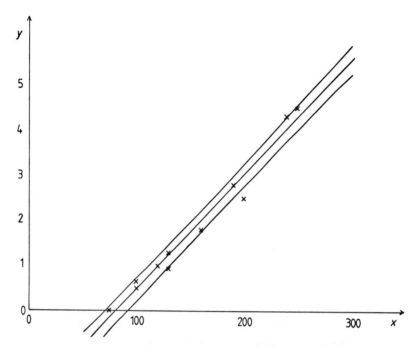

Fig. 9.4 The fitted regression line and 95% confidence limits for uptake by a single leaf, for the data of Fig. 9.3.

$$a = \frac{\Sigma(y) - b\Sigma(x)}{n} = \frac{33.00 - (0.0249)(2715)}{17} = -2.04$$

Hence the fitted regression line is, as shown in Fig. 9.4,

$$y = -2.04 + 0.0249x$$

The regression s.s. $= (S_{xy})^2/S_{xx} = (1078.21)^2/43224 = 26.90$

Analysis of variance

	s.s.	d.f.	m.s.	F
Regression	26.90	1	26.90	448
Residual	0.90	15	$s^2 = 0.060$	
Total variation about the mean	27.80	16		

S.E. of regression coefficient $= \sqrt{(s^2/S_{xx})} = 0.00118$

t-test of deviation or regression coefficient from zero:

$$t = \frac{0.0249}{0.00118} = 21.10***$$

The predicted mean value of y for a given concentration, $x = 150$ is

$$-2.04 + 0.0249(150) = 1.70$$

The standard error for this predicted mean value is

$$\sqrt{\left\{0.060\left[\frac{1}{17} + \frac{(150-159.7)^2}{43224}\right]\right\}} = 0.060$$

The predicted value of y for a single leaf for the same concentration, $x = 150$, is also 1.70.

The standard error for this predicted individual value is

$$\sqrt{\left\{0.060\left[1 + \frac{1}{17} + \frac{(150-159.7)^2}{43224}\right]\right\}} = 0.252$$

From such predictions and their standard errors we can obtain 95% confidence limits either for an individual leaf or for the mean value for many leaves, as

$$a + bx_0 \pm t(\text{s.e.})$$

where t is the 5% point of the t-distribution with 15 d.f.

Thus for an individual leaf 95% confidence limits are

x	Upper limit	Lower limit
50	−0.19	−1.40
100	1.01	−0.11
150	2.23	1.16
200	3.49	2.39
250	4.77	3.60
300	6.07	4.79

The curves for the 95% confidence intervals for individuals leaves are shown in Fig. 9.4.

EXERCISE 9.1 *Linear regression*

Direct measurement of the specific heat of apple twigs is difficult. An investigator proposes instead to predict specific heats of individual twigs from the more easily measured percentage of water in the twigs.

He first secures pairs of measurements of percentage water content (x) and specific heat (y) from 21 twigs. These are

x	y	x	y	x	y
49	46	57	100	65	131
53	57	52	53	61	113
62	119	51	65	52	69
58	90	53	89	52	66
50	44	51	70	56	96
63	131	56	85	58	111
59	104	60	96	54	69

The data have been simplified to make calculation easier. The x values have been rounded off to whole numbers and the specific heats coded so that

$$y = 1000 \text{ (specific heat} - 0.600000).$$

(a) Draw a graph of y against x.
(b) Calculate the regression line of y on x and enter on the graph.
(c) Complete the analysis of variance table:

	d.f.	s.s.	m.s.	F
Regression on x				
Residual			$s^2 =$	
Total				

(d) Estimate the mean specific heat for all twigs of 60% water content, and find the standard error of this estimate.
(e) If a single twig is found to contain 60% water, what is the estimate of its specific heat? Find the s.e. of this estimate and hence 95% confidence limits for the true value.
(f) You are given that the 95% confidence limits for the specific heat (coded) of a twig with water content

$$40\% \text{ are } -24 \text{ to } +25$$
$$45\% \text{ are } +5 \text{ to } +49$$
$$50\% \text{ are } +34 \text{ to } +75$$
$$55\% \text{ are } +62 \text{ to } +101$$
$$65\% \text{ are } +114 \text{ to } +157$$

Draw smooth curves through these upper and lower limits.

(g) A twig is known to have a specific heat (coded) of 90. Using the smooth curves, suggest how reasonable limits for its water content could be calculated.

9.5 CORRELATION

In regression the dependent variable, y, is the variable in which we are principally interested. The independent variable, x, is discussed only because of its possible effect on y. In some situation, however, we are interested not in the dependence of y on x but in the general problem of measuring the degree of association between the two variables x and y. The appropriate measure of such an association is the sample correlation coefficient

$$r = \frac{\Sigma(x - \bar{x})(y - \bar{y})}{\sqrt{[\Sigma(x - \bar{x})^2 \, \Sigma(y - \bar{y})^2]}}$$

or, in the notation of previous sections,

$$r = \frac{S_{xy}}{\sqrt{(S_{xx} S_{yy})}}$$

This definition limits r to values between -1 and $+1$. Positive values of r indicate that the two variables are positively associated, i.e. that large values of x occur with large values of y; in particular a value of $+1$ for r arises only when the values of x and y are exactly linearly related, the slope of the line being positive. Negative values show negative association and when $r = -1$ the graph of y against x gives points exactly on a straight line with a negative slope. A value of zero for r indicates that there is no linear association between the two variables. Some illustrations of typical patterns for various values of r are shown in Fig. 9.5(a), (b), (c), (d) and (e).

To test the significance of a particular value of r the simplest method, using the statistical tables we have already used, is to calculate

$$F = [(n - 2)r^2]/(1 - r^2),$$

and compare this with the F-distribution on 1 and $(n - 2)$ degrees of freedom, where n is the sample size. This test is equivalent to testing the significance of the regression of either variable on the other as described in section 9.2.

The use of the correlation coefficient is rather limited since the number of situations where we want to know whether two variables are associated, but are not interested in the equation of the relationship, is very small. Correlation techniques are useful in the preliminary examination of a large number of variables to see which variables are associated. However, even here there are regression techniques, which will be discussed in the next chapter, that are rather more effective than correlation methods.

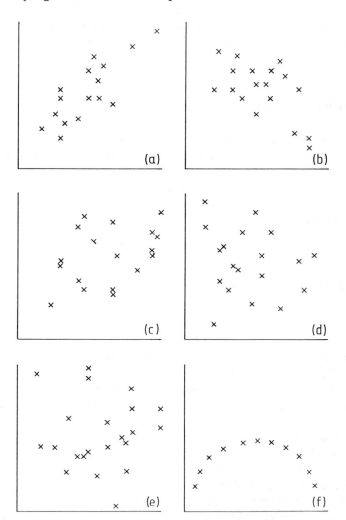

Fig. 9.5 Sample of x and y values for various correlations: (a) $+0.9$; (b) -0.8; (c) $+0.6$; (d) -0.4; (e) 0.0; (f) 0.0.

Although, as will be obvious, we do not consider the correlation coefficient of much practical use, the research worker will find frequent references to it in published research, more particularly in research in economics and in the social sciences. Consequently there are a few warnings on the interpretation of correlation coefficients which should be included here. First, the correlation coefficient measures only the linear association between two variables, so that even if there is an exact relationship between two variables, if that relationship is curved the correlation coefficient may be much smaller than 1. For example, an exact but not straight line relationship between x and y shown in Fig. 9.5(f) would give a value of $r = 0$. On the other hand, a high value of r does not necessarily give evidence of a linear relationship between the variables concerned. As an example, if x takes the values 1, 2, 3 up to n, and $y = x^2$, i.e. y takes the values 1, 4, 9 up to n^2, then the smallest possible value of r (whatever the value of n) is 0.968 and yet x and y are obviously not linearly related. Finally a significantly high value of r in no way implies that one variable is dependent on the other; correlation does not mean causation. If one examined the numbers of doctors and numbers of deaths in 1000 towns and villages in England there would undoubtedly be a high correlation between the two variables, because each tends to be large in the larger towns!

9.6 TESTING WHETHER THE REGRESSION IS LINEAR

The results for linear regression, like those for correlation, depend on the assumption that the true relationship between x and y is linear. In practice predicted values may be little affected if this assumption is untrue, because a straight line may be a very good approximation to the true relationship over the range of x values of interest. Nevertheless we obviously wish to be able to test this assumption, and this must be done by examining the residual mean square after fitting the regression. As was stated in section 9.2 the residual mean square gives an estimate of the variance of observations about the relationship provided the assumption of a linear relationship is correct. If the assumption is not correct then the residual sum of squares will include not only this random variation of y but also the systematic deviations of the true relationship about a straight line. Hence the residual mean square will tend to overestimate the variance. If we had another estimate of the variance which we knew to be unbiased then we could compare the ratio of the residual mean square to this unbiased estimate with the appropriate F-distribution and hence have a test of whether the residual mean square was significantly large.

It is not always easy to find an unbiased estimate of variance. One situation where such an estimate is available is when repeat observations of y have been recorded for some of the x values. The variation of y values for a given x value is clearly unaffected by the form of the relationship. Thus the sample variance of y values for a given value of x gives us an estimate of the variance of observations about the relationship. A better estimate is the pooled variance when there are repeat y values for several different x values (the same pooled variance as in section 3.5). As an example of this argument consider again Example 9.1. There are repeat values at: $x = 100, 130, 160$ and 200.

x	Values of y	$\Sigma y^2 - \dfrac{(\Sigma y)^2}{n}$
100	0.65, 0.50, 0.40	0.0317
130	0.95, 1.30	0.0612
160	1.80, 1.80, 2.10	0.1267
200	2.50, 2.90, 2.45, 3.05	0.2625
		0.4821

$$\text{Pooled variance} = \frac{0.0317 + 0.0612 + 0.1267 + 0.2625}{2 + 1 + 2 + 3} = 0.060$$

The residual sum of squares of 0.90 with 15 d.f. in Example 9.1 is made up of the total of the sums of squares for each of the set of repeated values above, 0.482, and a sum of squares measuring the non-linearity of the regression which by subtraction is $0.90 - 0.4821 = 0.42$ with $15 - 8 = 7$ d.f. The mean square for non-linearity $0.42/7 = 0.060$ can be compared with the pooled variance mean square of 0.060 using the F-test with 7 and 8 d.f., $F = 0.060/0.060 = 1.00$ which is clearly not significant (the value of exactly one is unusual but does occasionally occur). Clearly the linear model is adequate.

We have now introduced F-tests in two different ways in testing regression and it is important to distinguish between them since they are testing quite distinct aspects of regression. In section 9.2 we considered the analysis of variance which splits the total variation of y into that attributable to the regression on x and the residual variation. The F-test in section 9.2 tests whether the regression component of the variation is sufficiently large that we must accept that there is a tendency for y to increase or decrease with x. Since a linear trend is the only relationship we have considered we say that the linear regression is significant. However, this does not show that the relationship is in fact linear. This can only be tested by examining the variation about the fitted linear relationship and investigating whether this residual variation is too big. This is what we have done in this section. If the residual variation about the regression is large compared with the pooled variance then our relationship is not linear and the earlier testing of the significance of the linear regression is of less value because the assumption of a linear relationship on which the linear regression is based, is false.

9.7 REGRESSION ANALYSIS USING COMPUTER PACKAGES

All general statistical packages have routines for fitting regression lines using least squares as described in this chapter. They allow the data to be plotted easily to investigate the suitability of the proposed regression model and also have facilities to store and investigate the residuals from the model fitting. The residuals which are the differences between the observed dependent data values and the fitted points on the line contain all the information about the lack of fit of the model. If the model was a perfect fit all the observed points would lie exactly on the fitted line and the residuals would be zero; if the residuals are relatively large then the regression model does not explain much of the variability in the data. The facilities of a typical package are illustrated by analysing the data from Example 9.1:

The data is stored in two columns

x (concentration)	y (uptake)
75	0.00
100	0.65
100	0.50
120	1.00
130	0.95
131	1.30
160	1.80
190	2.80
200	2.50
240	4.30
250	4.50

The variable y is then specified as the dependent variable in the regression and x as the independent variable and the fitted values and residuals calculated and displayed. The typical form of output shown below includes the estimates of the intercept and slope with standard errors and an analysis of variance:

The regression equation is

$$y = -2.08 + 0.0254\, x$$

	Coeff	Std Error	t-ratio	p
Predictor	−2.0759	0.2220	−9.35	0.000
Constant	0.025448	0.001354	18.80	0.000
x				
R − sq = 97.5%				

Analysis of variance

Source	d.f.	s.s.	m.s.	F	p
Regression	1	22.304	22.304	353.29	0.000
Error	9	0.568	0.063		
Total	10	22.872			

The t-ratio tests the hypothesis that either of the model coefficients is zero. For example for the slope (x)

$$t\text{-ratio} = 0.025448/0.001354 = 18.80$$

This is highly significant (p is less than 0.0005) and so there is strong evidence of a relationship between y and x. As discussed in section 9.3 this is exactly equivalent to the F-test in the analysis of variance.

R-sq (or R^2) is commonly output as a summary measure of the regression. It measures the percentage of the total variability in the data (measured by the total sum of squares) which is explained by the regression line:

$$R^2 = \frac{\text{Regression Sum of Squares}}{\text{Total Sum of Squares}} = \frac{22.304}{22.872} \times 100 = 97.5\%$$

For a simple regression between two variables it is the square of the correlation coefficient between the two variables x and y.

The package also outputs the residuals:

Row	x	y	Fitted value	Residual	Standardized residual
1	75	0.00	−0.16726	−0.167256	0.78045
2	100	0.65	0.46895	0.181055	0.79374
3	100	0.50	0.46895	0.031055	0.13614
4	120	1.00	0.97791	0.022093	0.09398
5	130	0.95	1.23239	−0.282387	−1.18980
6	130	1.30	1.23239	0.067613	0.28488
7	160	1.80	1.99583	−0.195830	−0.81788
8	190	2.80	2.75927	0.040729	0.17362
9	200	2.50	3.01375	0.513752	2.22051
10	240	4.30	4.03167	0.268325	1.28116
11	250	4.50	4.28616	0.213844	1.06218

The fitted values are calculated from the fitted equation of the line

$$y = -2.08 + 0.0254\,x$$

This package outputs two forms of residual. The first residual is the difference between the observed value y and the fitted value on the regression line

$$\text{Residual} = y - \text{fitted value}$$

To decide if any of these residuals is unusually large they must be assessed relative to the underlying variability. This can be done using the standardized residuals which have been standardized to have unit variance. Using the properties of the standardized normal distribution we would expect 95% of residuals to be inside the range ± 2 if our model was adequately explaining the variation in the response y. For this data the ninth observation has a standardized residual outside the range ± 2 which indicates a large residual. However as we would expect 1 in 20 standardized residuals to be outside the range ± 2 this gives no strong cause for concern. The residuals can be used in this way to highlight any possible outliers in the data.

Since the residuals contain all the information about the lack of fit of the model they should always be examined critically. If the model is adequate the residuals should be a random sample from a normal distribution; any systematic patterns in the residuals may be indicating that a more complex model is required.

10

More complex relationships

10.1 MAKING THE CROOKED STRAIGHT

The last chapter was concerned solely with a relationship, assumed to be linear, between two variables one of which, y, was assumed to be dependent on the other, x. In practical situations we are likely to be concerned with more complex relationships estimated in different environments and with examining the possible dependence of y on several x variables. First, however, we shall consider the situation where the relationship between y and x is not linear, i.e. cannot be represented by a straight line.

Since we know how to deal with straight line relationships, the sensible approach when confronted with a curved relationship is to straighten it. Therefore we look for a simple function of y or x or both such that when we plot the graph of, say, some function of y against x then we have an acceptably straight line relationship. Of course we cannot hope to get an exact straight line because the observations will contain some random variation. Nevertheless, the eye is a remarkably good judge of whether a line is straight and it is a reasonable rule that what looks straight probably is sufficiently straight for our purposes. How do we know what functions or transformations of y and x to use? There are two answers to this problem. The first is that one learns from experience. Thus one learns fairly quickly what kinds of curves will be changed to reasonably straight lines if instead of plotting y against x one plots log y against x. However, this is not a very satisfactory answer and the real answer is that in any particular situation the experimental scientist who has collected the data should have a good idea of the kind of relationship between y and x which he would expect to hold either from theoretical considerations or from previous work. The transformations finally chosen should preferably not only achieve a linear relationship but also imply a relationship between the original variables y and x that is biologically meaningful.

Various examples of transformations will be discussed during the rest of the chapter, but we shall first consider briefly three typical situations where transformations are often needed.

(a) The early growth of a plant when there are no limiting constraints on growth. Here y is weight in grams and x is time in days from an arbitrary

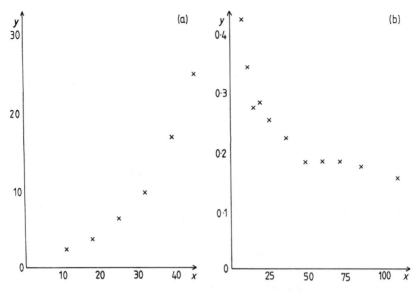

Fig. 10.1 (a) Relationship between the weight of a plant (y) in grams and time (x) in days. (b) Relationship between proportional rates of growth (y) in grams per gram per year and mean annual body weight (x) in grams.

date after sowing.

y	2.2	3.6	6.2	9.5	16.9	24.5
x	11	18	25	32	39	45

The data are plotted in Fig. 10.1(a)

(b) Growth rates of fish (dace). Here y is the proportional growth in grams per gram per year and x is the mean body weight of the fish over the year in grams.

y	0.42	0.34	0.27	0.28	0.25	0.22	0.18	0.18	0.18	0.17	0.15
y	7.4	10.8	14.6	19.3	25.1	36.4	49.2	59.2	71.1	85.0	109.2

The data are plotted in Fig. 10.1(b).

(c) The change in the proportion of deaths of insects sprayed with different concentrations of a given insecticide. Here y is the proportion of deaths out of 200 insects and x is the concentration.

y	5%	40%	55%	80%	90%	100%
x	0.0075	0.015	0.03	0.06	0.12	0.24

In the first situation the rate of increase of y with x is clearly larger for the larger values of y. To make the relaionship linear we require that the rate of increase of y with x is the same for all values of y. In other words we need to 'squash' up the top part of the y scale. Equivalently we could stretch out the higher values of the x scale. Is there any biological reasoning behind such transformations? Well, it is not unreasonable to assume that for small plants the rate of growth depends on the size of the plant because the dominant factor in plant growth is the rate of photosynthesis which depends on the leaf area. A rate of growth proportional to plant size would indicate that the rate of change of log weight of the plant should be constant. Hence, looking at log weight instead of weight should give a more linear relationship. Figure 10.2(a) shows what happens when we plot log y and against x.

If we now fit a regression of log y on x we should pause briefly and question whether the other assumptions involved in regression are likely to be true. We should consider whether it is reasonable to assume that the observations are independent and approximately normally distributed about the fitted line. However, the principal assumptions is that the amount of random variation about the true relationship is independent of the value of x (or equivalently of y). If we assume this is so for observations of log y, then we are saying that the variation of log y is much the same for low values of x (and y) as it is for high values. In terms of the variation of the original y, the plant weight, this means that the variation of weight should be larger for large values of x (and y) than it is for small values. This, fortunately, seems reasonable. We might well expect to get a variation from 20 to 30 g in weight after 45 days but only 5 to 7 g after 25 days. Certainly we would not expect the variation of y to be constant, i.e. the same for all values of x. Thus it would appear that the assumptions of

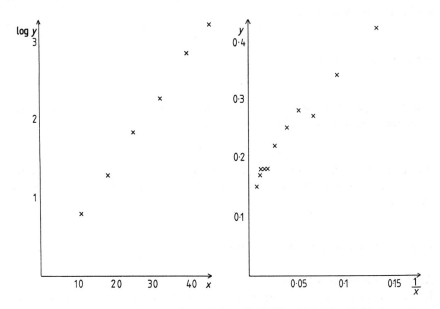

Fig. 10.2 (a), (b). Transformed data from Figs 10.1 (a) and (b).

uniformity of variation may be expected to be true for variation of log y and that the log transformation of y not only straightens the line but satisfies the other assumptions necessary for the regression analysis.

In the second situation, the growth rate of dace, a suitable form of transformation would seem to be one squashing up either the higher y values or the higher x values. The data suggest that the proportional growth rate y does not decrease right down to zero as body weight increases but has a lower limiting value. In such a case a log transformation is not usually so successful and an alternative is to consider the reciprocal of x, $1/x$ instead of x. Figure 10.2(b) shows what happens when we plot y against $1/x$. Again we must look at the assumptions, and in particular the assumption of homogeneity of variance. The graph in Fig. 10.2(b) does not give any great cause to worry about the assumption of constancy of variance so that we might well be prepared to accept the assumption and proceed as usual. On the other hand it could be argued that, as in the first situation, larger values of y might be expected to be more variable. In this case we cannot transform the values of y to make the variation more constant without losing the linearity achieved by transforming the value of x. The first aim of a transformation is always to achieve linearity. More complex regression techniques, known as weighted regression, can be used if the transformation to linearity has not also achieved a constant variance.

The third situation is a typical example of probit analysis and we shall discuss this form of model in Chapter 14. An alternative approach to describing and fitting curved relationships is to consider the polynomial family of response curves. The first members of this family of curves are the linear response, which we have already considered

$$y = a + bx,$$

the quadratic response

$$y = a + bx + cx^2$$

which is frequently used as an empirical description of a simple curved relationship, and the cubic response

$$y = a + bx + cx^2 + dx^3$$

To be able to fit these models we have to learn how to express the dependence of y on more than a single x variable, and this is the purpose of the rest of this chapter.

10.2 TWO INDEPENDENT VARIABLES

All the relationships we have considered so far have been between two variables. In most cases this will be a considerable simplification of the true situation and could be highly misleading. Take, for example, the following hypothetical data from a survey of fertilizer practice in agricultural crops:

Place	A	B	C	D	E
Yield (cwt per acre)	24	30	27	29	20
Fertilizer application	50	100	200	20	100

A glance at these figures or their graphical representation in Fig. 10.3(a) gives no suggestion of any relationship between yield and fertilizer level. However, suppose we now have further information on the rainfall at the five places during the growing season as follows:

Place	A	B	C	D	E
Rainfall (in)	3	5	1	6	0

If we make the assumption that there is a positive linear relationship between fertilizer and yield then the places where yields are 'too high' are B and D which had above average rainfall whereas yields at C and E with little rain are 'too low'. Hence it seems likely that if we assume that rainfall and fertilizer both simultaneously affect yield then we might well find a strong relationship of yield with fertilizer. Thus, for example, if we guessed that yield would increase by 2 cwt per acre for each extra inch of rain and on this basis corrected the yields to those which we would have expected at the average rainfall of 3 inches we would obtain the revised relationship of Fig. 10.3(b) by the following series of calculations.

	A	B	C	D	E
Rainfall	3	5	1	6	0
Difference from mean (= 3)	0	+2	−2	+3	−3
Yield	24	30	27	29	20

Correction to yield to allow for rainfall being higher or lower than average:

	A	B	C	D	E
− Difference × 2 cwt per acre	0	−4	+4	−6	+6
'Corrected' yields	24	26	31	23	23
Fertilizer level	50	100	200	20	100

If we were now to do a regression of corrected yield on fertilizer level we should clearly find a strong effect, significant at the 5% level. Note that we are not

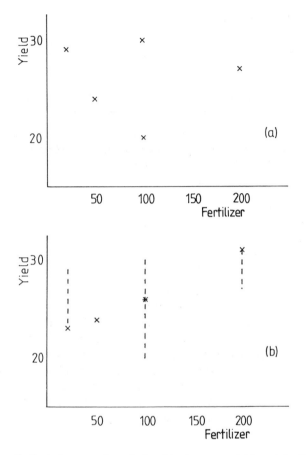

Fig. 10.3 Hypothetical data on the relationship between yield and fertilizer level, (a) with no adjustment for rainfall, (b) with yields adjusted for rainfall.

saying that the effect of rainfall is to increase yield by 2 cwt per acre for each inch of rain, merely that if we assume this is the effect then a clear relationship between yield and fertilizer level is suggested.

We could have tried different rates of increase of yield with rainfall, such as 1.5, 2.5 or more extreme values. For each set of 'corrected' yields allowing for a certain level of the effect of rain we could have fitted the regression of corrected yield on fertilizer level. In order to see which was the best of the effect of rainfall on yield we could argue as previously that it was the rate for which the residual sum of squares of the regression of corrected yield on fertilizer level was smallest. In other words we are using very much the same argument as when we originally tried to decide the best line for a simple linear relationship. Here again we can use mathematics to find the best value for the effect of rainfall.

To use mathematical techniques we must express the problem in a less 'wordy' fashion. If we call yield y, fertilizer level x_1, and rainfall x_2 then we want to find the rate at which yield increases with rainfall, say b_2 so that when we correct the yields from y to $y - b_2(x_2 - \bar{x}_2)$ we get the best regression on x_1. Equivalently we want to find the best line

$$y - b_2(x_2 - \bar{x}_2) = a_1 + b_1 x_1$$

or, adding $b_2(x_2 - \bar{x}_2)$ to both sides

$$y = a_1 + b_1 x_1 + b_2 x_2 - b_2 \bar{x}_2$$

and since $b_2 \bar{x}_2$ is the same for all values of x_2, we can again rewrite this as

$$y = a + b_1 x_1 + b_2 x_2$$

where $a = a_1 - b_2 \bar{x}_2$. The values b_1 and b_2 are the regression coefficients of y on x_1 and x_2 respectively. Sometimes b_1 and b_2 are called partial regression coefficients to distinguish them from ordinary regression coefficients obtained when y is regressed on only one variable.

Now this is a most interesting form because x_1, fertilizer level, and x_2, rainfall, appear in exactly the same way. Obviously we would have got the same form by considering a possible correction to the yields for the effect of fertilizer and then the regression of the corrected yields on rainfall. What we are in fact trying to do is to estimate the joint effects of both rainfall and fertilizer on yield. Obviously it would be easier to do this if we could control rainfall and fertilizer and change only one at a time. Equally obviously in survey data we cannot do this and it is fortunate that there is a statistical technique which will allow us to estimate both effects simultaneously. This general technique is called multiple regression and we are here concerned with the special case in which we have two x variables, x_1 and x_2.

The y variable is called the dependent variable and the x variables are called independent variables. The use of the word 'dependent' for the y variable is reasonable because we can generally think of the value of y as depending on the values of the x variables rather than the reverse. For example, the variable y may be the yields of a crop and the x variables, the amounts of various different fertilizers applied. However, unless the data come from carefully controlled experiments with adequate randomization, we must be careful not to argue that we are establishing causal relations linking y with the x. Providing that this is remembered, the use of the word dependent for the y variable is clearly not too misleading. On the other hand, describing the x variables as the independent variables is most unfortunate. In survey data, and sometimes in experiments, the values of the different x variables are certainly not independent of each other. This does not rule out the use of multiple regression techniques. Indeed, they are at their most powerful and useful when there are correlations among the x variables. Appreciable correlations obviously do cause problems in disentangling the effects of the x variables involved but multiple regression techniques are still applicable and are indeed the only techniques that take proper account of these correlations or dependencies.

Causal relations can be inferred from significant regression relationships when the values of the x variables are carefully controlled in an experiment with adequate randomisation. The randomization ensures that the values of the x variables are not systematically correlated with the values of any other variables or factors that may affect the dependent y variable. In surveys, this randomization of the correlations with other variables is not possible and so causal relations cannot be inferred except in the unlikely event that all possible important x variables are measured and included in the regression analysis.

Sometimes regression analysis is used not to establish causation but to select variables that can be used in the future as predictors of the values of the y variables. For example, the x variables may be easily measurable characteristics of patients thought to be suffering from a certain disease. The values of these x variables are thought to be correlated with a definitive characteristic of the disease, y, that is more difficult and more expensive to measure. The values of a selection of these x variables that are highly correlated with the y variable may then be used to screen the patients. Only those with, say, high values of the selected x variables will be measured for the y variable but it does assume that the correlation structure of the y and x variables will be the same in the future as it was in the data on which their selection was based. We will return to this point when discussing a multiple regression analysis of some plankton data in Example 10.1.

We can rewrite the criterion of minimizing the sum of squared deviations, or the principle of least squares, for this relationship as follows. The value of y expected on the basis of the relationship is

$$(a + b_1 x_1 + b_2 x_2)$$

The difference of the observed value, y, from this is

$$y - (a + b_1 x_1 + b_2 x_2).$$

The sum of squared deviations is

$$S = \Sigma [y - (a + b_1 x_1 + b_2 x_2)]^2$$

The principle of least squares means that we find the value of a, b_1, and b_2 which make S as small as possible. Some elementary calculus provides the following formulae for a, b_1 and b_2:

$$b_1 = \frac{S_{x_1 y} S_{x_2 x_2} - S_{x_2 y} S_{x_1 x_2}}{S_{x_1 x_1} S_{x_2 x_2} - (S_{x_1 x_2})^2}$$

$$b_2 = \frac{S_{x_2 y} S_{x_1 x_1} - S_{x_1 y} S_{x_1 x_2}}{S_{x_1 x_1} S_{x_2 x_2} - (S_{x_1 x_2})^2}$$

where $S_{x_1 x_2} = \Sigma(x_1 - \bar{x}_1)(x_2 - \bar{x}_2)$, the corrected sum of products of x_1 and x_2, and the other S quantities are the usual corrected sums of squares and products,

$$a = \frac{\Sigma(y) - b_1 \Sigma(x_1) - b_2 \Sigma(x_2)}{n}$$

The regression sum of squares (the variation of y accounted for by the relationship of y with x_1 and x_2) can be shown to be $b_1 S_{x_1 y} + b_2 S_{x_2 y}$ and hence the residual sum of squares is

$$S_{yy} - b_1 S_{x_1 y} - b_2 S_{x_2 y}$$

Because we have estimated one more quantity, the second regression coefficient, this r.s.s. has one fewer degrees of freedom than the residual sum of squares aftet regression on one variable, giving $(n - 3)$ d.f. Hence we have an estimate of the random variance

$$s^2 = \frac{\text{residual sum of squares}}{n - 3}$$

The standard errors of the regression coefficients are

$$\text{s.e.} = (b_1) = \sqrt{[s^2 S_{x_2 x_2}/(S_{x_1 x_1} S_{x_2 x_2} - (S_{x_1 x_2})^2)]}$$
$$\text{s.e.} = (b_2) = \sqrt{[s^2 S_{x_1 x_1}/(S_{x_1 x_1} S_{x_2 x_2} - (S_{x_1 x_2})^2)]}$$

We can use these standard errors to make inferences about the true values of b_1, or of b_2. However the information about b_1 and b_2 from the regression fitting is not independent because the two regression coefficients are estimated from a single set of data. (Note that this is different from information from a designed experiment structure where the estimates of treatment contrasts are independent (or orthogonal)). The interdependence of b_1 and b_2 can be expressed through the correlation of the two coefficients

$$\text{Corr}(b_1, b_2) = Sx_1 x_2/\sqrt{(Sx_1 x_1 \, Sx_2 x_2)}$$

Alternatively, we may describe the pattern of joint precision by the covariance

$$\text{Cov}(b_1, b_2) = s^2 Sx_1 x_2/(Sx_1 x_1 \, Sx_2 x_2 - (Sx_1 x_2)^2)$$

which will usually be used with the variances of b_1 and b_2 (variance = (standard error)^2)

The use of the covariance is mainly when a combination of b_1 and b_2 is considered. If we wish to predict y for a specific pair of values of x_1 and x_2, (x_{10}, x_{20}). The predicted value would be

$$a + b_1 x_{10} + b_2 x_{20}$$

The standard error of the predicted value derives from the variances and covariances of a, b_1 and b_2.

$$\text{s.e.}(a + b_1 x_{10} + b_2 x_{20})$$

$$= \sqrt{\left\{ s^2 \left(\frac{1}{n} + \frac{Sx_1 x_1 (x_{20} - \bar{x}_2)^2 + Sx_2 x_2 (x_{10} - \bar{x}_1)^2 - 2Sx_1 x_2 (x_{10} - \bar{x}_1)(x_{20} - \bar{x}_2)}{[Sx_1 x_1 \, Sx_2 x_2 - (Sx_1 x_2)^2]} \right) \right\}}$$

For the data on yield, fertilizer level and rainfall

$$\Sigma(y) = 130 \qquad \Sigma(y^2) = 3446 \qquad Syy = 66$$
$$\Sigma(x_1) = 470 \qquad \Sigma(x_1^2) = 62900 \qquad S_{x_1 x_1} = 18720$$
$$\Sigma(x_2) = 15 \qquad \Sigma(x_2^2) = 71 \qquad S_{x_2 x_2} = 26$$
$$\Sigma(x_1 y_1) = 12180 \qquad S_{x_1 y} = -40$$
$$\Sigma(x_2 y) = 423 \qquad S_{x_2 y} = 33$$
$$\Sigma(x_1 x_2) = 970 \qquad S_{x_1 x_2} = -443$$

$$b_1 = \frac{(-40)(26) - (33)(-440)}{(18720)(26) - (-440)^2} = \frac{13480}{293120} = 0.0460$$

$$b_2 = \frac{(33)(18720) - (-40)(-440)}{293120} = \frac{600160}{293120} = 2.047$$

$$a = \frac{130 - (0.0460)(470) - (2.047)(15)}{5} = 15.54$$

regression sum of squares $= (0.0460)(-40) + (2.047)(33) = 65.711$

$$\text{residual s.s.} = 0.289 \qquad s^2 = \frac{\text{residual s.s.}}{n\text{-}3} = 0.14$$

$$\text{s.e.}(b_1) = \sqrt{[(0.14)(26)/293120]} = 0.0035$$
$$\text{s.e.}(b_2) = \sqrt{[(0.14)(18720)/293120]} = 0.095$$
$$\text{Cov}(b_1, b_2) = 440/\sqrt{(18720)(26)} = -0.63$$

The precision of each regression indicates that both coefficients are well-determined. The b_1 value of 0.0460 has a standard error of 0.0035 and a 95% confidence interval of

$$0.0460 \pm (0.0035)(4.3) \text{ or } (0.031 \text{ to } 0.061),$$

using the 5% t value for 2 d.f. $(n - 3 = 2)$. The 95% confidence interval for b_2 is

$$2.047 \pm (0.095)(4.3) \text{ or } (1.64 \text{ to } 2.46)$$

The correlation of -0.63 shows that the uncertainty about the b_1 and b_2 values is quite strongly interdependent.

It is important to have a clear idea of the interpretation of the regression coefficients, b_1 and b_2. Each coefficients is an estimate of the increase of yield when the independent variable corresponding to that coefficients is increased by one unit, while the other independent variable is kept constant. Thus if we increase the fertilizer application by 10 units we expect an increase of 10×0.046 or 0.46 unit of yield. To increase yield by one unit we would therefore expect to have to increase the amount of fertilizer applied by about 20 units. If, on the other hand, we keep the level of fertilizer fixed then to increase yield by one unit we need about half an inch more rain.

10.3 TESTING THE COMPONENTS OF A MULTIPLE RELATIONSHIP

When we have two independent variables, we should be interested in assessing whether both or one or neither are important in determining yield. There are two methods available. The first is to use the standard errors of the regression coefficients to test the significance of each coefficient. This can lead to one of three results:

1. Both coefficients significant;
2. Only one coefficient significant;
3. Neither coefficient significant.

In the first situation the conclusion is clear; the dependent variable is related to both independent variables. It might appear that (2) and (3) also have clear-cut interpretations. However, there are some logical difficulties. With (2), suppose the coefficient of x_1 is significantly different from zero at, say, the 5% level, but the coefficient of x_2 is not. We may then reasonably accept the null hypothesis that y is not dependent on x_2. However, the coefficient of x_1 and the subsequent test of its significance are based on the assumptions that y is dependent on x_2. If we decide to omit x_2 then we must recalculate the regression of y on x_1 alone to be able to test the importance of x_1. With (3), when we find both coefficients non-significant we are saying

1. That if y depends on x_2 then the dependence of y on x_1 is not significant so that we can eliminate x_1, and
2. That if y depends on x_1 then the dependence of y on x_2 is not significant so that we can eliminate x_2.

Examination of these statements shows that they do not imply that both x_1 and x_2 should be ignored but rather that y is dependent on at most one of them. By attempting to examine the dependence of y simultaneously on both x_1 and x_2 we are asking the data to tell us about the effect on y of changing x_1 with x_2 held constant, and about the effect on y of changing x_2 with x_1 held constant. However, the data may not be sufficiently informative to apportion the separate effects of x_1 and x_2 so that we cannot recognize clearly which of them is the important variable. Hence to complete the investigation we must calculate the simple linear regressions of y on x_1 and of y on x_2. However, if both simple linear regressions are then found to be significant, as occurs quite commonly, then we still do not have a clear interpretation. The situation can be summarized as follows:

1. If we believe either one of the independent variables does have an effect on the dependent variable, then the other independent variable appears to be unimportant.
2. If we assume that one variable can be ignored as having no effect then the remaining variable clearly has a significant effect.

The most reasonable interpretation is that the dependent variable is affected by one or other of the independent variables but not by both. And the data is inadequate to tell us which independent variable is the important one.

An alternative way of presenting tests of significance of the independent variable is through the analysis of variance.

This time we consider from the start the four possible regressions

1. With both independent variables
2. With x_1 only
3. With x_2 only
4. With neither

and compare the four residual sums of squares. First consider the residual sums of squares for both variables included (R.S.S.$_{12}$) and for only x_1 included (R.S.S.$_1$). Because the model with both x_1 and x_2 included is a modification of that with only x_1, R.S.S.$_{12}$ must be an improvement on R.S.S.$_1$; in other words R.S.S.$_{12}$ must be smaller than R.S.S.$_1$. If x_2 is irrelevant to y then the reduction in the RSS will be small, about the size of the random variance. If, on the other hand, the difference R.S.S.$_1$ − R.S.S.$_{12}$ is large compared with the random variance then this indicates that a considerable part of the variation of y can be attributed to variation of x_2, **given that there is assumed to be a relationship between** y **and** x_1. This last assumption is important—we are comparing the residual variation about two assumed relationships both of which include x_1. Hence we are testing the effect of x_2 on y assuming that x_1 does have an effect on y. This situation can be compared with that in which we compare the residual for only x_2 included (R.S.S.$_2$) with the residual with neither x_1 nor x_2 included (R.S.S.$_0$). Again the difference between the two residuals tells us about the effect of x_2 on y. In fact the difference, R.S.S.$_0$ − R.S.S.$_2$ is the regression sum of squares for x_2 when we fit a simple linear regression of y on x_2. In this case there are no assumptions about x_1 and hence comparing the difference R.S.S.$_0$ − R.S.S.$_2$ with the random variance provides a test of the effect of x_2 on y, with x_1 assumed to be irrelevant.

The tests may be put formally in an analysis of variance as follows:

Source of variation	s.s.	d.f.	m.s.
Residual variation about full model (1)	$R.S.S._{12}$	$n-3$	s^2
Residual variation about model (2)	$R.S.S._1$	$n-2$	
Residual variation about model (3)	$R.S.S._2$	$n-2$	
Total variation = residual variation model (4)	$R.S.S._0$	$n-1$	

Effect of x_2, assuming x_1 in model $R.S.S._1 - R.S.S._{12}$ $F_a = \dfrac{R.S.S._1 - R.S.S._{12}}{s^2}$

Effect of x_1, assuming x_2 irrelevant $R.S.S._0 - R.S.S._1$ $F_b = \dfrac{R.S.S._0 - R.S.S._1}{s^2}$

Effect of x_1, assuming x_2 in model $R.S.S._2 - R.S.S._{12}$ $F_c = \dfrac{R.S.S._2 - R.S.S._{12}}{s^2}$

Effect of x_2, assuming x_1 irrelevant $R.S.S._0 - R.S.S._2$ $F_d = \dfrac{R.S.S._0 - R.S.S._2}{s^2}$

Comparison of F_a with the F-distribution on 1 and $n-3$ d.f. tests the hypothesis that y is dependent on x_2 given that y is dependent on x_1. Comparison of F_b with the same F-distribution tests the hypothesis that y is dependent on x_1 assuming that y is independent of x_2. We have a similar analysis of variance for the alternative sequence of models.

The whole pattern of testing can be represented pictorially as in Fig. 10.4.

Another way of stating the general problem of testing which x variables are important is that we are searching for a model with as few independent variables as is compatible with the requirement that no model including additional variables gives a significantly lower residual variation. In terms of Fig. 10.4, we are looking for the highest model in the tree which is not worse than any of the models below it. The language of these definitions of aims has been deliberately couched in terms of more than two independent variables and we shall return to this discussion of general multiple regression briefly in the next section.

Let us now look at the data on rainfall, fertilizer level and yield using this analysis of variance approach. The residual sums of squares are as follows:

$R.S.S._0 = S_{yy} = 66$

$R.S.S._{12}(\text{full model}) = 0.29$

$R.S.S._1(x_1 \text{ only}) = S_{yy} - \dfrac{(S_{x_1y})^2}{S_{x_1x_1}} = 66 - \dfrac{(-40)^2}{18720} = 65.91$

$R.S.S._2(x_2 \text{ only}) = S_{yy} - \dfrac{(S_{x_2y})^2}{S_{x_1x_1}} = 66 - \dfrac{(33)^2}{26} = 24.12$

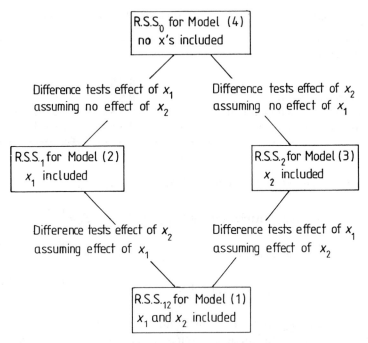

Fig. 10.4 Diagrammatic representation of the comparison of the regression R.S.S. for models 1 to 4.

Hence the analysis of variance:

Model	R.S.S.	d.f.	m.s.	F
(1) x_1 and x_2	0.29	2	0.14	
(2) x_1 only	65.91	3		
(3) x_2 only	24.12	3		
(4) neither	66.00	4		
Test of x_2 assuming x_1	65.62	1	65.62	469***
Test of x_1 ignoring x_2	0.09	1	0.09	0.6
Test of x_1 assuming x_2	23.83	1	23.83	170**
Test of x_2 ignoring x_1	41.88	1	41.88	299***

The logical deduction from these analyses is that both variables, rainfall (x_2) and fertiliser level (x_1), affect yield. For the effect of x_2 is significant whether or not we assume x_1 to have an effect, and if x_2 is assumed to have an effect then the effect of x_1 is significant. Hence both effects are significant. Again a diagrammatic representation as in fig. 10.5 my help.

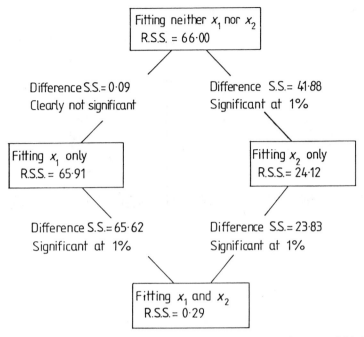

Fig. 10.5 Comparison of the different regressions fitted for the data on yield, fertiliser and rainfall.

Two final points on the technical side of these tests. It is possible to compare the residual sums of squares for the full model and for the model with neither x_1 nor x_2 and to test the hypothesis that x_1 and x_2 are both irrelevant to y. The F-statistic is

$$F = \frac{(\text{R.S.S.}_0 - \text{R.S.S.}_{12})/2}{s^2}$$

the divison 2 being the difference between the d.f. for R.S.S.$_{12}$ and R.S.S.$_0$. The degrees of freedom for this F-test are 2 and $(n-3)$. Second, it is not possible to test the difference between R.S.S.$_1$ and R.S.S.$_2$. This is essentially because neither regression model is a special case of the other as is the case in comparing

$$\text{Model (1)} \quad y = a + b_1 x_1 + b_2 x_2 \quad \text{with}$$

$$\text{Model (2)} \quad y = a + b_1 x_1$$

Example 10.1 Multiple regression with two variables

Part of an investigation of the factors affecting the productivity of plankton in the River Thames at Reading consisted of taking measurements at 17 monthly intervals of the production of oxygen ($y=$ in milligrams per cubic metre per

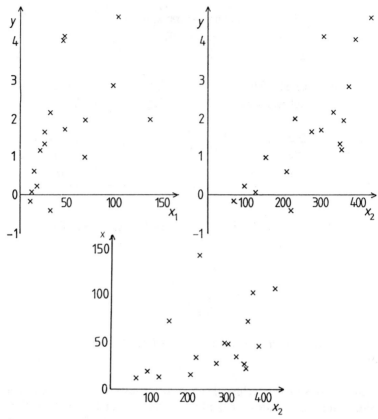

Fig. 10.6 Data on relationships for plankton of oxygen production (y) with chlorophyll (x_1) and light (x_2): (a) y plotted against x_1; (b) y plotted against x_2; (c) x_1 plotted against x_2.

day), the amount of chlorophyll present (x_1 in micrograms per cubic metre per day) and the amount of light (x_2 in calories per square centimetres per day). The data collected was as follows:

y	x_1	x_2	y	x_1	x_2
2.16	33.8	329.5	1.16	22.5	258.5
4.13	47.8	306.8	0.61	16.5	210.0
2.84	100.7	374.7	1.94	71.3	361.8
4.65	105.5	432.8	1.70	49.4	300.4
−0.42	33.4	222.9	0.2	19.3	96.9
1.32	27.0	352.1	0.98	71.6	151.8
4.04	46.0	390.8	0.06	13.4	126.0
1.97	139.5	232.6	−0.19	11.8	67.8
1.63	27.0	277.7			

Figure 10.6 shows plots of y against x_1(a), y against x_2(b) and x_1 against x_2 (c). These graphs are essential to check the various assumptions that are made in a regression analysis. Here we assume as always in regression analysis that the relationship of y to each of the x variables is linear, and that the deviations about the multiple regression equation are normally distributed, independently of each other, with the same variance for all values of the x variables. Not all these assumptions are easy to check but a few graphs can often highlights any serious failure there my be in the assumptions. Clearly there is some evidence that y increases with either x_1 or x_2. On the other hand the data are such that x_1 and x_2 themselves tend to increase or decrease together so that the apparent relationships of y with both x_1 and x_2 may be open to some suspicion. To isolate the separate effects of chlorophyll and light on oxygen production and to test their significance, we must use multiple regression techniques to allow for the correlation between x_1 and x_2. The calculations can be done by hand using the formulae presented in sections 10.2 and 10.3 but it is much more convenient to fit multiple regression models using a statistical package. The steps in the analysis and interpretation of the relationships between the variables using output from a typical package are as follows:

First we fit the simple regressions:

(a) Oxygen production (y) on chlorophyll (x_1).
A simple regression model is fitted relating y to x_1 (ignoring x_2)
The regression equation is

$y = 0.593 + 0.00224\, x_1$

Predictor	Coef.	Standard error	t-ratio	p
Constant	0.5927	0.5419	1.09	0.291
x_1	0.022372	0.008904	2.51	0.024

$R - sq = 29.6\%$

Analysis of variance

Source	d.f.	S.S.	m.s.	F	p
Regression	1	10.915	10.915	6.31	0.024
Error	15	25.933	1.729		
Total	16	36.848			

(b) Oxygen production on light
A simple regression model is fitted relating y to x_2 (ignoring x_1)
The regression equation is

$y = -1.18 + 0.0106x_2$

Predictor	Coef	Standard error	t-ratio	p
Constant	-1.1771	0.6596	-1.78	0.095
x_2	0.010625	0.002271	4.68	0.000

$R - sq = 59.3\%$

Analysis of variance

Source	d.f.	S.S.	m.s.	F	p
Regression	1	21.867	21.867	21.89	0.000
Error	15	14.981	0.999		
Total	16	36.848			

In both cases the regression is significant ($p = 0.024$ for x and $p = 0.000$ for x_2)
We next calculate the multiple regression:

The regression equation is
$y = -1.34 + 0.0118x_1 + 0.00908x_2$

Predictor	Coef.	Standard error	t-ratio	p
Constant	-1.3384	0.6289	-2.13	0.052
x_1	0.011776	0.006932	1.70	0.111
x_2	0.009077	0.002326	3.90	0.002

$R - sq = 66.3\%$

The variance-covariance matrix for the three parameter estimates is

Constant	0.3955		
x_1	-0.0006582	0.00004805	
x_2	-0.0011504	-0.000006209	0.000005410

Analysis of variance

Source	d.f.	S.S.	m.s.	F	p
Regression	2	24.427	12.214	13.77	0.000
Error	14	12.42	0.887		
Total	16	36.848			

Source	d.f.	Adjusted S.S.	Adjusted m.s.	F	d
x_1	1	2.56	2.56	2.88	0.112
x_2	1	13.512	13.512	15.23	0.002

The adjusted sum of squares for x_1 is the sum of squares for x_1 given that x_2 has already been included in the model. Similarly the adjusted sum of squares for x_2 is the sum of squares for x_2 given that x_1 has already been included in the model.

Adjusted S.S.x_1 = Residual S.S. full model − Residual S.S. regression on x_2 only and similarly:

Adjusted S.S.x_2 = Redidual S.S. full model − Residual S.S. regression on x_1 only

There is ample evidence that there is an effect of x_2 on y. However, the effect of x_1 on y is less well substantiated, the F-test of the hypothesis $b_1 = 0$ on the assumption that x_2 affects y not being significant at the 5% or 10% level ($p = 0.112$). If we are predisposed to expect an effect of x_1 then we would probably prefer the full model and write the relationship

$$y = 0.118x + 0.0091x_2 - 1.34$$

with standard errors of the coefficients, 0.0069 for x_1 and 0.0023 for x_2.

Alternatively we may prefer to accept the simpler model including only x_2,

$$y = 0.0106x_2 - 1.18$$

the standard error of the coefficient of x_2 being 0.0023.

The decision whether or not to include a variable in a regression equation will depend on many considerations, not all of them statistical. In a scientific investigation, the inclusion of a variable will depend on *a priori* considerations and on the information provided by the data about the values of the corresponding regression coefficient that are compatible with those data. For example, the largest value contained in a 95% confidence interval for the regression coefficient may be sufficiently small, in scientific or in commercial terms, for the effects of the variable, over some defined range, to be judged unimportant. If variables are being chosen more pragmatically to be used in predicting, for example in medical diagnosis, then the effects of any increased precision of prediction may have to be set against the costs of measuring the extra variable.

The overall adequacy of the full multiple regression model can be assessed by examining the standardized residuals calculated as deviations of observed from fitted values, standardized by the residual standard deviation, as defined in section 9.7.

Row	y	x_1	x_2	Standardized residual	Fitted values
1	2.16	33.8	329.5	0.12291	2.05055
2	4.13	47.8	306.8	2.33193	2.00936
3	2.84	100.7	374.7	−0.48279	3.24865
4	4.65	105.5	432.8	1.00982	3.83256
5	−0.42	33.4	222.9	−1.65416	1.07821
6	1.32	27.0	352.1	−0.98665	2.17562
7	4.04	46.0	390.8	1.48791	2.75065
8	1.97	139.5	232.6	−0.71120	2.41569
9	1.63	27.0	277.7	0.14424	1.50027
10	1.16	22.5	358.5	−1.19570	2.18072
11	0.61	16.5	210.0	−0.17125	0.76210
12	1.94	71.3	361.8	−0.95006	2.78534
13	1.70	49.4	300.4	−0.29646	1.97011
14	0.21	19.3	96.9	0.52965	−0.23156
15	0.98	71.6	151.8	0.11621	0.88266
16	0.06	13.4	126.0	0.11371	−0.03689
17	−0.19	11.8	67.8	0.49185	−0.58403

A useful visual check can be made by plotting the residuals against the fitted values from the model, if the assumptions underlying the regression model hold

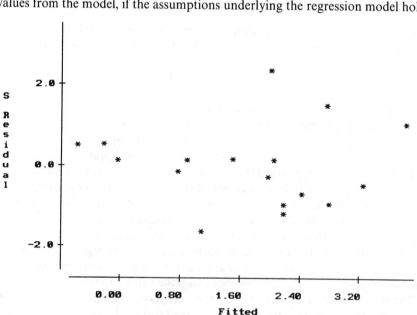

Fig. 10.7 Residual values plotted against fitted values for the joint regression of Example 10.1.

then the residuals will be a random sample from a normal distribution. If any systematic patterns are apparent in the residual plot then this may indicate a violation of the model assumptions and a more complex model is required. Further details on residual plots may be found in Draper and Smith (1981). A residual plot for this data is shown in Fig. 10.7.

EXERCISE 10.1   *Multiple regression*

In an investigation into the interdependence of water uptake, food intake and egg production the following records were obtained from a ten-day period of observation of 12 birds.

Water uptake (y) (ml per bird/day)	Food intake (x_1) (g per bird/day)	Egg production (x_2) (eggs/10 days)
175	90	2
342	150	5
252	114	6
362	152	9
284	122	6
219	117	7
229	114	4
260	117	6
88	55	0
132	72	0
254	106	2
199	93	0

Clearly all three variables increase and decrease together. The particular question for which the data were obtained was 'Does water consumption depend on food intake or on egg production or both?' Fit the separate regressions of water uptake (y) on food uptake (x_1) and egg production (x_2) and also the multiple regression of y on x_1 and x_2 and give your answer to the question

$$\Sigma(y) = 2796 \qquad \Sigma(x_1) = 1302 \qquad \Sigma(x_2) = 47$$
$$\Sigma(y^2) = 720080 \qquad \Sigma(x_1^2) = 150052 \qquad \Sigma(x_2^2) = 287$$
$$\Sigma(x_1 y) = 327374 \qquad \Sigma(x_2 y) = 13051$$
$$\Sigma(x_1 x_2) = 5903$$

$$S_{yy} = 68612 \qquad S_{x_1 x_1} = 8785 \qquad S_{x_2 x_2} = 102.92$$
$$S_{x_1 y} = 24008 \qquad S_{x_2 y} = 2100$$
$$S_{x_1 x_2} = 803.5$$

10.4 MULTIPLE REGRESSION

Multiple regression is a large subject and in an introductory text we can only give a brief outline of the possibilities. For a more extensive description the reader may find Draper and Smith (1981) appropriate. Multiple regression is a very powerful tool for investigating relationships and here we shall describe the principles and illustrate the use of multiple regression. The calculations for fitting multiple regression will inevitably involve the use of a computer package.

The extension from the situation of the last two sections, where we considered the possible dependence of y, the dependent variable, on two independent variables, x_1 and x_2, to the more general situation where we allow y to be dependent on however many variables we think may be important, is difficult only in that the calculations become much more complex. The principle of testing the significance of particular regression coefficients and the logical approach to selecting the best set of independent variables, where best is to be interpreted as 'including all variables which really affect y, and only those', remain the same. Thus with three variables x_1, x_2 and x_3 we test the significance of the effect of x_1 by comparing the residual sums of squares (a) for the regression on all three variables, and (b) for the regression on x_2 and x_3 only. The difference between the residual sums of squares is compared with the residual mean square for the full model and the F-test is used. Alternatively, if we decide that x_2 is irrelevant, then we test the effect of x_1 by comparing the residual sums of squares for (c) the regression on x_1 and x_3, and (d) the regression on x_3 only. And so on.

In some applications there is a natural ordering of the independent variables. The best example of this is the fitting of a polynomial response function of x. By taking $x_1 = x$, $x_2 = x^2$, $x_3 = x^3, \ldots, y = a + bx + cx^2 + dx^3 + \ldots$, we can use multiple regression to fit to the values of the independent variables y, polynomials in x of any chosen degree. Obviously we would try fitting polynomials of increasing degree and so there is no question of the order in which to fit the various independent variables. First we fit (a, b); then (a, b, c); then (a, b, c, d) and so on. The fact that the independent variables are all functions of the same variable, $x, x^2, x^3, \ldots$ means that they will inevitably be quite highly correlated so that it may be difficult to identify the precise form of the relationship. The use of polynomials to describe a response to a quantitative factor is very common and may be justified as a short-term expedient to interpolate between a few data points. There is however generally very little, if any, biological justification for fitting polynomials and they should be used with caution. There are particular dangers in using polynomials to extrapolate beyond the range of the data from which they have been calculated.

In the more general problem of selecting the best multiple regression complications arise from the number of possible regressions to be calculated. The actual time to calculate the necessary regressions is not usually probibitive. The difficulties arise in sorting through the calculated regressions in a logical fashion while at the same time trying to build in the prior knowledge and expertise of the investigator. Many statistical packages offer a range of options for the selection of subsets of independent variables. The general problem of finding the set of independent variables which gives the 'best fit' is a subjective matter in which statistical tests should be combined with the researcher's knowledge about the variables concerned. Any computer program which purports to provide the 'best fit' must therefore be treated with caution, not least because different methods of finding the 'best fit' can give different answers.

The methods of selecting the best fit which may be encountered include

1. Finding the best single-variable regression, then the best two-variable regression, then the best three-variable regression and so on. These methods are based on complete surveys of all possible regressions or on logical elimination of regressions which cannot, because of results for the regressions already considered, be as good as the best so far found.
2. Selection of variables sequentially, at each stage adding the variable which most reduces the residual sum of squares.
3. The gradual elimination of variables starting from a complete model, at each stage eliminating that variable which least increases the residual sum of squares.

An important precaution when using a selection of variables procedure in a multiple regression package is to find out precisely what the procedure does and how it is programmed to stop. In particular any procedure labelled 'stepwise regression' must be identified more precisely as it may be (2) or (3) above or most probably, a modification of (2) in which variables are added sequentially, but at each stage the current set of variables are tested to see whether any variable can be eliminated. To illustrate the selection of a best multiple regression we return to Example 10.1 on plankton to illustrate how results from a multiple regression package can be interpreted.

Example 10.2 Multiple regression

In addition to the variables x_1 (chlorophyll) and x_2 (light) considered earlier, two other x variables were measured. These were temperature (in degrees Celcius), and a quantity called the 'compensation point' which effectively measures the opacity of the water in which the plankton are found. This gives us four possible independent variables but some thought about the nature of the problem suggested that to consider only these four might be an over-simplification. If we have a model of the form

$$y = a + b_1x_1 + b_2x_2 + b_3x_3 + b_4x_4$$

we are assuming that if we keep, say, x_2, x_3 and x_4 at fixed levels the effect on y of increasing x_1 by one unit is to increase y by b_1 units, regardless of the particular levels at which x_2, x_3 and x_4 are fixed. In the language of factorial experiments we are assuming that there is no interaction between the effects of

x_1, x_2, x_3, and x_4. Now it seems reasonable that the effect of light on production of oxygen by the plankton which are found some way below the surface of the water will depend to some extent on the level of the compensation point. Hence instead of assuming that our initial model for $x_2 =$ light is $y = a + b_2x_2$, and the effect of the variation of the compensation point (x_3) is to increase or decrease the oxygen production by a given amount, which leads to the model $y = a + b_2x_2 + b_3x_3$, we would like to say that the effect of light on oxygen production, measured by b_2, is itself dependent on x_3

$$b_2 = b + cx_3.$$

This gives as our new model

$$y = a + (b + cx_3)x_2 + b_3x_3$$

$$= a + bx_2 + b_3x_3 + c(x_3x_2)$$

If we define a new variable x_5 to be x_2x_3 then we can write this model as

$$y = a + bx_2 + b_3x_3 + cx_5$$

and we see that this is just another multiple regression model with x_5, our 'interaction' variable, considered in just the same way as any other independent variable. Remember that interrelations among the independent variables do not invalidate the use of multiple regression; rather multiple regression is the only technique which can manage such situations. In the analysis of the plankton it was thought that there were three possible interactions, those of light and compensation point, light and chlorophyll, and chlorophyll and compensation point. Thus the starting point of the multiple regression analysis was the possible dependence of y, the oxygen production, on seven independent variables:

$$x_1 = \text{chlorophyll}$$
$$x_2 = \text{light}$$
$$x_3 = \text{compensation point}$$
$$x_4 = \text{temperature}$$
$$x_5 = x_2x_3$$
$$x_6 = x_1x_2$$
$$x_7 = x_1x_3$$

There are a number of systematic approaches to choosing the 'best' regression but these ignore the knowledge of the system being investigated and therefore none of these was used when the data were originally examined. The account here follows that original examination closely and chronologically. Initially a set of eight regressions was tried. The sets of independent variables were

1. x_1 only (the linear regression on chlorophyll)
2. x_2 only
3. x_1 and x_2
4. x_1, x_2 and x_3

5. x_1, x_2 and x_4
6. x_1, x_2, x_3 and x_4
7. x_1, x_2, x_3, x_5, x_6 and x_7 (all except temperature)
8. all variables.

The reasons for this choice were that chlorophyll and light were the first two variables to come to the researcher's mind; temperature appeared unimportant; and it was desirable to see how much of the variation of y oxygen production, could be explained by all the variables considered. Obviously the results from the regressions contained a lot of information, but the single most informative quantity from each regression was the residual sum of squares and these are tabulated in Table 10.1.

As more variables are included in the regression the residual sum of squares decreases. Since the residual s.s. measures the unexplained variation this is a predictable pattern. Even adding a variable which is quite unrelated to the dependent variable will produce some small reduction in the residual sum of squares. The problem is to distinguish reductions in residual s.s. due to adding important variables from reductions which are inevitable because variables are being added. For this we need to know how big a reduction we should expect if the variables we are adding are irrelevant.

First consider the complete regression (h) in which all seven variables are included. The residual s.s. is 1.64 on 9 d.f. The residual mean square, $s^2 = 1.64/9 = 0.18$ gives an estimate of the variation of observations about the fitted relationship. To compare two different regressions it is necessary that one 'includes' the other in the sense that the set of variables for one includes all the variables for the other plus some additional variables. Thus regression (e) can be compared with (a), (b), (c), (f) or (h) but not with (d) or (g). To assess the effect of the additional variables we consider the difference between the residual sums of squares for the two regressions. If the additional variables are worthless then this difference should be about the same as (the number of additional variables) $\times$ (the residual m.s. for the full regression). For the special case of a single additional variable the difference between the two residual sums of squares

Table 10.1

Variables included	Residual sum of squares	Degrees of freedom
(a) 1	25.93	15
(b) 2	14.97	15
(c) 1, 2	12.42	14
(d) 1, 2, 3	7.38	13
(e) 1, 2, 4	10.65	13
(f) 1, 2, 3, 4	6.24	12
(g) 1, 2, 3, 5, 6, 7	1.71	10
(h) 1, 2, 3, 4, 5, 6, 7	1.64	9

should be about the same as the overall residual m.s. Thus if we compare regressions (c) and (e), the additional variable 4 appears to have an important effect since the difference in residual s.s. $12.42 - 10.65 = 1.77$, is large compared with the residual mean square of 0.18. In fact all the differences which can be compared in Table 10.1 seem to be important except for the difference between (g) and (h). This suggests that when all six other variables are included, variable 4, temperature, may be unimportant.

Having calculated a multiple regression equation, more detailed information about the contributions of the individual variables can be obtained by comparing each partial regression coefficient with its standard error. Examination of the standard errors of the regression coefficients of the other six variables when x_4 is omitted shows that the coefficients for x_1 and x_7 are large compared with their standard errors but the other four coefficients have large standard errors. These results suggest that x_1 and x_7 are probably important and that some of x_2, x_3, x_5 and x_6 may also be needed. The next group of regressions tried reflects the questions arising from these tentative conclusions. The variables included in the regression and the resulting residual sum of squares are shown in Table 10.2.

These results confirm that the regression of y on the two variables x_1 and x_7 is an excellent one, giving a residual sum of squares of 2.28, only 0.64 greater

Table 10.2

Variables included	Residual s.s.	d.f.
1, 7	2.28	14
2, 7	6.59	14
1, 5	5.30	14
1, 2, 5	4.86	13
1, 2, 6	11.70	13
1, 2, 7	2.02	13
1, 2, 3, 5	4.40	12
1, 2, 3, 7	1.85	12

Table 10.3

Variables included	Residual s.s.	d.f.
3	16.13	15
7	9.55	15
1, 3	8.43	14
3, 7	5.03	14
1, 3, 7	2.22	13
1, 4, 7	2.23	13

than the residual s.s. using all seven variables. All regressions not including both x_1 and x_7 are substantially worse than the (x_1, x_7) regression. Inclusion of further variables with x_1 and x_7 does not give any marked improvement. As a final check a further group of regressions was tried with results shown in Table 10.3. These merely confirmed the previous findings.

A simple illustration of the arguments involved in comparing the residual sums of squares may be helpful in explaining the logic of the method. The method of illustration has already been used in Figs. 10.4 and 10.5 for two x variables. Figure 10.8 shows the relevant part of the corresponding picture for the seven variables of this example. Each regression is shown as a box with the variables included and the resulting residual s.s. shown in the box. Lines are shown connecting those regressions which can be compared; regressions can only be compared if one 'includes' the other. The difference between the residual sums of squares can be read off easily and compared with the residual mean square for the full regression.

If we look at the regressions either with one x variable or with two we see that whenever a comparison can be made the difference between the residual sums of squares is large compared with 0.18 and hence the regression with any single

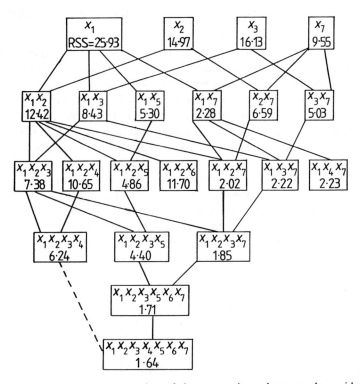

Fig. 10.8 Diagrammatic representation of the comparisons between the residual sums of squares for 22 fitted multiple regressions for the plankton data.

x variable must be rejected as inadequate. For example comparing (x_1, x_5) with (x_1) the difference in r.s.s. is $25.93 - 5.30 = 20.63$. If we now compare the two variable regressions with those with three variables then when the two variables are (x_1, x_7) the three variable regressions are little improvement. If we move on down from these three variable regressions including x_1 and x_7 we continue to find only small difference in r.s.s. showing that the (x_1, x_7) regression is not improved upon by adding other variables. It is, however, still possible that there are other regressions unrelated to (x_1, x_7) which are as good or better. However, all those tried give higher r.s.s. than (x_1, x_7) and if in fact we try regressions including all variables except either x_1 or x_7 then these also give a larger r.s.s.

Our conclusion then in this particular example, which it should be admitted is more clear-cut than is usual, is that the important variables are x_1 and x_7. Writing the final regression out in full, remember that x_7 was defined as $x_1 x_3$ we have

$$\text{oxygen} = 0.27 - 0.038(\text{chlorophyll})$$
$$+ 0.050(\text{chlorophyll} \times \text{compensation point})$$

or

$$\text{oxygen} = 0.27 + 0.050(\text{compensation point} - 0.76) \times \text{chlorophyll}$$

The interpretation is that when the compensation point is greater than 0.76, which occurs in 14 our of the 17 observations, then the more chlorophyll there is the greater is the oxygen production; further for a given amount of chlorophyll the oxygen production is greater when the compensation point is higher. On the other hand when the compensation point is low, less than 0.76, the effect of increasing the level of chlorophyll is to reduce the oxygen production, even to the point of producing negative oxygen production. This is not merely a figment of the imagination; in fact 'oxygen production' is really the balance between production and absorption; two of the oxygen production values in the original data (Example 10.1) are negative. The critical values of 0.76 probably corresponds to some real physical phenomenon since it does occur within the range of our data and is therefore not an extrapolation. We might summarize the results by saying that oxygen production appears to be dependent on the amount of chlorophyll present but the rate at which oxygen production increases with increasing chlorophyll depends on the level of the compensation point, the rate being high when the compensation point is high and negative when the compensation point is low or zero.

One aspect of multiple regression which the novice often finds surprising is the way in which the apparent importance of a variable changes according to the other variables included with it in the set of independent variables. Thus our final selected relationship in Example 10.2 includes x_1 and x_7 as the two important variables for predicting oxygen production. But in Example 10.1 variable x_1 (chlorophyll) appeared relatively unimportant compared with variable x_2 (light) which is ultimately omitted from the fitted relationship. The explanation of this phenomenon lies in the interrelationship of the observed values of the set of independent variables. Because x_1, x_2, x_3 and also x_5, x_6, x_7 which are derived from x_1, x_2, x_3 are to some degree interrelated, when we omit the important explanatory variable $x_7 (= x_1 x_3)$ the effect of x_1 may be masked by the variation of x_3, and x_2 may more nearly resemble x_7 than does x_1. Consequently x_2 acts (in Example 10.1) as a surrogate variable for the combination of x_1 and x_7.

Another example of a surrogate variable which is apparent on an initial examination of the data is the 'obvious' decline of oxygen production with time. This is a common phenomenon when data relating to periodic annual patterns are recorded for an incomplete number of years (17 months in this example). There is little doubt from the biologist's viewpoint that the final fitted model is a realistic description of the oxygen production process.

10.5 POSSIBLE PROBLEMS IN COMPUTER MULTIPLE REGRESSION

Even the computer can experience computational difficulties in multiple regression fitting and the users of multiple regression programs should beware of the computational problems which may arise. The principal source of numerical difficulties is the inclusion, in the set of independent variables, x_1, $x_2, \ldots, x_7$, of variables which are closely interrelated. Sometimes it may be known on biological grounds that two or more variables will be highly correlated; for example, if we are trying to relate crop yields to environmental factors we may consider both maximum daily temperature and total daily radiation as potential independent explanatory variables and these will tend to vary similarly over a range of different locations. Alternatively, we may construct variables for inclusion in a multiple regression equation knowing that they will be related. In the plankton example, we used $x_1 =$ chlorophyll, $x_2 =$ light and $x_7 = x_1 x_2$. Here there is an exact relationship and yet the relationship

$$y = a + b_1 x_1 + b_2 x_2 + b_7 x_7$$

is perfectly sensible since it says that a linear relationship between production (y) and chlorophyll (x_1) depends on the amount of light, both for the intercept and the slope of the production–chlorophyll relationship. An apparently even more extreme case is when we consider a quadratic relationship between y and x;

$$y = a + bx + cx^2$$

If we write $z = x^2$ this becomes a typical multiple regression relationship

$$y = a + bx + cz$$

and yet x and z are exactly related.

Provided the calculations are made with care all these multiple regression relationships with approximate or exact relations between the independent variables can be fitted except for one special case and that is when there is an exact **additive** relation between the x variables, of the form

$$x_1 = k_1 x_2 + k_2 x_3,$$

where any number of x variables may be involved in the relation. This occurs for example if x_1 = total monthly rainfall, and x_2, x_3, x_4, x_5 are the four weekly rainfalls for weeks $1, 2, 3$ and 4 of the month, when we would have the relation $x_1 = x_2 + x_3 + x_4 + x_5$. It is clearly meaningless to ask how y is related to x_1 as well as to x_2, x_3, x_4 and x_5. If a multiple regression program is required to fit the multiple regression of y on x_1 to x_5 then it will either produce a message indicating a numerical failure within the program, or it may be able to 'recognize' the impossibility of the request and 'refuse' to attempt the calculations. Users should accept that avoiding this situation is their responsibility.

11
Linear models

11.1 THE USE OF MODELS

So far in this book we have said very little about the use of models, either biological or mathematical, for the processes being investigated. Nevertheless, the analysis and interpretation of data which we have discussed in the preceding chapters have been based on implicitly assumed or in some cases, explicitly stated, models. As emphasized in Chapter 8, most of our analyses of experimental data assume that each yield from an experimental unit is the sum of a unit effect and the effect of the treatment applied to that unit. In addition, the set of unit effects is assumed to be, approximately, normally distributed. When units are grouped into blocks the unit effect is decomposed into a block effect and the residual unit, or error, effect. This assumption can be written formally as a model,

$$y_{ij} = b_i + t_j + e_{ij}.$$

If the treatments have a factorial structure, being combinations of fertilizer level (f_j) and variety (v_k), then if we assume that treatment effects are due simply to the main effects of fertilizer and variety our model for the analysis becomes

$$y_{ljk} = b_i + f_j + v_k + e_{ijk}$$

If, in addition, we allow for interaction between fertilizer and variety, so that different varieties react differently to changes of fertilizer, then the model becomes

$$y_{ijk} = b_i + f_j + v_k + (fv)_{jk} + e_{ijk}$$

This sequence of implied models had provided a general structure for the analysis of experimental data and thinking about the assumed models should have clarified the assumptions on which our conclusions are based. If we believe that the model adequately represents the biological situation then we shall be confident that the conclusions from the analysis are valid.

We may also formulate and use models more explicitly. Explicit models may be used to provide predictions, or to express and increase our understanding of underlying biological processes.

By choosing an appropriate model to describe the data, results may be predicted for situations not directly investigated. The linear regression and multiple linear regression models of Chapters 9 and 10 are examples of this second kind of model. If a simple linear regression model,

$$y_i = a + bx_i + e_i$$

is fitted to a set of data then from the resulting numerical form of the model we can predict the value of y for some particular value of x. This prediction may be the purpose of the analysis or it, and other predictions, may be tested against new observations and a modified model may be developed such as

$$y_i = a + bx_i + cz_i + e_i$$

where z is a second independent variable or

$$y_i = a + bx_i + cx_i^2 + e_i$$

Predictions from the new models can similarly be proposed and tested, hopefully leading to an improved predictive model.

A related use of models is to try to increase our understanding of the underlying biological process. The philosophy is that of the classical scientific method. By attempting to express the behaviour of a system in terms of a precise and quantitative model we can identify the form of experimental data which would provide the most critical test of our understanding. We then test the adequacy of our model (as in the lack of fit test in section 9.6) to describe the data we have collected. Since a model is always a simplification we will often detect a failure of the model. The data will provide some information on the way in which the model is inadequate and this should lead to an improved model and thence to further data, and the cycle continues.

In this chapter and in Chapters 12 and 14 we shall develop, more formally, the ideas of sequences of models for describing, interpreting and assessing data. The power of the model-based approach is substantially due to the computer facilities now available for fitting more complex models to data. We shall therefore assume the use of computer packages for all calculations and this will allow us to concentrate our attention on conceptual ideas of the models.

11.2 MODELS FOR FACTORS AND VARIABLES

There are several similarities and one major difference between the class of models implied in Chapters 5 to 8 for the analysis of data from designed experiments and those considered more explicitly for regression relationships in Chapters 9 and 10. The first similarity is that all the models assume that the observed yield or output (y) can be expressed as the sum of a systematic component and a random component. The systematic component represents an explanation of the major variation between yields, while the random component represents the uncontrolled, inherent, variation between the observed units.

Systematic	Random
$y_{ij} = b_i + t_j +$	e_{ij}
$y_i = b_0 + b_1 x_1 + b_2 x_2 + b_3 x_3 +$	e_i

Further similarities are that the systematic and random components are assumed to combine by simple addition, and that the systematic component may be composed of several subcomponents added together. A final similarity is that the error term, which constitutes the random component, is assumed to be approximately described by a normal distribution and this random variation of each y value about its systematic component value is assumed to be uneffected by the systematic component of the model. Note that these characteristics are precisely those identified in Chapter 8 for experimental data.

The major difference between experimental design models and the regression models is in the form of the individual components of the systematic component of the model.

In the experimental design model the systematic variation is described in terms of factors, with each factor including a set of possible levels. Every y observation includes exactly one level from each factor in its systematic part. The 'block' factor comprises (block 1, block 2, ...) and each observation occurs in exactly one block. Similarly, if we are considering a single, unstructured, set of treatments, the treatment factor comprises (treatment 1, treatment 2,...) and contributes one level to the systematic component of each observation. When the treatment for each unit is described in terms of several factors, then the effect of each treatment combination is represented by a combination of main effect factors and interaction factors.

Thus, for a trivial experiment, with single observations for six factorial combinations (2 genotypes $\times$ 3 chemicals) in two laboratories (block), the 12 observations are modelled in terms of two block effects (b_1, b_2), two genotypes (g_1, g_2), three chemical effects (c_1, c_2, c_3) and six genotype $\times$ chemical interaction effects $((gc)_{11}, (gc)_{12}, (gc)_{13}, (gc)_{21}, (gc)_{22}, (gc)_{23}$, as follows,

$$y_{111} = b_1 + g_1 + c_1 + (gc)_{11} + e_{111}$$
$$y_{112} = b_1 + g_1 + c_2 + (gc)_{12} + e_{112}$$
$$y_{113} = b_1 + g_1 + c_3 + (gc)_{13} + e_{113}$$
$$y_{121} = b_1 + g_2 + c_1 + (gc)_{21} + e_{121}$$
$$\vdots \qquad \vdots$$
$$y_{223} = b_2 + g_2 + c_3 + (gc)_{23} + e_{223}$$

For each observation the systematic component of the model includes exactly four terms. We have used a total of 13 factor level parameters to describe the systematic component of the model and the 12 observed data values clearly cannot provide sufficient infomation to identify appropriate numerical values for all 13 parameters. However, this superfluity causes no computational problems either for the analysis of variance or for the calculation of treatment means or of differences between means.

Note that we can write a general form of model to summarize the set of 12 models as follows,

$$y_{ijk} = b_i + g_j + c_k + (gc)_{jk} + e_{ijk}$$

In contrast the regression models are described in terms of explantory variables, with exactly the same parameters appearing as coefficients in the systematic component of the model for each observation. Thus, if growth is determined by temperature and rainfall then the general form of model would be

$$g_i = a + b_1 t_i + b_2 r_i + e_i,$$

and the models for individual observations

$$g_1 = a + b_1 t_1 + b_2 r_2 + e_1$$
$$g_2 = a + b_1 t_2 + b_2 r_2 + e_2$$

and so on.

Now the underlying arithmetical procedure for fitting either form of model to data, that is determining the most appropriate estimates for the parameters, is the same in both experimental data and regression data. Both forms of model are particular cases of the General Linear Model. The thinking which leads to a recognition that the two forms can be treated as one stems from the idea that we can define a 'dummy' variable for each level of an experimental factor in such a way that the parameter for that level is included in the model only for those observations for which it is required. A dummy variable for a particular level of a factor has only two possible values;

1 for observations for which the factor level is included, and
zero for all other observations.

Consider again the simple experimental structure with 2 blocks, 2 genotypes and 3 chemicals. We could define dummy variables

x_1 representing block 1
x_2 representing block 2
x_3 representing genotype 1
x_4 representing genotype 2
x_5 representing chemical 1
x_6 representing chemical 2
x_7 representing chemical 3
x_8 representing the (11) genotype–chemical combination
x_9 representing the (12) genotype–chemical combination
x_{13} representing the (23) genotype–chemical combination
Variable x_1 has values of 1 for $y_{111}, y_{112}, y_{113}, y_{121}, y_{122}, y_{123}$, and zero values for all other y's
Variable x_2 has values of 1 for $y_{211}, y_{212}, y_{213}, y_{221}, y_{222}, y_{223}$, and zero values for all other y's

Variable x_5 has values of 1 for $y_{111}, y_{121}, y_{211}, y_{221}$, and zero values for all other y's

Variable x_{13} has values of 1 for y_{123} and y_{223} and zero values for all other y's.

These definitions display the superfluous nature of the factor information, mentioned earlier. The two variables x_1 and x_2 represent exactly the same information since x_1 has value 1 whenever x_2 has the value zero, and **vice versa**. Thus we need only one of x_1 and x_2 to define the split of the data into two groups corresponding to the two blocks. In a similar way only one of x_3 and x_4 is needed, only two of x_5, x_6 and x_7, and only two of x_8 to x_{13}. Note that the number of dummy variables required to represent differences between the levels of a factor, or combinations of factors, is equal to the number of degrees of freedom. Hence in our example we require one for blocks, one for genotypes, two for chemicals and two for the genotype–chemical interaction.

Once we have represented factor levels by dummy variables we can analyse the results from a designed experiment by using a multiple regression program. When sets of dummy variables contain superfluous variables we include the minimum number necessary. Suppose we have the following set of yields:

Block	Genotype	Chemical	Yield
1	1	1	13
1	1	2	16
1	1	3	12
1	2	1	5
1	2	2	13
1	2	3	5
2	1	1	13
2	1	2	16
2	1	3	15
2	2	1	7
2	2	2	15
2	2	3	6

The analysis of variance for this (artificial) set of yields is

	s.s.	d.f.	m.s.
Blocks	5.33	1	5.3
Genotypes	96.33	1	96.3
Chemicals	80.67	2	40.3
G × C	20.67	2	10.3
Error	3.67	5	0.73
Total	206.67	11	

Analysing the data through a multiple regression program would require the following array of yields and dummy variables.

13	0	0	0	0	0	0
16	0	0	1	0	0	0
12	0	0	0	1	0	0
5	0	1	0	0	0	0
13	0	1	1	0	1	0
5	0	1	0	1	0	1
13	1	0	0	0	0	0
16	1	0	1	0	0	0
15	1	0	0	1	0	0
7	1	1	0	0	0	0
15	1	1	1	0	1	0
6	1	1	0	1	0	1
y	x_2	x_4	x_6	x_7	x_{12}	x_{13}

The analysis of variance can be constructed through fitting an appropriate sequence of models, successively including additional terms and noting the values of the residual sum of squares for each model. For this example, the residual sums of squares after fitting sets of dummy variable regression are

Fitting	Residuals s.s.	d.f.	Reduction in s.s.	
Mean (= Total s.s.)	206.67	(11)		
x_2	201.33	(10)	5.33	(1)
x_2, x_4	105.00	(9)	96.33	(1)
x_2, x_4, x_6, x_7	24.33	(7)	80.67	(2)
$x_1, x_4, x_6, x_7, x_{12}, x_{13}$	3.67	(5)	20.67	(2)

All the sums of squares required to complete the analysis of variance can be deduced from the set of residual sums of squares from this set of five regression results. The regression coefficients for the final model provide estimates of the block and treatment effect parameters from which the treatment mean yields can be calculated.

Note that this approach to the analysis of variance for experimental data enables us to handle missing data very simply. If an observation is missing it is simply not included in the analysis. Thus, if the last observation in the example is omitted, the resulting residual sums of squares are:

Fitting	Residual s.s.	(d.f.)	Reduction in s.s.
Mean	174.73	(10)	
x_2	158.13	(9)	16.60
x_2, x_4	91.59	(8)	66.54
x_2, x_4, x_6, x_7	21.26	(7)	70.33
$x_2, x_4, x_6, x_7, x_{12}, x_{13}$	3.67	(6)	17.59

However in this situation, where different sets of treatments occur in different blocks, the order in which terms are fitted modifies the sums of squares attached to each factor, as we would expect from multiple regression. The testing of an effect must use an order of terms in which the effect to be tested is the last term added to the model. To test all the effects of interest may require several orders of fitting. It is important when choosing different orders of fitting to remember that interactions are interpretable only when fitted with at least one of the related main effects. The special case with all treatment combinations occurring in each block is a special case of multiple regression where order of fitting is immaterial.

Note that with this approach to fitting models and calculating the analysis of variance we can analyse data from any of the incomplete block designs discussed in Chapter 5. In considering the analysis of incomplete block designs we must recognize that the precision of estimates of treatment differences will not usually be identical for all pairs treatments. The standard error of the estimate of a treatment difference will depend on the variances and covariance of the estimates of treatment effects. As with other uses of multiple regression, to calculate the standard errors of differences between estimates of treatment effects we need the variance–covariance matrix of the estimates. This should be available from the statistical package used for the multiple regression fitting. If the variances for treatment effects t_2 and t_3 are V_1 and V_2, and the covariance is V_{12}, then the standard error of the difference, $t_2 - t_3$, will be

$$\sqrt{(V_1 + V_2 - 2V_{12})}$$

When treatment effects are fitted using a general linear model package the treatment effects are often provided in the form of differences from an 'origin' treatment usually taken as treatment 1. The standard error for $t_2 - t_1$ is then given directly, and the standard error of $t_2 - t_3$ calculated as above from the variances and covariance of $t_2 - t_1$ and $t_3 - t_1$.

The purpose of this demonstration that a multiple regression approach with dummy factors can provide the analysis of variance for data from a designed experiment is not, of course, to suggest that this would usually be an appropriate method for analysing such data. For complete block designs and many incomplete block designs we would obviously expect to use factor models. The intention is to demonstrate that the apparent distinction between models expressed in terms of factors and those expressed in terms of variables is no barrier to analysis through a common method.

The benefit of this conclusion is that we can now consider models in which the systematic component includes both factor and variable terms. For example suppose we are investigating the regeneration of heathland managed (treated) in different ways, and that we also expect that rainfall, altitude and mean summer temperature affect regeneration rates, we could postulate a model

$$y_{ij} = a + b_1 r_{ij} + b_2 a_{ij} + b_3 t_{ij} + m_i + e_{ij}$$

when i represents different management methods and j represents replicate samples of heathland for a particular management method.

The effects of all the components of the model could be estimated simultaneously by fitting the complete model. The importance of rainfall, altitude, temperature on imposed management systems could be assessed by fitting various subsets of the terms.

In the next four sections we shall examine different ways of using models with a mixture of factors and variables for interpreting data involving structured relationships and for improving the interpretation of experimental data.

11.3 COMPARISON OF REGRESSIONS

In Chapters 9 and 10 we have used linear regression models for a single set of data. In biological research it is rare to have information about a relationship between variables from only a single set of data. More commonly there are several similar sets of data, possibly from widely spread locations, or obtained under a number of different environmental or applied conditions. For any fitted relationship to have biological relevance there must be some degree of repeatability. It is therefore important to be able to compare the fitted relationships for different sets of data. The models we shall use for this comparison will be the basic model for linear regression with additional components to represent the variation between the different data sets.

As an example consider the relationship between the weight of part of an animal and the weight of the whole animal. If we use logarithms of the weights then this kind of relationship is commonly found to be reasonably linear for animals at a particular time of the year or in a particular environment. However, the relationships for records taken at different times of the year may be markedly different. Another example, mentioned in Chapter 9, is the relationship between leaf area and leaf weight. For a particular species the relationship between log leaf area and log leaf weight is almost invariably found to be linear for a given variety or strain of the species grown in a particular environment. With other strains or in other environments the fitted relationship may still be linear but with different intercept or slope.

In these situations we are interested in comparing the fitted relationships for different times, environments or strains with the hope of finding a pattern in the relationships. The motivation for looking for such a pattern is twofold. First, if we are using the fitted relationship for prediction of the dependent variable it will be of little value if we have to re-estimate the relationship for every new strain or environment. Ideally the relationship we wish to use for prediction would be exactly the same for all strains for all environments, in which case, having examined the relationship for a representative set of strains and environments we should be able to use the fitted relationship from the entire set of data to make predictions for any strain in any environment. Even if the relationships are not identical there may be a certain degree of consistency which can be used, such as the slope of the line remaining the same while the intercept changes systematically with, say, the maximum temperature in the particular environment. A second advantage of finding pattern in fitted relationships is the increased validity of the relationship. In the words of the Queen to Alice (Carroll, 1971), 'What I tell you three times is true'. If the relationship between leaf area and leaf weight is the same if the plant species is grown in England or in Nigeria or in Thailand, then the fitted regression may be argued to have some biologically meaningful interpretation.

Suppose we have two sets of data each consisting of pairs of observations of two variables, y and x, the two sets of data arising from two different environments. Assume that there are good reasons for accepting that the relationship between y and x is linear. We are interested in comparing the two lines. The three possible cases are shown in Fig. 11.1. If the relationships for the two sets of data are genuinely different then we have the situation shown in Fig. 11.1(a) where the two lines have different slopes and different intercepts. At the other extreme the relationships are identical in Fig. 11.1(b) and one line describes all the data quite adequately. An intermediate situation is shown in Fig.11.1(c). Here the lines are clearly distinct but the slopes of the lines are the same. This is not an improbable situation since the slope of the line is the rate of increase of y as x increases. If x is a time scale then the slopes of the lines will be the same if the growth rate of the variable y with time is the same in the two environments the only effect of environmental change being to increase all y values by a constant amount.

First we consider the choice between (a) and (b). We need to fit linear regressions to each set of data separately and to the combined set of data. The method for fitting a straight line regression, described in Chapter 9, can be used three times to produce the required values of slope, intercept and residual sum of squares.

Data set 1	a_1	b_1	$RSS(a_1, b_1)$
Data set 2	a_2	b_2	$RSS(a_2, b_2)$
Combined data set	a	b	$RSS(a, b)$

To calculate the results using a computer program we shall need three columns

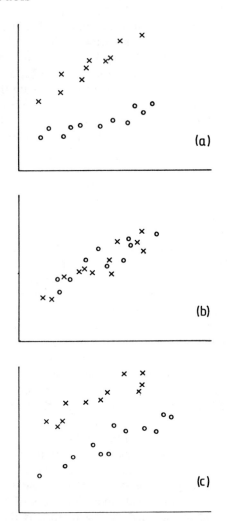

Fig. 11.1 Two sets of data showing the linear relationships between two variables y and x: (a) when the relationships are different for the two sets; (b) when the relationships are identical; (c) when the relationships are different but parallel.

of data, one for y, one for x, and one to indicate the data set (1 or 2) to which each observation belongs. The model for fitting the separate lines to the two data sets includes a factor, a_i, to represent the variation of the intercept term between the two data sets, and another factor, b_i, to represent the variation of the slope, and the full model can be written

$$y_{ij} = a_i + b_i x_{ij} + e_{ij}$$

where $i = 1$ or 2, and (y_{ij}, x_{ij}) are the pairs of observations in set i.

For fitting a single line to the combined data sets the model is reduced to the normal linear regression form, since the intercept and slope do not vary between data sets,

$$y_{ij} = a + bx_{ij} + e_{ij}$$

The computer output for each model will include the fitted parameter values and the RSS for the model.

Now we know several things about these values of b, a and RSS. First b_1 and b_2 are almost certain to be different since even if there is only one true relationship, b_1 and b_2 are calculated from two different samples and therefore are no more likely to be identical than are two yields of wheat from two different plots. The same applies to a_1 to a_2 and also to b and b_1 or b and b_2 (after all b must be some kind of 'average' slope and if two values are different the 'average' will not be equal to either). Also, although the RSS for the separate lines is the smallest possible sum of squares of deviations about a single line it must be larger than the sum of the RSS for the separate lines because they in their turn are the minimum sums of squares of deviations about a line for the two sets separately. In other words the single fitted line is by definition worse than the individual fitted line for each set of data separately since the sum of squares of deviations about the single line is simply the sum of squares of deviations of points in set 1 about the single line, and the sum of squares of deviations of points in set 2 about the single line (this argument is illustrated in Fig. 11.2).

The fact that RSS(a, b) must be greater than RSS(a_1, b_1) + RSS(a_2, b_2) suggests a way of testing whether the two relationships are the same. If the two lines are very different then each will be considerably different from the single line and RSS(a, b) for the single line will be substantially greater than RSS(a_1, b_1, a_2, b_2). On the other hand if there is really only one relationship then the variation of the two individual lines about the single line, and hence the excess of RSS(a, b) over RSS(a_1, b_2, a_2, b_2), will be simply random sampling variation. Now we already have two estimates of the variance of observations about the fitted relationship from the residual mean squares about the two individual regressions and the best estimate will be obtained by pooling these two estimates. It is useful now to write the calculations in an analysis of variance.

	s.s.	d.f.	m.s.
Residual variation for set 1	RSS(a_1, b_1)	$n_1 - 2$	
Residual variation for set 2	RSS(a_2, b_2)	$n_2 - 2$	
Sum of residual variations	RSS(a_1, b_1, a_2, b_2)	$n_1 + n_2 - 4$	s^2
Residual variation about a single line	RSS(a, b)	$(n_1 + n_2) - 2$	
Difference	RSS$(a, b) - RSS(a_1, b_1, a_2, b_2)$		2

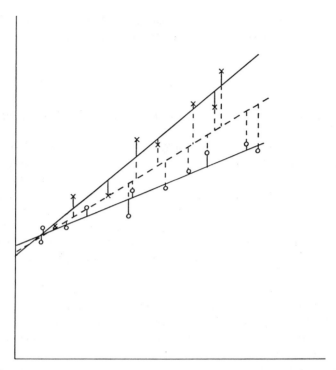

Fig. 11.2 Comparison of the deviations about a single line with the deviations about separate lines for the two data sets.

We now calculate

$$F = \frac{\text{difference m.s.}}{s^2} = \frac{[\text{RSS}(a, b) - \text{RSS}(a_1, b_1, a_2, b_2)]/2}{s^2}$$

and compare it with the F-distribution on 2 and $(n_1 + n_2 - 4)$ d.f. to test the hypothesis that there is no difference between the two relationships. If there is no difference then the difference mean square, like the residual mean square, is an estimate of random variation: if there is a difference then $\text{RSS}(a, b) - \text{RSS}(a_1, b_1, a_2, b_2)$ will become large and the value of F will be large and hence extreme and significant. Note the parallel with the test of the difference between the means of two samples (section 4.2). In both cases we have an estimate of sampling variation with which we compare the variation due to the difference between the two samples, either of the mean value or of the fitted line. A further parallel is that we have to make the assumption that the variance of observations about the fitted line is the same for both samples. For the regression situation this means that the residual mean squares about the two separate fitted lines should be similar. The judgement of whether the residual mean squares are acceptably similar can be based on an F-test comparison of the two mean squares, but will also take into account the biological reasonableness of the assumption that the variances are the same.

Example 11.1 Comparison of regressions

In a study on the growth of early summer cauliflowers the numbers of leaves on cauliflower plants were counted at various dates in each of two years, the sample size being 10 plants per date. The data are $y =$ mean leaf number and $x =$ accumulated day degrees above $32°$ F, divided by 100. From much previous work we can assume that there is a linear relationship between mean leaf number and accumulated day degrees above $32°$ F, and the object is to compare the relationships for the two years. Fig. 11.3 shows y plotted against x with the various fitted lines discussed.

1956/7		1957/8	
y_1	x_1	y_2	x_2
3.8	4.5	6.0	4.5
6.2	7.5	8.5	8.0
7.2	9.5	9.1	9.5
8.7	10.5	12.0	11.5
10.2	13.0	12.6	13.0
13.5	16.0	13.3	14.0
15.0	18.0	15.2	16.5

The computer input is

y	x	year
3.8	4.5	1
6.2	7.5	1
7.2	9.5	1
8.7	10.5	1
10.2	13.0	1
13.5	16.0	1
15.0	18.0	1
6.0	4.5	2
8.5	8.0	2
9.1	9.5	2
12.0	11.5	2
12.6	13.0	2
13.3	14.0	2
15.2	16.5	2

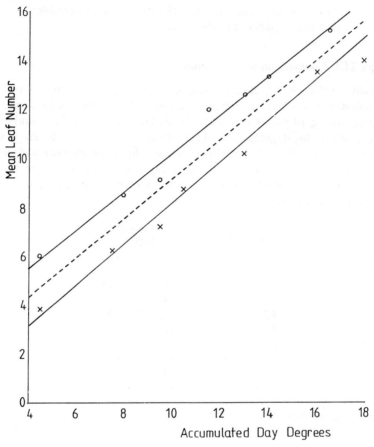

Fig. 11.3 Data on the relationship between leaf number (y) and accumulated day degrees (x) with fitted lines for each data set separately (———) and for the combined data set (----).

The results are

Model 1 $b_1 = 0.840,$ $a_1 = -0.25,$ $RSS(a_1, b_1) = 0.73(5 \text{ d.f.})$
           $b_2 = 0.789,$ $a_2 = \quad 2.28,$ $RSS(a_2, b_2) = 0.93(5 \text{ d.f.})$

and the sum of the separate line residuals $RSS(a_1, b_1, a_2, b_2) = 1.66(10 \text{ d.f.})$

Model 2 $b = 0.810$ $a = \quad 1.07$ $RSS(a, b) = 15.27(12 \text{ d.f.})$

The analysis of variance for assessing the variation about the alternative fitted models is

Analysis of variance	S.S.	d.f.	m.s.	F
Residual variation about a single line	15.27	12		
Sum of residual variations about individual lines	1.66	10	0.17	
Difference (variation of individual lines about a single line)	13.61	2	6.80	40***

Hence the use of single line is significantly worse than the individual fits and we conclude that the relationships are different for the two years.

Note that the residual mean squares about the individual regression lines are very similar (0.146 and 0.186) and that the pooling of the residual mean squares is clearly reasonable.

11.4 FITTING PARALLEL LINES

If we conclude that a single relationship holds for both sets of data then, of course, there is no need to consider any other possibility. But suppose that, as in the example, we find conclusive evidence that the relationships for the two sets of data are not identical. Then we shall want to investigate the possibility that the slopes of the two relationships are the same and that the two sets of data differ only in the height of the linear relationship [Fig. 11.1(c)].

The model for this intermediate situation has a factor term for variation between the intercepts and a single, b to represent the common slope of the two parallel lines, and is

$$y_{ij} = a_i + bx_{ij} + e_{ij}$$

The computer program produces the fitted values of b, a_1 and a_2 and the RSS for the parallel line model, $RSS(a_1, a_2, b)$.

The fitted parameter values for the parallel line model can also be calculated without the computer.

The estimate of the common slope is

$$b_c = \frac{(S_{xy})_1 + (S_{xy})_2}{(S_{xx})_1 + (S_{xx})_2}$$

a weighted average of the individual slopes. The estimates of the intercepts are

$$a_i = \frac{\Sigma(y_i) - b_c \Sigma(x_i)}{n_i}$$

and the residual sum of squares is

$$\text{RSS}(a_1, a_2, b) = (S_{yy})_1 + (S_{yy})_2 - \frac{[(S_{xy})_1 + (S_{xy})_2]^2}{(S_{xx})_1 + (S_{xx})_2}$$

The $\text{RSS}(a_1, a_2, b)$ has $(n_1 + n_2 - 3)$ degrees of freedom. This can be justified in two ways. First the RSS is the difference between the sum of the two corrected sums of squares of y [based on $(n_1 - 1)$ and $(n_2 - 1)$d.f.] and the sum of squares for fitting the common regression coefficient (1 d.f.). Alternatively, it can be argued that from $(n_1 + n_2)$ observations, three parameters, a_1, a_2 and b, have been estimated leaving $(n_1 + n_2 - 3)$ degrees of freedom for the residual variation.

Now we can test whether the slopes of the two relationships are the same by examining whether $\text{RSS}(a_1, a_2, b)$ is substantially greater than $\text{RSS}(a_1, b_1, a_2, b_2)$.

If the difference is small compared with s^2 then we can accept that the slopes are the same. We can again set the calculations in the structure of an analysis of variance as follows:

	s.s.	d.f.	m.s.
Residual variation about parallel lines	$\text{RSS}(a_1, a_2, b)$	$n_1 + n_2 - 3$	
Sum of individual residuals	$\text{RSS}(a_1, b_1, a_2, b_2)$	$n_1 + n_2 - 4$	s^2
Difference of slopes	$\text{RSS}(a, a_2, b) - \text{RSS}(a_1, b_1, a_2, b_2)$	1	

And again we test the hypothesis that the slopes are equal by comparing

$$F = \frac{\text{difference m.s.}}{s^2} = \frac{\text{RSS}(a_1, b_1, a_2, b_2)}{s^2}$$

with the F-distribution on 1 and $(n_1 + n_2 - 4)$ d.f.

For the example on growth rates of cauliflower seedlings,

$$b = 0.819, \qquad a_1 = -0.01, \qquad a_2 = 1.95, \qquad \text{RSS}(a_1, a_2, b) = 1.80,$$

and the analysis of variance is

	s.s.	d.f.	m.s.	F
Residual variation about parallel lines	1.80	11		
Sum of individual residuals	1.66	10	0.17	
Difference of slopes	0.14	1	0.14	0.84

There is clearly no evidence that the two slopes are different. We have therefore found a point of consistency in the growth rates of cauliflower and obviously can have greater confidence in the generality of the value obtained for the rate of increase of leaf number with temperature, when temperature is measured as accumulated day degrees above 32° F.

Example 11.2 Comparison of regressions to compare treatments

To illustrate another aspect of these methods we consider data from an experiment to investigate the factors which affect the time taken for food particles to pass through the guts of **Nassarius**. Two groups of foods were distinguished, organic and inorganic, and various different types of food particles from each group were fed to the animals. For each type of particle the time in seconds, and the particle size in thousandths of a centimetre, were recorded.

Preliminary investigation suggested that an approximately linear relationship exists between $y = \log$ time and $x = \log$ size. The basic records were therefore transformed to a logarithmic scale and were as follows:

(a) Inorganic foods (9 types)

Log time	y	1.25	1.18	0.88	1.99	1.72	1.86	2.28	1.70	2.97
Log size	x	1.17	1.39	1.53	1.94	2.18	2.30	2.54	2.63	3.77

(b) Organic foods (7 types)

Log time	2.81	2.47	3.50	3.87	3.36	3.39	3.57
Log size	2.48	2.31	2.55	3.49	3.45	3.53	4.07

The primary object of the investigation at this stage was to compare the two groups of foods and to estimate the difference between the organic and inorganic foods. It is difficult to define the difference between organic and inorganic foods simply. As can be seen from the data which are plotted in Fig. 11.4 the particle sizes are generally larger for the organic foods. Also the time increases with increasing particle size. Therefore a simple comparison between the mean times for organic and inorganic foods would not be useful because if a large difference was found it would not be possible to say whether it represented the difference between organic and inorganic or was due to the difference between the particle sizes for the two groups. Equally, if no difference was found, a genuine difference between the groups could have been obscured by the size differences. Hence to estimate the difference between the groups we fit the regression lines of log time on log size and compare the lines, theus eliminating the effects of size differences.

Fitting separate linear regressions gives

Inorganic:	$b_1 = 0.718,$	$a_1 = 0.208,$	$\mathrm{RSS}(a_1, b_1) = 0.599$
Organic:	$b_2 = 0.516,$	$a_2 = 1.669,$	$\mathrm{RSS}(a_2, b_2) = 0.655$

For any particular particle size we have estimates of the log time for the two groups

$$y_1 = 0.208 + 0.718(\log \text{ lize})$$

$$y_0 = 1.669 + 0.516(\log \text{ size})$$

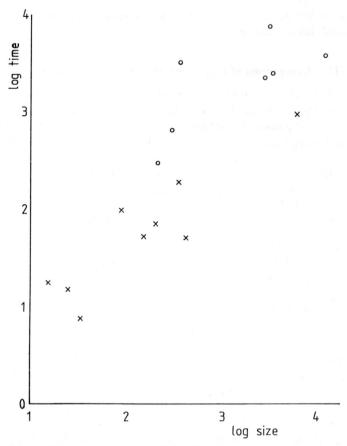

Fig. 11.4 Relationship between the log of digestion time and the log of particle size for inorganic foods (x) and organic foods (o).

and hence an estimate of the difference

$$a = y_o - y_1 = 1.461 - 0.202(\log \text{ size})$$

This result is not particularly useful since the estimated difference depends on the particular particle size chosen. It would be far more useful to find an estimate which did not depend on particle size. This is possible only if the rate of change of log time with log particle size is the same for both groups of particles—in other words, if the slopes of the two regression lines are the same. An alternative argument is that if the difference between the two lines is to be constant, whatever the particle size, then the lines must be parallel. Either way we come back to the need to fit parallel lines.

The results from fitting the parallel line model shown in Figure 11.5

$$b = 0.646, \qquad a_1 = 0.365, \qquad a_2 = 1.265 \qquad \text{RSS}(a_1, a_2, b) = 1.326$$

For completeness we also fit the common line model

$$b = 0.947, \qquad a = 0.022, \qquad RSS(a, b) = 3.480$$

Hence we have the anlysis of variance:

	s.s.	d.f.	m.s.	F
Residual variation about organic line	0.599	7	0.086	
Residual variation about inorganic line	0.655	5	0.131	
(1) Sum of residual variation for individual lines	1.254	12	0.105	
(2) Residual variation about Parallel lines	1.326	13		
(3) Residual variation about a single line	3.480	14		
Difference (2) − (1)	0.072	1	0.072	0.69
Difference (3) − (2)	2.154	1	2.154	20.5
Difference (3) − (1)	2.226	2	1.113	10.6

Note that in (1), we have pooled the two residual sums of squares minimizing the variation about the two separate lines. As discussed in section 11.3, this is only justifiable if the residual variations about the two lines are of the same order of magnitude. In this example, the two residual mean squares do differ by about 50% but their ratio is not significantly different from one as judged by an F-test with 5 and 7 degrees of freedom.

The tests to compare the different regressions can be approached in two equally logical ways. One approach is to test first, as in section 11.3, whether a single line is as good a fit as the two separate lines ($F = 10.6$, significant at $p < 0.01$, therefore single regression line not acceptable), and then to test whether the slopes of the two relationships may be assumed to be the same ($F = 0.69$, not significant, therefore accept parallel lines). The second approach which seems more natural in this example is to test first whether the slopes can be assumed the same, and then, given that the slopes are the same, whether there is **any** difference between the two parallel lines by comparing the parallel lines RSS with the single regression RSS($F = 20.5$, significant at 0.1%) which, naturally leads to the same conclusion as before that the slopes may be the same but the lines are certainly not identical.

EXERCISE 11.1 *Comparison of regression*

The rates of uptake of rubidium (Rb) and bromide (Br) for potato discs immersed in a solution of rubidium bromide were measured, the following data

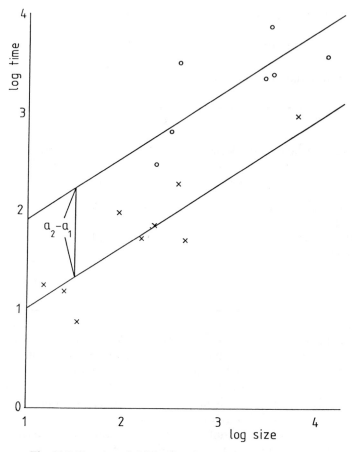

Fig. 11.5 Two parallel lines fitted to the data of Fig. 11.4.

being obtained:

Time (minutes) (x)	mg equivalents per100 g of water in the tissue Rb(y_R)	Br(y_B)
22	7.2	0.7
46	11.4	6.4
67	14.2	9.9
90	19.1	12.8
96	20.0	15.8
$\Sigma(x) = 321$	$\Sigma(y_R) = 71.9$	$\Sigma(y_B) = 45.6$
$\Sigma(x^2) = 24\,405$	$\Sigma(y_R^2) = 1148.25$	$\Sigma(y_B^2) = 552.94$
	$\Sigma(xy_R) = 5273.2$	$\Sigma(xy_B) = 3641.9$

Assuming that the rate of uptake is constant for each ion find the fitted lines and residual sums of squares for the two ions.

Test the hypotheses that

1. The relationships for the two ions are identical;
2. The rates of uptake for the two ions are the same, i.e. the regression lines are parallel.

11.5 COVARIANCE ANALYSIS

In the previous section we introduced two factor terms into a linear regression model to represent variation of the intercept and slope of the linear relationship between different data sets. We now consider the addition to an experimental design factor model of a regression term to improve the precision of treatment comparisons.

Even when experimental units have been grouped into blocks to try to make the within-block variation small there may still be substantial variation within a block. Sometimes we can recognize that an initial measurement, x, could be made on each experimental unit which, in the absence of treatment effects, might be expected to be strongly correlated with the yield variable, y. The accuracy of the experiment can then be improved by adjusting the value of the y variable by this initial variable, x, often referred to as a concomitant variable. As an example consider an experiment on pigs in which litters were used as blocks. Although pigs in the same litter would be expected to react similarly to a particular treatment by virtue of their similar genetic make-up, there may well be considerable differences in the initial weights of pigs in a litter which might affect the final weights, or whatever yield variable is being considered. If we are prepared to assume that there is some relationship between initial weight and final weight then we could adjust the final weights to reduce the effects of the differences between initial weights. Such an adjustment would be expected not only to provide a more valid estimate of the treatment effects but also to reduce the random variation in the experiment and hence the standard error of treatment means and differences between treatment means.

Adjusting final weights in this way is obviously similar to using the weight gain as the variable to be analysed. However, it is quite possible that weight gains will themselves be related to initial weights and that an adjustment by initial weight would also improve the analysis of weight gains. The use of adjustment methods is more general than simply analysing weight gain. Other examples where such adjustments might be valuable are small-scale trends in light intensity across a glasshouse, field fertility trends not taken into account in blocking, variation in past history of trees and differential germination rates in a drilled crop.

Suppose we took measurements of our yield variable, y, and our adjusting variable, x, for a number of units all receiving the same treatment and found an approximately linear relationship between y and x. By fitting a regression of y on x we could obtain a prediction of y for a further unit for which $x = x_0$. If a different treatment was applied to this further unit then we could compare the actual yield of the unit with the yield that we predict would have been achieved if the first treatment had been applied. Hence we would have a comparison between the treatments adjusted for the different values of x for the particular units to which the two treatments were applied. In an experiment we have allocated the units to different blocks and treatments so that we do not have a group of units from which we can calculate the regression. However, if we are prepared to assume that the relationship between the yield variable and the adjusting variable is approximately linear then we can make exactly the same kind of adjustment. The technique we use is called **covariance**.

The basic model is the usual model for the particular experimental design, with an additional term to represent the regression dependence on the covariate. Thus, if the experimental design is a randomized block design, the basic model would be

$$y_{ij} = b_i + t_j + e_{ij}$$

The expanded model includes an additional term for the regression on the covariate, x_{ij}, with regression coefficient, c.

$$y_{ij} = b_i + t_j + cx_{ij} + e_{ij}$$

We can fit both models and obtain residual sums of squares

$$\text{RSS}(b_i, t_j) \text{ for the first model, and}$$
$$\text{RSS}(b_i, t_j, c) \text{ for the second.}$$

The change in the residual sum of squares,

$$\text{Covariance s.s.} = \text{RSS}(b_i, t_j) - \text{RSS}(b_i, t_j, c),$$

allows, us to test whether including the covariance term produces a useful reduction in the residual sum of squares. We can write the analysis of variance

Source	s.s.	d.f.	m.s.	F
Blocks	usual	$b - 1$	usual	
Treatments	usual	$t - 1$	usual	
Covariance	$\text{RSS}(b_i, t_j) - \text{RSS}(b_i, t_j, c)$	1		
Error	$\text{RSS}(b_i, t_j, c)$	$bt - b - t$	s^2	

The F ratio: Covariance s.s./s^2 tests the hypothesis that the true covariance coefficient is zero. Our main interest, of course, is in the treatment variation and we now encounter a problem that we met previously in this chapter and in Chapter 10, namely that the order of fitting terms changes the sum of squares for the terms. The values of the covariate, x_{ij}, are not identical for the different treatments and hence inclusion of the covariate in the model will change the apparent pattern of treatment differences. To assess the significance of the overall treatment variation, allowing for the covariance effect, we must also fit the model without the treatment effects,

$$y_{ij} = b_i + cx_{ij} + e_{ij}$$

for which the residual s.s. will be

$$\text{RSS}(b_i, c)$$

The treatment s.s. after allowing for the covariance effect is
Treatment (adjusting for covariate) $= \text{RSS}(b_i, c) - \text{RSS}(b_i, t_j c)$ and we have the analysis of variance

Source	s.s	d.f	m.s.	F
Blocks	usual	$b - 1$	usual	
Covariance	$\text{RSS}(b_i) - \text{RSS}(b_i, c)$	1		
Treatments	$\text{RSS}(b_i, c) - \text{RSS}(b_i, t_j, c)$	$t - 1$		
Error	$\text{RSS}(b_i, t_j, c)$	$bt - b - t$	s^2	

The significance of the overall treatment variation is tested in the usual way, using the adjusted treatment mean square,

$$F = \text{Treatment m.s.}/s^2$$

The comparison of residual sums of squares for different models enables us to assess the significance of the covariance term and of the treatment differences, adjusted for the effect of covariance. However, our primary interest is, as always, in comparing treatment means. The effects of including the covariance term in the model are hopefully to reduce the random variation and hence the standard errors for differences between treatment means, and to correct the comparisons between treatments to allow for any 'bias' arising from uneven distribution between treatments of 'good' and 'bad' covariate values.

To understand the argument for adjusting treatment means by the covariate, suppose that the fitted covariance coefficient is positive. This means that experimental units with large values of the covariate, x, will tend to have larger values of the yield, y. Correspondingly, treatments whose units have above average values of x will tend to have above average values of y quite independently of whether the treatment tends to increase or decrease yields. Hence a more valid comparison of treatments will be achieved by adjusting treatment means down for those treatments with high values of x and adjusting up those treatments with low values of x.

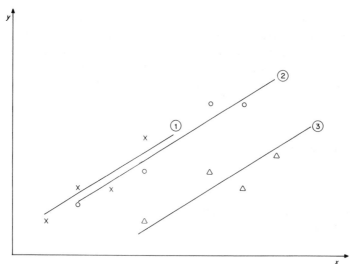

Fig. 11.6 Effects of covariance adjustment for dependence of variable y on the covariate x. Treatments 1 (x), 2 (o), 3 (Δ).

The procedure is illustrated in Fig. 11.6 in which each treatment shows an upward trend of y as x increases. Clearly for comparable values of x treatment 1 is superior to treatment 2, which is superior it treatment 3. However, the unadjusted treatment means show an apparent advantage for treatment 2 and suggest little difference between treatments 1 and 3. The parallel lines, superimposed on the figure, show the estimated covariance trend and the vertical differences between the lines provide estimates of the differences between treatment means allowing for that covariance trend.

The output from a computer program should give the differences between treatment means adjusted for the covariance effect. Usually we adjust the treatment means to the value to be expected for the overall mean value, $\bar{x}$, of the covariate. The adjusted means can also be calculated directly, given the treatment means for both y and x, and the covariance coefficient. If the treatment means of y and x for a particular treatment are $\bar{y}_T$ and $\bar{x}_T$, the overall mean of x is $\bar{x}$ and the covariance coefficient is c, then the adjusted treatment mean is

$$\bar{y}_T - c(\bar{x}_T - \bar{x})$$

The standard errors for differences between adjusted treatment means are composed from the variance of the difference between the unadjusted treatment means and the variance of the covariance coefficient, c, multiplied by the square of the difference between the treatment means of the covariate, $\bar{x}_1 - \bar{x}_2$. The variance of the covariance coefficient is in the same from as that of any other regression coefficient, s^2/S_{xx}. In this case the S_{xx} is the error sum of squares of the covariate (from an analysis of variance of the covariate, x). The formula for the standard error of the difference between two adjusted treatment means is

$$\sqrt{\left\{ s^2 \left[\frac{2}{r} + \frac{(\bar{x}_1 - \bar{x}_2)^2}{S_{xx}} \right] \right\}}$$

This will vary according to which pair of treatments are being compared but often the variation will not be substantial and an average standard error may be used for all comparisons. Computer packages will usually provide standard errors for comparing adjusted treatment means but the user should check carefully what particular forms of standard errors are offered by the package.

As usual, in making the calculations of an analysis of covariance we make a number of assumptions and while checking the validity of the assumptions is not always easy they should always be borne in mind so that the user can convince himself that they are not unreasonable. First, we have assumed that the slope of the regression of y and x is the same for all the experimental treatments and for all the blocks. Second, we have assumed that the variance of the adjusted y values is the same for all treatments and blocks, the assumption of homogeneity of variance. Third, when interpreting the adjusted treatment means we usually assume that the treatments have no effect on the x values. If we consider again some of the examples at the beginning of this section we can see that in four out of five cases this last assumption is reasonable. Thus we would not expect the initial weights of pigs to be affected by treatments applied afterwards. The lack of any treatment effects on the x values should always be checked by calculating, from the analysis of covariance of the covariate, the F-value for treatments versus error. A significant or near-significant F-value is a warning sign that all may not be well. The one example at the beginning of this section where the treatments may well affect the values is where germination rate in a drilled crop is used as the covariate x. However, the use of covariance in this case is still valid if it is used to predict the yield difference that would be observed if the drilling rates of different treatments could be adjusted to give the standard density. This is a rather different use of covariance from the usual intention simply to reduce variation. The assumption that the slope of the relationship is the same for all the treatments is clearly vital whatever the purpose of the covariance adjustments.

Another common application of covariance analysis is to correct for accidents which have occurred to some experimental units. An example of this might be bird damage in pea crops, with the amount of damage assessed subjectively on a predetermined scale and used as the covariate, x. In using covariance analysis for this kind of situation one has to be particularly careful that the amount of damage is unrelated to the particular treatments. The case of bird damage might seem a reasonable premise but suppose that the covariate was frost damage in a trial of cauliflower varieties. Then one would have to be convinced that the different varieties were equally susceptible to frost and that differences in frost damage between varieties were purely random. Such an argument is obviously less easy to maintain and this is a good example of a situation where covariance analysis might be positively harmful since the result of the adjustments to treatment means would be to give falsely high means for varieties susceptible to frost damage.

Example 11.3 The analysis of covariance

Sixteen steers were weighted to the nearest 10 kg(x). They were then used in an experiment to compare 4 different hormone treatments A, B, C, D, in 4 randomized blocks. After treatment the weight of kidney fat in grams to the nearest 10 g(y) was measured on the carcasses.

Block	A x	A y	B x	B y	C x	C y	D x	D y	Total x	Total y
I	56	133	44	128	53	129	69	134	222	524
II	47	132	44	127	51	130	42	125	184	514
III	41	127	36	127	38	124	43	126	158	504
IV	50	132	46	128	50	129	54	131	200	520
Totals	194	524	170	510	192	512	208	516	764	2062

We fit four models

$$y_{ij} = b_i + e_{ij}$$

$$y_{ij} = b_i + t_j + e_{ij}$$

$$y_{ij} = b_i + t_j + cx_{ij} + e_{ij}$$

$$y_{ij} = b_i + cx_{ij} + e_{ij}$$

The residual sums of squares for these four models are

$$\text{RSS}(b_i) = 71,$$

$$\text{RSS}(b_i, t_j) = 42,$$

$$\text{RSS}(b_i, t_j c) = 12$$

$$\text{RSS}(b_i, c) = 36$$

from which we construct two analysis of variance:

Source	s.s.	d.f.	m.s.	F
Blocks	57	3	19	
Treatments	29	3	9.7	
Covariance	30	1	30	37
Error	12	14	0.8	

showing that the covariance effect is very substantial, and

Source	s.s.	d.f.	m.s.	F
Blocks	57	3	19	
Treatments	35	1	30	
Covariance	24	3	8	10
Error	12	14	0.8	

showing that, after adjustment for the covariate effect, the variation between the treatment means is clearly significant.

The treatment means (y) adjusted for the differences in the x values are calculated as follows:

Treatment	Unadjusted y means $(\bar{y}_\mathrm{T})$	Deviation of x mean from overall mean $\bar{x}_\tau - \bar{x}$	Adjusted mean $\bar{y}_\tau - b(\bar{x}_\tau - \bar{x})$
A	131.0	0.75	130.7
B	127.5	− 5.25	129.3
C	128.0	0.25	127.9
D	129.0	4.25	127.5
		0.00	

s.e. for comparing any two unadjusted means $= \sqrt{[(42/9)(2/4)]} = 1.53(9\text{ d.f.})$

Comparison	s.e. of comparison (8 d.f.)
A–B	0.99
A–C	0.87
A–D	0.91
B–C	0.97
B–D	1.15
C–D	0.93

Conclusion

The adjustment for covariance reduced the value of s^2 from 4.7 to 1.5 and consequently the standard errors of comparisons are reduced. It has also altered the interpretation of the results. Before adjustment only the difference between treatments A and B is significant. After adjustment this difference is not

significant but treatment A has a significantly higher mean than treatments C
and D.

EXERCISE 11.2 *The analysis of covariance*

The data in this exercise are taken from a cover crop experiment with apples
at East Malling in which y, the crop in pounds over a four-year period, was
adjusted by x, the crop in bushels over the immediately preceding four-year
period during which no differential treatments had been applied. The experiment
was designed in randomized blocks in the orchard.

	Blocks							
	I		II		III		IV	
Treatment	x	y	x	y	x	y	x	y
A	8.2	287	9.4	290	7.7	254	8.5	307
B	8.2	271	6.0	209	9.1	243	10.1	348
C	6.8	234	7.0	210	9.7	286	9.9	371
D	5.7	189	5.5	205	10.2	312	10.3	375
E	6.1	210	7.0	276	8.7	279	8.1	344
F	7.6	222	10.1	301	9.0	238	10.5	357

Carry out the analysis of covariance. Calculate the adjusted treatment means
and their appropriate standard errors. Discuss the conclusions to be drawn
from the results and the effect the covariance analysis has had on the treatment
means and standard errors.

11.6 REGRESSION IN THE ANALYSIS OF TREATMENT VARIATION

One other use of regression in the analysis of experimental data is when one or
more of the treatment factors are of a quantitative nature, e.g. amount of protein
in animal feed, quantity of fertilizer applied. In such situations the treatment
differences are often large and the main interest lies in whether the response to
the changing level of the factor is linear over the range of levels used. If the
response is linear then it can be very simply summarized by quoting the values
of a and b in the fitted linear relationship $y = a + bx$. For a curvilinear response
pattern the quadratic relationship, $y = a + bx + cx^2$ may provide an adequate,
empirical, description. Alternative non-linear model response functions are
discussed in Chapter 12. Factors for which this approach may be useful are
harvest time, fertilizer levels, concentration of a drug or level of protein in diet.
For any of these factors we may be interested to know the mean rate of increase

of yield with, say, harvest time, but may also be interested in detecting departures from a linear relationship since these will normally take the form of falling below the linear relation for high levels of the factor and would indicate that the 'law of diminishing returns' applies.

The details of the calculations for the linear regression term are simple, being almost exactly the same as the linear regression calculations already discussed. Since treatment totals will already have been calculated for the treatment sum of squares we use these rather than the treatment means. As before we write $T_1, T_2, \ldots,$ for the treatment totals and $x_1, x_2, \ldots,$ for the corresponding levels of the treatment factor. The regression coefficient of yield on x is

$$b = \frac{\Sigma Tx - \Sigma T\Sigma x/t}{r[\Sigma x^2 - (\Sigma x)^2/t]}$$

In this formula r is the number of blocks and t is the number of treatment levels. The r in the denominator effectively converts the treatment totals to treatment means so that b is an estimate of the increase of yield per plot for a unit increase of x. The component of the treatment sum of squares attributable to the regression of yield on x is

$$\text{regression s.s.} = \frac{(\Sigma Tx - \Sigma T\Sigma x/t)^2}{r[\Sigma x^2 - (\Sigma x^2/t)]}$$

This is a single degree of freedom component and the difference between the full treatment s.s and this regression s.s. is the residual s.s. measuring the deviations of treatment means from a linear relationship. The standard error of the regression coefficient is

$$\text{s.e. } (b) = \sqrt{(s^2/rS_{xx})}$$

where s^2 is the error mean square from the initial analysis of variance. These formulae are for the simple situation of a single-factor experiment with r blocks. If the experiment is a factorial and we are considering the regression on a single quantitative factor then r is replaced in the formulae by the number of plots at each level of the factor concerned. Thus in an experiment with 4 blocks × 5 protein levels in diet × 3 grazing methods, the effective replication of each protein level is $12(= 4 \times 3)$ and so $r = 12$ in all the regression calculations.

To fit and assess regression components in the usual model for experimental data we compare two alternative models. These are the usual experimental design model, with a complete set of treatment effects.

$$y_{ij} = b_i + t_j + e_{ij}$$

and the regression model where the set of treatment effects, t_j, are replaced by a linear regression dependence on the quantitative treatment levels, x_j,

$$y_{ij} = b_i + cx_j + e_{ij}$$

Note that here the x values are values of a variable, whereas the t_j values are levels of a factor. The second model can be extended to include quadratic or other curvilinear terms.

The second model, which attempts to describe the variation of yield by a single parameter, c, is a special case of the first, in which there are no patterns assumed for the variation amongst the treatment levels, the total variation of which is represented by the $(t - 1)$ degrees of freedom. Comparing the residual sums of squares for the two models shows how much of the treatment variation is attributable to the linear trend, and also how much treatment variation remains unexplained by the linear trend.

Example 11.4 Regression on treatment levels in analysis of variance

Consider again the data from the factorial experiment in example 6.2 on survival of *Solmonella typhimurium*. Both factors, water activity and sorbic acid, are

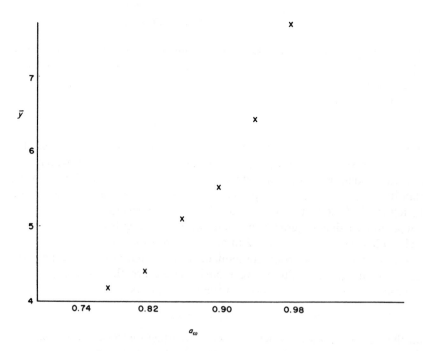

Fig. 11.7 Relationship between mean log salmonella density and water activity level, data from Example 11.4.

quantitative with six and three equally spaced levels respectively. We therefore consider the possibility of fitting regression trends to explain the yield variation between water activity levels and between sorbic acid levels. Salmonella density clearly increases with higher values of water activity, as shown in Fig. 11.7, and there is also a lesser trend for salmonella density to increase at lower sorbic acid levels. Since the relationship of salmonella density with water activity has a suggestion of a curve we shall fit a quadratic term in addition to the obvious linear trend.

The computer input consists of

y_{ijk} yield
b_i block level (factor)
w_j water activity level (factor)
s_k sorbic acid level (factor)
x_j amount of water activity (variable)
z_k amount of sorbic acid (variable)

The first two rows and last row of the data appear as

| 8.19 | 1 | 1 | 1 | 0.98 | 0 |
| 8.37 | 2 | 1 | 1 | 0.98 | 0 |
| .... |
| 4.15 | 3 | 6 | 3 | 0.78 | 200 |

The usual model for the experimental data is

$$y_{ijk} = b_i + w_j + s_k + (ws)_{jk} + e_{ijk}$$

The model for fitting a regression trend on the water activity variable, including the block differences, is

$$y_{ijk} = b_i + w_1 x_j + w_2 x_j^2 + e_{ijk}$$

If the quadratic term, $w_2 x_j^2$ is omitted we have the linear regression. Similarly the model for fitting a regression trend on the amount of sorbic acid is

$$y_{ijk} = b_i + sz_k + e_{ijk}$$

We could combine the two regression effects, as in multiple regression, and also include the possibility of a product term which would represent the linear change of the sorbic acid regression across the water activity levels. This model would be

$$y_{ijk} = b_i + w_1 x_j + w_2 x_j^2 + sz_k + (ws)x_j z_k + e_{ijk}$$

It would be possible to define other models with a mixture of factor and variable terms, such as

$$y_{ijk} = b_i + w_j + sz_k + e_{ijk}$$

which allows for variation between all six water activity levels, with a linear regression on sorbic acid level. In this data set such models would provide no additional information because the data set includes all 54 possible combinations of water activity level, sorbic acid level and block. Consequently the information about regression effects is a subset of the full treatment factor information, independently of the effect of other factors.

Model	RSS	Change in RSS
Mean	87.03	
b_i	87.02	0.01
b_i, w_j	5.46	81.56
b_i, w_j, s_k	2.71	2.75
$b_i, w_j, s_k, (ws)_{jk}$	1.38	1.33
b_i, w_1	11.19	75.83
b_i, w_1, w_2	5.48	5.71
b_i, s	84.35	2.67
b_i, w_1, s	8.52	
$b_i, w_1, s, (ws)$	07.98	0.54

The analysis of variance, summarizing the variation attributable to all the different components of the models is

Source	s.s		d.f.	m.s.	F
Blocks	0.01		2		
Water activity	81.56		5		
Linear regression		75.83	1	72.33	1390
Quadratic regression		5.71	1	5.71	140
Residual		0.02	3	1.68	0.5
Sorbic acid	2.75		2		
Linear regression		2.67	1	2.67	65
Residual		0.08	1	0.08	2.0
W × S	1.33		10		
Linear × linear		0.54	1	0.54	13
Residual		0.79	9	0.009	2.2
Error	1.38		34	0.041	

For water activity the linear regression sum of squares is (of course) vastly significant, and the quadratic regression accounts for almost all the remaining regression.

The linear regression for the dependence on sorbic acid is also very significant, and again the residual sum of squares is not large, suggesting that the linear regression provides a reasonable explanation of the effect of sorbic acid; note, however, that there is only 1 d.f. for this residual.

Finally, the linear × linear product term accounts for a large, and clearly significant, portion of the interaction s.s. but the residual interaction variation is not wholly negligible.

The interpretation from the regression models fitted would be that the treatment variation can be explained fairly satisfactorily by using linear and quadratic terms for the water activity effects, a linear term for sorbic acid and a linear product term for the interaction effect. The fitted model is then

$$\log(\text{density}) = 5.121 + 17.35(aw - 0.88) + 81.5(aw - 0.88)^2$$

$$- 0.00272(sa - 100) - 0.018(aw - 0.88)(sa - 100)$$

The quality of the fit can be assessed by noting that the treatment regression s.s. is 84.75 ($= 75.83 + 5.71 + 2.67 + 0.54$) on 4 d.f. while the treatment residual s.s. is 0.89 ($= 0.02 + 0.08 + 0.79$) on 13 d.f. or by comparing the observed and fitted treatment means.

Observed

sa/wa	0.98	0.94	0.90	0.86	0.82	0.78
0	8.30	6.53	6.00	5.14	4.56	4.28
100	7.67	6.41	5.36	4.70	4.30	4.23
200	7.08	6.31	5.24	5.53	4.30	4.07

Fitted

8.12	6.84	5.81	5.04	4.54	4.29
7.67	6.46	5.50	4.81	4.37	4.20
7.22	6.08	5.19	4.57	4.21	4.11

Treatment Residuals

$+0.18$	-0.31	$+0.19$	$+0.10$	$+002$	-0.01
0	-0.05	-0.14	-0.11	-007	$+0.03$
$--0.14$	$+0.23$	$+0.05$	-0.04	$+0.09$	-0.04

The fit is generally good, though a few treatment combinations behave in slightly eccentric manner.

EXERCISE 11.3 *Regression on treatment levels in a randomized block*

The yields below are extracted from an experiment on the timing of harvesting rhubarb for canning. The experiment consisted of four blocks for seven harvesting dates at four-day intervals

Date	I	II	III	IV	Total
3 May	21.2	21.4	12.0	17.2	71.8
7 May	19.3	17.4	24.5	30.2	91.4
11 May	22.8	29.0	18.5	24.5	94.8
15 May	26.0	34.0	33.5	30.2	123.2
19 May	43.5	37.0	25.1	23.4	129.0
23 May	32.1	30.5	35.7	32.3	130.6
27 May	33.0	32.2	35.4	35.4	136.0
Total	197.9	201.1	184.2	193.2	776.8

Calculate the simple analysis of variance, fit a linear regression of yield on harvesting time and divide the treatment s.s. into that due to the linear regression, and the residual variation. Summarize the results. (The uncorrected sum of squares is 23020.58.)

12
Non-linear models

12.1 ADVANTAGES OF LINEAR AND NON-LINEAR MODELS

In Chapters 9 and 10 we discussed the use of various regression models when fitting data on relationships. All the models of those chapters are in the class of models called 'linear models', because the parameters in the models occur in a simple linear form. That is, in the three examples from Chapters 9 and 10 shown below, the yield, y, is a linear combination of the parameters a, b, c and d:

$$y = a + bx$$
$$y = a + bx + cx^2$$
$$y = a + bx + cz + dxz$$

It is important in discussing such models to be clear that, for example, the second model is a linear model because the parameters occur in a linear fashion, though the relationship between y and x is definitely not linear, but quadratic.

Linear models have dominated the statistical methods for investigating relationships, not because such models are always the most appropriate, but because the theory of fitting such models to data is very simple. The calculations involved in obtaining estimates of parameters in linear models require only the solution of a set of simple simultaneous equations. These calculations can be done without the aid of computers, though they can be done quicker with computers.

In contrast other, possibly more realistic, forms of model which involve parameters in non-linear fashion cannot be so simply fitted without the use of a computer. Some forms of non-linear model were investigated before the development of modern computers, and complicated methods of fitting them devised, but these models inevitably had little appeal to statisticians or research workers because of their complexity, and they were not widely used, with the exception of probit analysis. With the availability of high-speed computers the fitting of non-linear models should be no more difficult than that of linear models. The complicated methods of fitting have been replaced by simpler methods, which require a large amount of computation. It is therefore important that the research biologist is aware that there should be no difficulties in fitting these models.

To see why non-linear models should be useful, it is necessary to consider why linear models are inadequate to model biological situations. If we are considering a relationship between a stimulus variable, x, and a resulting yield variable, y, then the three simplest forms of linear model are the straight line,

$$y = a + bx$$

the quadratic,

$$y = a + bx + cx^2$$

and the cubic,

$$y = a + bx + cx^2 + dx^3$$

The three models are displayed in Fig. 12.1. The straight line is obviously a very restricted relationship. Very few biological relationships are even approximately straight for a reasonable range of x values, the most common form of straight line relationship being, perhaps, the allometric relationship between the logarithm of weight of a plant or animal part, and the logarithm of the whole plant or whole animal weight.

The quadratic model allows for curvature but is restricted in two critical ways. First, it is symmetric, the rise of y with increasing x to a maximum being of exactly the same form as the subsequent decline of y with further increase in x. The symmetry can be avoided by considering not x but a power of x.

$$y = a + b(x^\theta) + c(x^\theta)^2$$

which is often a useful form of model, but θ is now a non-linear parameter. The second disadvantage of a quadratic model is that the value of y must become negative when x is either large or small and this will usually be biologically unreasonable.

The cubic polynomial, and polynomials of yet higher degree, overcome the disadvantages of symmetry but not those of producing unrealistically negative or very large values of y for large or small values of x. In addition, they have a very rigid structure of maximum and minimum values.

None of the curves in the polynomial family of models allows for a relationship which tends to an asymptotic level of y as x becomes large, or for relationships where y is necessarily positive. In contrast, most of the non-linear models in common use do allow such biologically realistic forms of behaviour. In addition, many of the commonly used non-linear models can be derived from simple biological concepts which, to the extent that they are appropriate, 'justify' the use of the non-linear model.

Three particular types of non-linear model are discussed later in this chapter. The first uses exponential terms to represent a tendency to an asymptote. Two models of this form are:

(i) $y = A(1 - e^{-kx})$

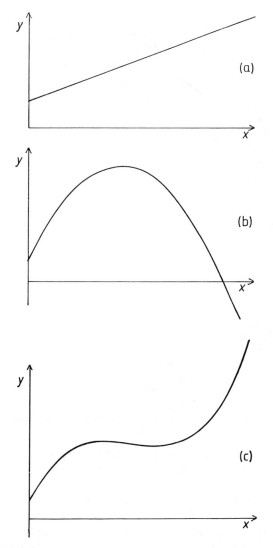

Fig. 12.1 Polynomial relationships; (a) $y = a + bx$, (b) $y = a + bx + cx^2$, (c) $y = a + bx + cx^2 + dx^3$.

which represents a gradual approach of yield to an upper limit, A, imposed by the environment and which is illustrated in Fig. 12.2(a) for various values of k. The stimulus, x would typically be the concentration of a chemical applied to living material in laboratory experiment, or the level of fertilizer applied to field plots in an agronomic crop experiment;

(ii) $y = y_0 e^{-kt}$

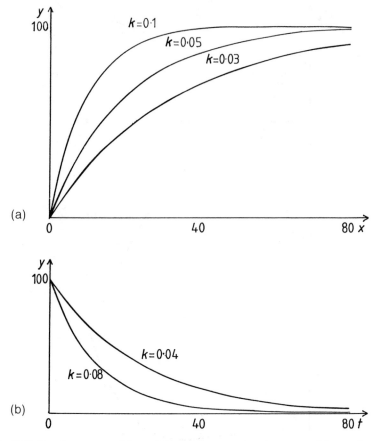

Fig. 12.2 Simple exponential models: (a) asymptotic response, with varying rates of approach to an upper limit on yield, (b) decay curves.

which represents a decay with time (t) from an initial value y_0 and which is illustrated in Fig. 12.2(b).

A second type of non-linear model which is commonly used in based on reciprocal relationships, typically the rectangular hyperbola. Simple enzyme and chemical kinetic relationships can often be expressed in the forms

$$y = \frac{Ax}{k + x}$$

where k would be the Michaelis–Menten constant, or

$$\frac{1}{y} = \frac{1}{A}\left[1 + \frac{1}{k(x - x_0)}\right]$$

used to relate photosynthesis to light intensity. These models are illustrated in

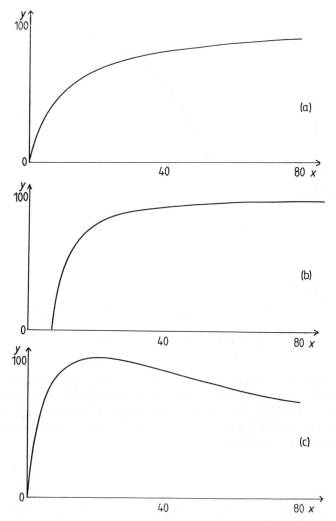

Fig. 12.3 Reciprocal models; (a) and (b) asymptotic, (c) with an optimal stimulus level.

Fig. 12.3(a) and (b). An extension of this reciprocal relationship leads to a group of response curves used in expressing agronomic response to crop density, fertilizer or chemical input (x),

$$\frac{x}{y} = a + bx + cx^2$$

When $c = 0$, this model reduces to the asymptotic curve of Fig. 12.3(a), but for non-zero c, the model, illustrated in Fig. 12.3(c), produces a non-symmetric response curve which, if a, b and c are positive, permits only positive values of y.

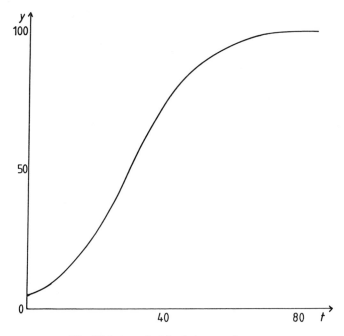

Fig. 12.4 A typical logistic growth curve.

A third type of non-linear model is the logistic curve, extensively used to represent the growth of organisms from a very small initial stage, during which growth is proportional to size, to later stages when size approaches an asymptote. The underlying idea is that the rate of change is related to current size by a differential equation

$$\frac{dy}{dt} = ky\left[\frac{A-y}{A}\right]$$

and in terms of the size at different times, we obtain the non-linear model

$$y = \frac{y_{max}}{1 + e^{(a-kt)}}$$

This is illustrated in Fig. 12.4

12.2 FITTING NON-LINEAR MODELS TO DATA

The philosophy of fitting non-linear models is the same as that for linear models. In Chapter 9, we discussed and used the principle of least squares, that if we are fitting a straight line to a set of observations then we choose that straight line which makes the sum of squares of deviations of y from the line as small as possible. Formally, this is expressed as choosing the parameters a and b of the line $y = bx$ so that

$$S = \Sigma[y - (a + bx)]^2$$

is minimized where the summation, Σ, is over the set of observations. As mentioned in the previous section, the fact that a and b are linear parameters leads to very simple estimates for a and b.

In considering the fitting methods of non-linear models, we shall use some artificial data on the change of activity of micro-organisms with time:

$y =$ activity	10	13	22	24	29	35	32	36
$t =$ time		2.5	5	10	20	30	40	60	80

The data are plotted in Fig. 12.5, and a curve of the form

$$y = \frac{at}{k + t}$$

seems a reasonable suggestion.

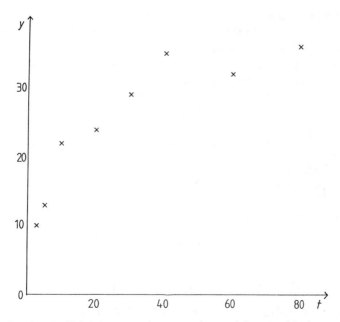

Fig. 12.5 Artificial data on micro-organism activity (y) with time (t).

The least squares principle requires that we choose a and k so that the sum of squares,

$$S = \Sigma\left(y - \frac{at}{k+t}\right)^2$$

is minimized. The mathematical equations for the values of a and k which minimize S are not directly solvable, but there is no difficulty in principle in finding the minimizing values of a and k, essentially by a process of organized, and improving, guesswork. The major problem in such a method of direct search for minimizing values is the initial guesswork. For the data of Fig. 12.5, a reasonable guess for a, which is the upper limit to which y tends, would be between 40 and 45. To obtain an initial guess for k we might consider an intermediate value of y at $t = 20$, from which we could choose k so that

$$\frac{20a}{k+20} = \text{observed } y = 24$$

for $a = 42$, this gives $k = 15$.

An alternative approach is to rewrite the model in the form

$$\frac{1}{y} = \frac{k+t}{at} = \frac{k}{a}\left(\frac{1}{t}\right) + \frac{1}{a}$$

If we now plot $1/y$ against $1/t$ as in Fig. 12.6, then we should obtain a straight line with intercept $1/a$ and slope k/a. The points in Fig. 12.6 appear reasonably linear, except perhaps for the highest value, and the dotted line drawn by eye gives an intercept of 0.026 and slope of 0.22, yielding initial guesses of $a = 38.5$ and $k = 8.5$. There are rather different from the earlier guesses, so we shall try both sets of initial guesses ($a = 42$, $k = 15$) and ($a = 38$, $k = 9$).

For $a = 42$, $k = 15$, the predicted values are:

$$t = 2.5 \quad 5 \quad 10 \quad 20 \quad 30 \quad 40 \quad 60 \quad 80$$

$$\frac{42t}{15+t} = 6.0 \quad 10.5 \quad 16.8 \quad 24.0 \quad 28.0 \quad 30.5 \quad 33.6 \quad 35.4$$

compared with observed values

$$y = 10 \quad 13 \quad 22 \quad 24 \; .29 \quad 35 \quad 32 \quad 36$$

giving deviations

$$-4.0 \quad -2.5 \quad -5.2 \quad 0.0 \quad -1.0 \quad -4.5 \quad +1.6 \quad -0.6$$

and a sum of squared deviations 73.4. Clearly the fitted curve values are generally too low.

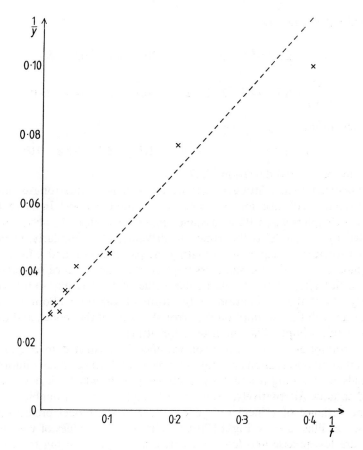

Fig. 12.6 Transformed relationship for the data of Fig. 12.5.

For $a = 38$, $k = 9$

$t =$	2.5	5	10	20	30	40	60	80
$\dfrac{38t}{9+t} =$	8.3	13.6	20.0	26.2	29.2	31.0	33.0	34.2

giving deviations

$$-1.7 \quad +0.6 \quad -2.0 \quad +2.2 \quad +0.2 \quad -4.0 \quad +1.0 \quad -1.8$$

and the sum of square deviations 32.4. This second curve is a much better fit, but is still a bit low, particularly for the higher t values.

Before leaving the reader to continue searching for the best a and k values, we will try one further set of values, keeping k at 9 but increasing a to 40. The predicted values are now

$$t = 2.5 \quad 5 \quad 10 \quad 20 \quad 30 \quad 40 \quad 60 \quad 80$$

$$\frac{40t}{9+t} = 8.7 \quad 14.3 \quad 21.1 \quad 27.6 \quad 30.7 \quad 32.7 \quad 34.8 \quad 36.0$$

giving deviations

$$-1.3 \quad +1.3 \quad -0.9 \quad +3.6 \quad +1.7 \quad -2.3 \quad +2.8 \quad 0.0$$

and a sum of squared deviations 33.2.

It is important that, in fitting models, we always make quite strong assumptions about the pattern of variation of observations about the model. In the regression models of Chapters 9 and 10, we assumed that the variance of y about the value predicted by the model is the same for all values of x. Similarly, in defining the least squares principle for estimating the parameters a and k for the non-linear model, we make the same assumption that the variance of y about the model is not affected by the particular value of x. It is important to think carefully about the assumption of the form of variation about the model. Statisticians call this assumption the error structure of the model, and different error structures imply different criteria for fitting.

The assumption of uniform error variance is illustrated in Fig 12.7(a); it implies that all observations are equally precise and hence equally informative, and leads to choosing a and k by minimizing S, in which all deviations are treated equally. Alternatively, it might be thought more reasonable to assume that the variance of y increased very rapidly as the model prediction for y increases, as illustrated in Fig. 12.7(b); this implies that values of y for larger x values are less precise and less informative and that, in fitting the curve, we should put less reliance on these values. Equivalently, we could express this in the assumption that the variance of $1/y$ was constant for all values of x. If this assumption is considered reasonable then, by writing the models in the form

$$\frac{1}{y} = \frac{1}{a} + \frac{k}{a}\frac{1}{x}$$

and by writing b for $1/a$ and c for k/a, we obtain a simple version of the model

$$\frac{1}{y} = b + c\frac{1}{x}$$

and can choose b and c to minimize S, where

$$S = \left[\frac{1}{y} - \left(b + c\frac{1}{x}\right)\right]^2$$

Clearly we are now fitting the linear regression of $1/y$ on $1/x$.

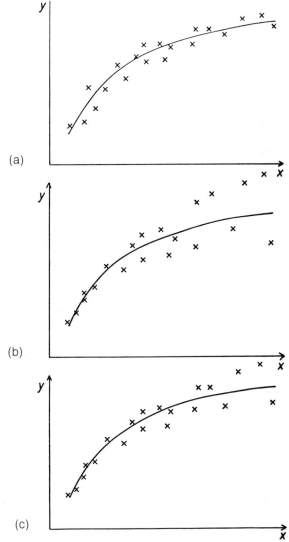

Fig. 12.7 Examples of data for which different error structures assumptions are appropriate: (a) uniform error variance, (b) variance of y increasing rapidly as y increases, (c) variance of y increasing steadily as y increases.

Notice that, for our data example, the assumption of constant variation appears quite reasonable in Fig. 12.5, but Fig. 12.6 suggests strongly that it would be wrong to assume that the variation of $1/y$ was constant, since it clearly increases with $1/x$.

The assumptions that we make about the error structure of the model are critical and influence both the method of estimation and the inferences we make about the estimates of the parameters. We must not choose our assumptions about error structure simply in order to make the estimation procedure simple, though if such an error structure seems valid then we should certainly use it. To illustrate the problem further, consider a third alternative for this error assumption. This is an intermediate assumption, that the variance of y does increase as the model prediction for y increases, but not so rapidly as in Fig. 12.7(b). Such a pattern of error variation is illustrated in Fig. 12.7(c), and would be reasonably summarized by assuming that the relative variance of y was constant. Equivalently we could assume that the variance of $\log y$ was constant and hence write the model in the form

$$\log y = \log a + \log x - \log(k + x)$$

and choose a and k to minimize S_2 where

$$S_2 = \Sigma \{\log y - [\log a + \log x - \log(k + x)]\}^2$$

This is again a non-linear estimation problem. In practice, it is very probable that one of the three error assumptions we have discussed here will be acceptable. To decide which is acceptable the most effective aids are to plot the data for y against x, $\log y$ against x, and $1/y$ against x or, in this particular model, $1/x$, and also to give some thought to the biological properties of the model.

12.3 INFERENCES ABOUT NON-LINEAR PARAMETERS

In linear regression, after obtaining the least squares estimates of the parameters we considered statistical conclusions about them and about the fitted relationship. In particular, we considered the standard error of the regression coefficient, the standard error of a predicted value for y and its dependence on the value of x, the use of the analysis of variance to obtain an estimate of the variance of observations about the fitted line, and the comparison of regression lines for different sets of data. For non-linear models, the corresponding results are not so easy to obtain and are generally approximate. Nevertheless, results similar to those for linear regression can be obtained and interpreted in the same way, but with caution.

First, consider the analysis of variance. The residual sum of squares is the minimized value of S. There are no formulae for the values of a and k which minimize S, and consequently no formula for the residual s.s., but, with the aid of computer programs, we can obtain the minimizing values of a and k and consequently the minimized value of S. The total sum of squares about the mean is, as before

$$\Sigma(y^2) - [\Sigma(y)]^2/n$$

and we can calculate the sum of squares due to the fitted relationship as the difference between this total sum of squares and the minimized value of S.

For the artificial data on micro-organism activity, the minimizing values for a and k are $a = 39.25$, $k = 9.25$, for which the value of S is 28.7. The total sum of squares is

$$(10^2 + 13^2 + ... + 36^2) - (201)^2/8 = 664.9$$

Hence we can write the analysis of variance

Source	s.s	d.f.	m.s.
Fitted relationship	636.2	1	$4.45 = s^2$
Residual	28.7	6	
Total	664.9	7	

For simple linear regression models, mathematical theory can be used to show that the residual mean square, s^2, provides an unbiased estimate of the variance of observations about the linear relationship. Part of the justification for asserting that the degrees of freedom for the residual s.s should be $(n - 2)$ is this unbiased property. For non-linear models there is no corresponding mathematical theory and the justification of the use of six degrees of freedom in the above analysis is by analogy with the results for linear regression models. The intuitive justification, that we are estimating two parameters from eight observations and that the deviations of observations about the fitted relationship are subject to two relationships and therefore include only 6 independent deviations, applies for non-linear models just as for linear models in section 9.2. Investigation of simulation results for non-linear models such as that discussed here, show that s^2 provides a very good estimate of the variance of observations about the non-linear relationships.

Instead of using the standard errors of the estimates of the parameters, we now have to use the possible values of S to derive confidence intervals for the parameters. Consider again the 'trial' sets of a and k and the corresponding values of S the sum of squared deviations

$$a = 42 \qquad k = 15 \qquad S = 73.4$$
$$a = 38 \qquad k = 9 \qquad S = 32.4$$
$$a = 40 \qquad k = 9 \qquad S = 43.2$$

Compared with the best fit of

$$a = 39.25 \qquad k = 9.25 \qquad S = 28.7$$

the first trial seems a very poor fit, the second quite good and the third a bit dubious. We shall measure quantitatively whether a pair of (a, k) values is within the confidence region of parameter values by calculating the increase in the value of S compared with the minimum. If a particular pair of (a,k) values gives a value S, then we shall accept the pair of values as being inside the confidence region if

$$F = \frac{(S - S_{min})}{2} \bigg/ \frac{S_{min}}{(n-2)}$$

is less than $F_{2,n-2}$. Effectively, we are testing whether the increase in the sum of squared deviations is significantly large. For our example,

$$S_{min} = 28.7, \ n - 2 = 6 \quad \text{and} \quad S_{min}/(n-2) = s^2 = 4.45$$

For $a = 42$, $k = 15$,

$$F = \left(\frac{73.4 - 28.7}{2}\right) \bigg/ 4.45$$

$$= \frac{22.35}{4.45} = 5.02$$

which is significant at the 5% level. Hence the pair $(42,15)$ is not in the 95%

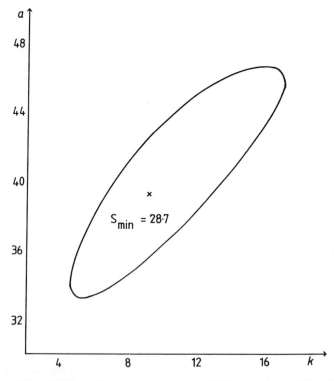

Fig. 12.8 A 95% confidence region for a and k for the data of Fig. 12.5.

confidence region. For $a = 40$, $k = 9$,

$$F = \left(\frac{43.2 - 28.7}{2}\right) \Big/ 4.45 = 1.63$$

which is not significant at the 5% level. Hence (40, 9) and also (38, 9) are within the 95% confidence region. In fact, we can turn the equation round and argue that, for a pair of values (a, k) to be in a 95% confidence region, we must have the corresponding S value to be less than

$$S_{min} + 2\frac{S_{min}}{(n-2)} F_{2, n-2}, 5\%$$

The 95% confidence region for a and k for the data example is shown in Fig. 12.8. Again, the construction of the region is laborious by hand, but quite simple by computer.

Frequently we wish to compare fitted models for different sets of data. Exactly as for linear regression, comparisons are made in terms of the residual sums of squares about the fitted models for each set of data separately, and for the combined set of data. Thus, if the non-linear model

$$y = \frac{ax}{k+x}$$

is fitted to three data sets, the minimized residual sums of square are S_1, S_2 and S_3 respectively, and when the model is fitted to the combined data set, the resulting residual sum of squares is S_c. Then we can construct an analysis of residual variation:

Source	s.s	d.f.	m.s.
Residual for set 1	S_1	$n_1 - 2$	
Residual for set 2	S_2	$n_2 - 2$	
Residual for set 3	S_3	$n_3 - 2$	
Σ residual	$S_1 + S_2 + S_3$	$n_1 + n_2 + n_3 - 6$	s^2
Residual for combined data	S_c	$n_1 + n_2 + n_3 - 2$	
Difference between fitted models	$S_c - (S_1 + S_2 + S_3)$	4	Difference m.s.

To test (approximately) the difference between the models fitted to the three sets of data we compare

$$F = \frac{\text{difference m.s}}{s^2} \quad \text{with } F_4, n_1 + n_2 + n_3 - 6$$

One final comment before we consider particular models and examples of their use in more detail. An alternative approach to the analysis of non-linear models, which is beyond the scope of this book, is through the analysis of likelihood. All the methods used so far in this book are maximum likelihood methods, that is they have the intuitively appealing property of estimating the values of unknown parameters so that the probability of observing the data actually obtained is maximized. Many of the more powerful statistical packages have routines for fitting non-linear models using maximum likelihood methods. To use them successfully the user must have a good idea of the mathematical form of the class of curves appropriate to the data and must also be able to supply appropriate starting values for the model parameters, using empirical procedures as discussed in this chapter. The rather easier analysis of residual variation discussed in this chapter is generally close to the analysis using maximum likelihood methods.

12.4 EXPONENTIAL MODELS

The simplest form of exponential model is that for the decay of some variable with time. The model for such a response pattern,

$$y = y_0 e^{-kt}$$

was illustrated in Fig. 12.2(b). The one assumption of the model is that the rate of decay is proportional to the current value of y. In the model, k represents the proportionality factor, and y_0 represents the initial value of the variable before decay begins.

A particular case is when y is the size of a population of individuals, and there is a fixed probability of death in a unit time interval, the probability, k, being the same for each individual in the population, and unaffected by the past history of the population. This model is closely related to the Poisson model of Chapter 14.

Exponential decay models have proved useful in studying the survival of medical and other biological populations at risk—for example, the survival of transplant patients or patients with cancer. In the laboratory, the models are used for populations of micro-organisms or of insects treated with chemicals. In agricultural investigations, the decay of chemicals in the soil or in animals have been modelled by exponential functions.

The other area where exponential models have been extensively used is where the response variable either increases with time, t, or as a result of an increasing level of a stimulus variable, x. The change of y is initially fast, but gradually declines until y reaches an asymptotic level. Such a model was illustrated in Fig. 12.2(a). A more general form is

$$y = A[1 - e^{-k(t - t_0)}]$$

illustrated in Fig. 12.9. The three parameters each have a simple biological

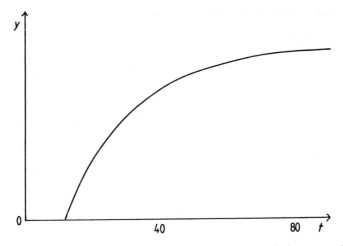

Fig. 12.9 Yield response curve with an asymptote approached exponentially.

interpretation, A being the asymptotic level which y approaches as t increases, k the rate of approach and t_0 the time at which y begins to increase from zero.

For each model, there are methods of rewriting the model to produce linear relationships which can be used to plot the data graphically, to see whether the data follow a linear pattern. Thus the model $y = y_0 e^{-kt}$ can be rewritten

$$\log y = \log y_0 - kt$$

and hence plotting $\log y$ against t, should produce an approximately straight line relationship. For the model

$$y = A[1 - e^{-k(t - t_0)}]$$

the alternative form is

$$\log(A - y) = -kt + \alpha$$

where $\alpha = \log A + kt_0$. This alternative form is not immediately useful for plotting the data since we can only calculate, and plot, the data if we assume a value for A, which is one of the model parameters which we require to estimate. However, A must be bigger than any observed y value but, provided a plot of y against t indicates that y has almost reached its asymptotic level, not much bigger. Hence by considering the larger y values, we can make a reasonable guess of A and plot the data. It may be useful to plot the data for several values of A to see for which, if any, a straight line relationship emerges. In considering different forms of writing a particular relationship, it must be remembered that, in fitting the model to the data, the choice of the correct error structure assumption is critically important.

Example 12.1 Fitting an exponential response model

An experiment to investigate the effect of nitrogen fertilization on sugar cane yields included five nitrogen levels (0, 50, 100, 150, 250 kg/hectare) in four randomized blocks of five plots each. The yields were as follows:

		Block				
		I	II	III	IV	Total
Nitrogen	0	0	73	77	72	282
	50	125	144	145	116	530
	100	152	154	160	141	607
	150	182	167	181	185	715
	200	198	188	189	182	757
Total		717	726	752	696	2891

The analysis of variance for this data is:

Source	s.s	d.f	m.s.	F
Blocks	323	3	108	
Treatments	35393	4	8848	109
Error	967	12	81	
Total	36683			

Clearly there is a very substantial effect of nitrogen, but little variation between blocks. The mean yields for the nitrogen levels are:

N	0	50	100	150	200
Yield	70.5	132.5	151.8	178.8	189.2

with standard error for comparing any two mean values: $\sqrt{[(2 \times 81)/4]} = 6.4$.

The mean yields are plotted against N in Fig. 12.10, with the individual yields also shown

A model which is often used for describing fertilizer response is the Mitscherlich response function which is most simply written

$$y = a[1 - e^{-b(x+c)}]$$

As discussed earlier, a represents the asymptotic level of yield, c the amount of fertilizer already available in the soil, and b the rate at which yield tends to its asymptotic level. A reasonable rough guess for a might be 200 and we therefore try plotting $\log(200 - y)$ against x. This is shown in Fig. 12.11 and obviously gives a reasonable straight line. Fitting a slope and an intercept for this transformed

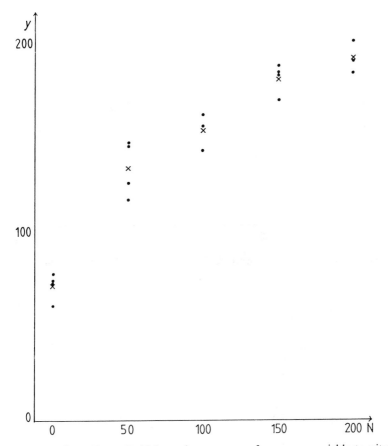

Fig. 12.10 Data from Example 12.1 on the response of sugar cane yields to nitrogen fertiliser.

data we get a slope of -0.01224 and an intercept of 4.90. The transformed model is

$$\log(a - y) = -bx - bc + \log a$$

Hence a first estimate of b is 0.01224 and a first estimate of c is

$$[(\log 200) - 4.90] \div 0.01224, \text{ or } 32.5$$

To obtain the best estimates of a, b and c, we search for those values which minimize the sum of squares

$$S = \Sigma\{y - a[1 - e^{-b(x+c)}]\}^2$$

For this particular model, however, there is a slightly simpler version of this

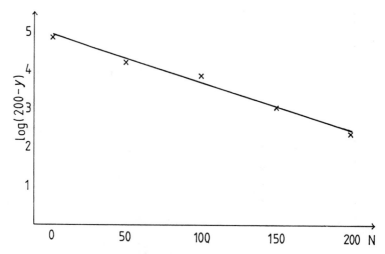

Fig. 12.11 Transformed mean yields for sugar cane data of Example 12.1.

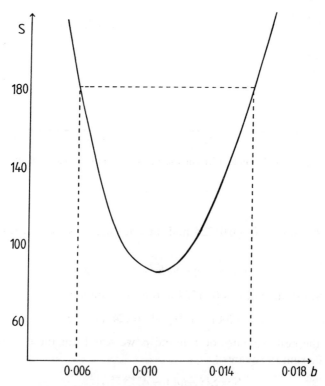

Fig. 12.12 Variation of the sum of squared deviations about the fitted model, as the parameter b takes different values.

search, since the model can be written

$$y = a - (ae^{-bc})e^{-bx}$$

If we write z for e^{-bx}, then this model has the form as the simple linear regression model $y = \alpha + \beta z$. Hence, if we search over possible values of b and calculate $z = e^{-bx}$ for each of the five x values for each possible value of b, we can find that value of b which gives the smallest residual sum of squares from the regression of y on z. The values of a and c can be deduced from the intercept and slope of the regression.

The estimates resulting from this procedure are

$$b = 0.0109$$
$$a = 203$$
$$c = 40$$

and the minimized value of S is 86.0. The degrees of freedom for S_{min} are $5 - 3 = 2$, and so the residual mean square for the fitted model is $86/2 = 43$. From the analysis of variance, the variance of a treatment mean is $81/4 = 20.25$, and this is the value we would expect for the residual m.s. An approximate F-test of whether the residual m.s. is significantly large gives

$$F = \frac{86/2}{81/4} = \frac{43}{20.25} = 2.13$$

which is not significant when compared with the F-distribution on 2 and 12 degrees of freedom.

To construct a confidence interval for the parameter b, we consider the way in which S varies with b, in all cases using the best estimates of a and c for the value of b being considered. The way S varies with b is shown in Fig. 12.12. For values of b between 0.009 and 0.012, S is not much greater than 86, indicating that the estimate of b is not very precise. Since the residual m.s. from the fit to the five mean values has only 2 d.f., it is preferable to use the estimate of variance from the analysis of variance, which has 12 d.f. The estimate of the variance of mean yields about the fitted line is then $s^2 = 20.25$ on 12 d.f. We examine the values of S and b to discover for what b values S is not greater than $S_{min} + s^2 F_{1,12,5\%}$. The critical value is $86.0 + 20.25 \times 4.75 = 182.19$. We can thus obtain an approximate 95% confidence interval for b of (0.0063, 0.00158) which shows clearly that the data do not provide enough information to estimate b very precisely.

12.5 INVERSE POLYNOMIAL MODELS

We have already discussed models based on the reciprocals of variables in the first three sections of the chapter. Now we shall reconsider these models in their formulation as inverse polynomials, examining the properties of the models briefly and then using them to analyse data in an example which illustrates many of the techniques of analysis for non-linear models.

Inverse polynomial models were first used for yield–density relationships for agricultural crops in the form

$$\frac{1}{w} = \alpha + \beta\rho$$

where w is the yield per plant and ρ the plant density per unit area. The curve is illustrated in Fig. 12.13(a) and the corresponding curve for the yield per unit

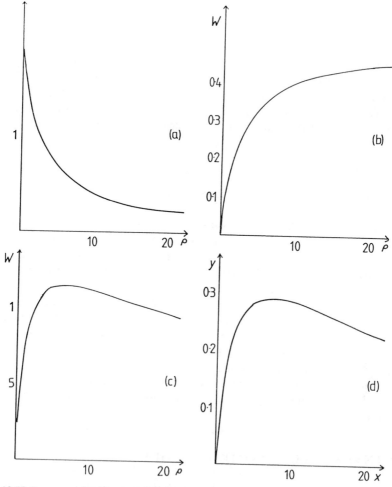

Fig. 12.13 Inverse polynomial models; (a) yield per plant (w) against plant density (ρ); (b) yield per area (W) against density; (c) powered inverse linear polynomial; (d) inverse quadratic polynomial.

area. $W = w\rho$, is shown in Fig 12.13(b),

$$\frac{\rho}{W} = \alpha + \beta\rho$$

The asymptotic yield of W is $1/\beta$ and the yield per plant, w, when the plant density becomes very small, so that the plants are effectively grown in isolation, is $1/\alpha$.

A modification of the relationship to allow for the possibility that W, as a function of increasing ρ, reaches a maximum and then declines, is

$$\frac{1}{w^\theta} = \alpha + \beta\rho,$$

illustrated in Fig. 12.13(c).

Another generalization, also allowing a maximum, has been more extensively used than the power modification with w^θ, and this is

$$\frac{x}{y} = \alpha + \beta x + \gamma x^2$$

where y is the yield per unit area and x is the plant density, or some nutrient stimulus level. The relationship is illustrated in Fig. 12.13(d). The maximum value of y occurs when

$$x = \sqrt{(\alpha/\gamma)}$$

Essentially, the ratio of α to β dictates the form of the rising portion of the curve, and the ratio of β to γ the form of the falling portion. The individual parameters do not have simple interpretations. One restriction on the curves is that they pass through the origin, i.e. $y = 0$ when $x = 0$. One particular flexibility is that the rising and falling portions of the curve are not symmetrical as they are in a simple quadratic relationship. This flexibility is a considerable advantage for fertilizer response curves. The restriction to pass through the origin is often unrealistic and so a further modification allows for an origin for x in the form

$$\frac{x + x_0}{y} = \alpha + \beta(x + x_0) + \gamma(x + x_0)^2$$

where the right hand side of the equation is still a simple quadratic polynomial.

Further generalizations allow y to be expressed as an inverse polynomial in several x variables. One general point to note is that, in the normal use of these models and in the illustrations in Fig. 12.13, the coefficients are all assumed to be positive. The algebraic relations do exist for negative parameter values, but it is not apparent that such relations are biologically useful.

As with all models, it is necessary to consider the error structure assumptions for each use of the model to fit data. However, in many yield–density and fertilizer response sets of data, it has been found reasonable to assume that the relative variability of y remains constant. That is, the standard deviation of y increases with y, or the logarithm of y is uniformly variable. When this assumption is true, there is a particularly simple form of parameter estimates which gives approximately the same value as the general estimation procedure. The form of the sum of squares to be minimized to give this approximation is

$$S = \Sigma \left[\frac{y}{x}(\alpha + \beta x + \gamma x^2) - 1 \right]^2$$

for the inverse polynomial, or

$$S = \Sigma [w^\theta(\alpha + \beta x) - 1]^2$$

for the power version. In the example, we shall use this approximate method of estimation.

Example 12.2 Invariance fits for a non-linear model

Three onion varieties were grown at ten different plant densities. Because of variable germination rates, it was decided that the observed plant density was a more appropriate measure of density than the intended density. The mean yields per plant, w (kg), and the plant densities, ρ (plants per ft^2), are given below. Examination of the yield variation between replicates for each density confirmed that the variation of $\log w$ was acceptably constant and the means

Variety 1		Variety 2		Variety 3	
w	ρ	w	ρ	w	ρ
0.1056	3.07	0.1316	2.14	0.1168	2.48
0.0894	3.31	0.1091	2.65	0.0916	3.53
0.0710	5.97	0.0937	3.80	0.0727	4.45
0.0603	6.99	0.0722	5.24	0.0528	6.23
0.0476	8.67	0.0531	7.83	0.0488	8.23
0.0377	13.39	0.0497	8.72	0.0391	9.59
0.0303	17.86	0.0378	10.11	0.0303	16.87
0.0242	21.57	0.0333	16.08	0.0242	18.69
0.0208	28.77	0.0245	21.22	0.0200	25.74
0.0185	31.08	0.0183	25.71	0.0163	30.33

were in fact averaged on the log scale. An analysis of variances of $\log w$ gave an estimate of the variances of each mean as 0.00395 on 52 degrees of freedom.

The data are plotted in two forms. In Fig. 12.14(a) the yield per area, $W = w\rho$, is plotted against ρ to show the asymptotic form of the relationship, and in Fig. 12.14(b) $1/w$ is plotted against ρ to examine the linearity of the relationship. Examination of the fitted model for w^θ does not produce evidence that θ is

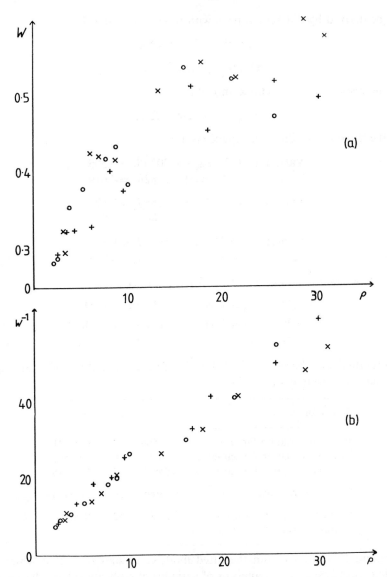

Fig. 12.14 Data for Example 12.2 for varieties 1(x), 2(o) and 3(+): (a) yield per area (W) against plant density (ρ); (b) reciprocal of yield per plant (w^{-1}) against plant density.

different from 1, and we shall therefore consider only the model $\dfrac{1}{w} = \alpha + \beta\rho$ and the sum of squares to be minimized

$$S = \Sigma[w(\alpha + \beta\rho) - 1]^2$$

The values of α and β which minimize S are given by two equations, which may be derived by differentiating S with respect to α and β,

$$\alpha\Sigma w^2 + \beta\,\Sigma w^2\rho = \Sigma w$$

$$\alpha\Sigma w^2\rho + \beta\Sigma w^2\rho^2 = \Sigma w\rho$$

and the residual s.s of the fitted model is

$$S_{min} = n - \alpha\Sigma w - \beta\Sigma w\rho$$

For the three sets of data the equations are:

Variety 1 $0.0338\alpha_1 + 0.2070\beta_1 = 0.505$
$0.2070\alpha_1 + 2.2282\beta_1 = 4.619$

Variety 2 $0.0520\alpha_2 + 0.2265\beta_2 = 0.623$
$0.2265\alpha_2 + 1.7180\beta_2 = 4.062$

Variety 3 $0.0362\alpha_3 + 0.1834\beta_3 = 0.513$
$0.1834\alpha_3 + 1.6785\beta_3 = 4.015$

From these equations, we obtain the fitted parameters:

$\alpha_1 = 5.24$ $\beta_1 = 1.59$ residual s.s $= 0.02566$
$\alpha_1 = 3.96$ $\beta_2 = 1.84$ residual s.s $= 0.04655$
$\alpha_3 = 4.58$ $\beta_3 = 1.89$ residual s.s $= 0.05877$

Each residual s.s is based on eight degrees of freedom, and we can summarize the results for residual s.s as follows:

Source of variation	s.s	d.f.	m.s.
Residual variation for variety 1	0.02566	8	0.00321
Residual variation for variety 2	0.04655	8	0.00582
Residual variation for variety 3	0.05877	8	0.00735
Combined residual variation	0.13098	24	.000546
Estimate of residual variation from analysis of variance	0.20607	52	0.00396

The residual variations from the fitted models are similar to that which we should expect based on the analysis of variance of replicate values, the F-ratio of $0.00546/0.00396 = 1.38$ being not significant at the 5% level and certainly not large enough to make us worry about the fit of the model.

We can analyse the results further by comparing the fitted models in the same way that we compared linear regressions (section 11.2). Thus, to fit a single curve to all three sets of data, we simply combine the three sets of equations, getting:

Combined fit $0.1220\alpha + 0.6169\beta = 1.641$
$0.6169\alpha + 5.6246\beta = 12.696$

and fitted parameters are

$$\alpha = 4.58 \qquad \beta = 1.75 \qquad \text{residual s.s} = 0.20802 \text{ (28 d.f.)}$$

If we wish to test whether the α values are the same, which implies that the mean weight of an isolated plant is identical for the three varieties, then we need equations for α, β_1, β_2 and β_3 and these are derived from the separate pairs of equations for the three varieties by combining the first equations from each pair.

$$
\begin{aligned}
0.1220\alpha + 0.2070\beta_1 + 0.2265\beta_2 + 0.1834\beta_3 &= 1.641 \\
0.2070\alpha + 2.2281\beta_1 &= 4.619 \\
0.2265\alpha \qquad\qquad + 1.7180\beta_2 &= 4.062 \\
0.1834\alpha \qquad\qquad\qquad\qquad + 1.6785\beta_3 &= 4.015
\end{aligned}
$$

The fitted parameters are:

$$
\begin{aligned}
\alpha = 4.51 \quad \beta_1 &= 1.65 \quad \text{residual s.s} = 0.14676 \text{ (26 d.f.)} \\
\beta_2 &= 1.77 \\
\beta_3 &= 1.90
\end{aligned}
$$

Similarly, to examine whether the β values are the same, which would mean that the maximum achievable yields per unit area are identical for the three varieties, we can derive a corresponding set of equations and obtain fitted parameters:

$$
\begin{aligned}
\alpha_1 &= 4.20 \quad \beta = 1.76 \quad \text{residual s.s.} = 0.18035 \text{ (26 d.f.)} \\
\alpha_2 &= 4.34 \\
\alpha_3 &= 5.26
\end{aligned}
$$

The residual sums of squares can be summarized in an analysis of variance as follows:

Curve fitted	r.s.s	d.f.	r.m.s
(1) Combined fit (α, β)	0.20802	28	
(2) α invariant $(\alpha, \beta_1, \beta_2, \beta_3)$	0.14676	26	
(3) β invariant $(\alpha_1, \alpha_2, \alpha_3, \beta)$	0.18035	26	
(4) Separate fit $(\alpha_1, \beta_1, \alpha_2, \beta_2, \alpha_3, \beta_3)$	0.13098	24	0.00546

Source of variation	s.s.	d.f.	m.s.	F
(2) − (4) variation of α	0.01578	2	0.00789	1.45
(3) − (4) variation of β	0.04937	2	0.02468	4.52*
(1) − (2) variation of β assuming α invariant	0.06126	2	0.03063	5.61*

Each F-ratio, testing the variation of the α values or of the β values, compares the appropriate difference mean square with the mean square from the residual variation about the three fitted lines. Thus, to test whether the α values show significant variation, we compare

$$F = \frac{0.00789}{0.00545} = 1.45$$

with $F_{2,24}$ the value of 1.45 being far from significant.

The conclusions from this analysis are in two stages. First, the yield–density relationships appear to give an acceptable fit to the data and we should be happy to use the model for describing the effects of variation of crop density over a wide range of densities. Second, the three varieties appear to be similar in their yields at very low densities, but different in their asymptotic yields for high densities.

12.6 LOGISTIC MODELS FOR GROWTH CURVES

Many biological investigations are concerned with the growth of organisms with time. Extensive studies have been made of the growth of whole plants, or of the growth of individual leaves, or of the growth of animals. Qualitatively, the growth of a biological organism can be thought of in four stages. Early growth, starting from a very small initial size, is relatively very rapid; we can express this quantitatively by saying that the rate of growth is proportional to the size of the organism. This form of growth is often called 'exponential growth' because, if the rate of growth is exactly proportional to size,

$$\frac{dy}{dt} = ky$$

then the size of the organism is described by the exponential function,

$$y = y_0 e^{kt}$$

The second stage of growth is relatively less rapid, as more of the energy of the organism is devoted to maintaining its current size, and during this stage the growth of the organisms may be well approximated by the linear relationship

$$y = \alpha + \beta t$$

In the third stage, the organism's growth diminishes further as a balance between the energy of the organism and the maintenance requirements is approached. The size of the organism in this stage is asymptotic, tending to an upper limit on size, in the same form as has been illustrated, for other biological situations in the previous sections.

The fourth, and the final stage of growth, with which we shall not be concerned here, is the antithesis of growth, through senscence.

The first three stages can be thought of together in terms of the rate of growth

$$\frac{dy}{dt} = ky\left[\frac{A-y}{A}\right]$$

When y is small, the rate of growth of y is approximately proportional to y, since the expression in brackets is close to 1. As y increases, the relative growth rate declines, and as y approaches A, dy/dt approaches zero. The equation for y implied by this form of growth rate is the logistic curve:

$$y = \frac{A}{1 + e^{(a-kt)}}$$

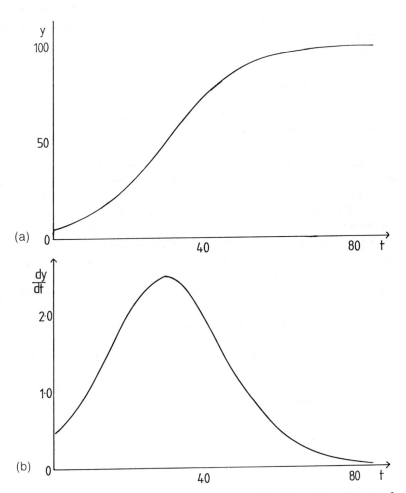

Fig. 12.15 Logistic growth functions: (a) size (y) as a function of time (t); rate of growth (dy/dt) as a function of time.

The curves showing the behaviour of the rate of growth dy/dt, and the size, y, with time are shown together in Fig. 12.15.

Generalizations of this basic form of curve are developed by Richards (1959) and Nelder (1961) and there is an extensive monograph on the subject by Causton and Venus (1981).

There are two practical difficulties which are likely to beset the use of these models. The first is that it is difficult to obtain data over a sufficient range of times to be able to obtain good information about all the parameters of the response function. This problem is more acute when a generalization of the function involving a fourth parameter is used, but is still important for the basic logistic. To see how the data might fail to be adequate, consider again the three stages of growth represented by the logistic. The estimation of k is based on the first stage of growth when the rate of increase of y with time is increasing sharply. The observations must therefore begin at an early stage to cover the exponential growth stage. By contrast, the estimation of A is based on the third stage when y is approaching its asymptotic value. There must therefore be observations as y approaches its limiting size and at least one observation when y has effectively ceased to increase. The third parameter, a, has no simple direct interpretation but is essentially measuring the size of the plant at the time from which the time scale is measured: provided that both the two extreme portions of the curve are well defined there should be no difficulty in estimating a.

In many situations, repeat measurements for the same experimental unit will be available. The logistic growth model is then fitted separately to the data for each unit and the fitted growth curve for different units compared through a comparison of the fitted parameter values. In some situations, of which the most common is the growth analysis of whole plants, the measurement of plant weight at a particular time involves the destruction of the plant or of the plant's growing environment. To obtain information about the growth curve, different samples of plants are required at each of a number of different times. The fitted growth curve is then appropriate to a population of plants rather than an individual plant. In this case, we have to decide when to sample plants as well as how many plants to harvest and measure on each sampling occasion. These problems will be discussed in Chapters 16 and 17. However, one general piece of advice is appropriate here and this applies to other forms of analysis of growth data. For a given total of resources, that is a given total number of plants which can be sampled, more information about the form of the growth curve is obtained by taking larger samples at a relatively small number of well chosen occasions than by taking smaller samples at very frequent occasions.

13

The analysis of proportions

13.1 DATA IN THE FORM OF FREQUENCIES

Suppose you wish to compare the size of the ant populations in two different regions of regenerated forest which have been treated in different ways. In each region you have a number of areas available. The standard method of comparison would be to define small sample areas in each region, count the number of ants in those areas, calculate the mean and variance of the sample of counts for each region in turn and then do a t-test to compare the means. If you were thinking carefully about the data you might transform the counts to try to satisfy the assumption that the counts come from a normal distribution. Alternatively, if you were seriously worried about the assumptions required for a t-test, you might decide not to count the ants in each sample area, but instead simply categorize each area as sparse or abundant in respect of the ant population being considered. Clearly by using such a crude form of assessment you are discarding information, but the actual assessment will be much quicker so that you may be able to assess many more areas from each region, and thus improve the total information.

The results obtained using the two different approaches might be:

(a) Forest region 1: ant counts 26, 8, 0, 2, 43, 19, 1, 6
    Forest region 2: ant counts 5, 14, 32, 1, 1, 0, 6, 8
(b) Region 1: 6 areas classified as sparse, 14 as abundant
    Region 2: 12 areas classified as sparse, 8 as abundant

We know how to analyse data from the first approach, but not yet data from the second. Before going on to the appropriate method of analysis it is interesting to consider some other situations which may give rise to the same kind of data.

(i) Two groups of cattle are given, respectively, a standard vaccination against brucellosis, and a new form of vaccine. Of 14 animals given the standard vaccine 10 contact brucellosis, 4 do not. Of the 16 animals who received the new vaccine 5 contract brucellosis, 11 do not. Is the difference strong evidence in favour of using the new vaccine?
(ii) Two different drying processes are used in preparing seed. When seeds from each process are tested it is found that 70 out of 100 germinate for seeds

from process A, but only 62 from process B. How strong is the evidence that the two processes affect germination differently?

(iii) In storing fruit the time of picking may be crucial. Of 30 apples picked at the normal time 8 are classified as 'bad' after 4 months storage, whereas of 20 apples picked a week earlier only 2 are 'bad'. Again, how much evidence do the data provide on the importance of the time of picking?

13.2 THE 2 × 2 CONTINGENCY TABLE

In all these situations the basic question is 'How strong is the evidence that different treatments alter the proportion of units belonging to a particular category?' The statistical approach we have used so far in this book is to assume that differences are purely due to chance until the probability of getting such a large difference purely by chance is too small to be believed. In other words we set up a null hypothesis that the proportions of 'abundant' and 'sparse' areas are the same in the populations of possible areas which we could have examined in the two different forest regions. The data are set out formally in a 2 × 2 contingency table showing the frequencies of all four possible contingencies:

Classification of areas in each region as abundant or sparse:

	Abundant areas	Sparse areas	Total number of areas
Region 1	14	6	20
Region 2	8	12	20
Combined regions	22	18	40

Now if the proportions of abundant to sparse are really the same in the two regional populations then the best estimate of the common proportions of abundant areas is $(14 + 8)/(20 + 20)$, i.e. the total number of areas classified as abundant out of the total number of areas sampled. Thus if we suppose that the ant populations are not differentially affected by the two regeneration methods then we would expect 22 areas out of 40 to be classified as abundant whichever cultivation method had been applied. So in each region we would expect 11 areas out of 20 to be classified as abundant and 9 out of 20 as sparse. If we now compare the observed values with our expected values we get:

	Observed	Expected	Difference
Region 1, abundant	14	11	+3
Region 1, sparse	6	9	−3
Region 2, abundant	8	11	−3
Region 2, sparse	12	9	+3

Note the pattern in the differences. Had you expected it? To assess whether the differences are substantial or negligible, we appeal to mathematical theory which says that if we calculate

$$X^2 = \frac{(\text{observed} - \text{expected})^2}{\text{expected}}$$

for each of the four lines in the table and sum these four quantities then this sum should have approximately a χ^2 (chi-squared) distribution. We shall meet this

$$X^2 = \Sigma \frac{(\text{observed} - \text{expected})^2}{\text{expected}}$$

many times more.

The principle of the use of the X^2 statistic is simple. The bigger the difference (observed − expected) the stronger the evidence against our null hypothesis. We square the difference so that we are summing only positive numbers and we divide by the expected number for each group because a difference of 3 from an expected value of 9 is of greater importance than a difference of 3 from an expected value of 90. Before we start calculating X^2 for these particular data, notice that we said only that we knew the approximate distribution of X^2 to be the chi-squared distribution. The reason why it is only approximate is essentially that we are dealing with counts rather than continuous measurements like weights or lengths or time. We could have 14 areas out of 20 classified as abundant, or 13, or 12 or 0, but not 13.5 or 12.13. The chi-squared distribution is, like t- and F-distribution, a continuous distribution and so we are comparing the statistic X^2 which can take one of a limited set of values with a distribution including all positive values as possible. The approximation involved in using the continuous χ^2 distribution, instead of a distribution based on only integer values, is slight unless some of the expected frequencies are very small. There is no absolute limit on the permissible smallness of expected frequencies but conclusions based on expected frequencies which included a value of less than 1 would be very dubious and expected frequencies between 1 and 5 should be treated with some caution.

There are in fact many different χ^2 distributions just as there are many different t- and F-distributions. With a 2 × 2 contingency table we use the χ^2 distribution with 1 degree of freedom. This distribution is summarized in the first line of Table A.6. We use the term 'degrees of freedom' in very much the same way as we did when discussing, in section 2.3, the degrees of freedom associated with a sample variance, s^2, as an estimate of the population variance, σ^2. There is said to be only one degree of freedom because, if the row and column totals remain unaltered, the number in any one of the four cells of the table immediately determines the other three numbers. The other χ^2 distributions will be needed when we consider contingency tables with more than two rows or columns.

To illustrate the use of the χ^2 test we consider again the data for the comparison of the two forest regions.

	Observed	Expected	Differences	$(O - E)^2/E$
Region 1, abundant	14	11	+3	0.818
Region 1, sparse	6	9	−3	1.000
Region 2, abundant	8	11	−3	0.818
Region 2, sparse	12	9	+3	1.000
				3.636

The observed value of X^2 is 3.64. To determine the significance of this evidence that there is a difference in ant density between the two regions we are interested in the probability of obtaining, by chance, a value bigger than 3.64. Only the values bigger than 3.64 are of interest because they will correspond to larger deviations between observed and expected frequencies than the deviations we have actually observed and, in general, the larger the deviation the stronger the evidence against the null hypothesis. Now from Table A.6 we find that 5% of the values in a χ^2 distribution with 1 d.f. would be larger than 3.84. Our calculated value is less than 3.84 so that in our example the evidence for different ant densities in the two regions is not significant at the 5% level, although the calculated value of 3.64 is near enough to the 5% significance value for us to have doubts about the null hypothesis of no difference.

One final complication in calculating the X^2 statistic for a 2 × 2 contingency table concerns a modification known as Yates' correction. The purpose of this correction is to improve the approximation of the continuous χ^2 distribution to the discrete distribution of possible contingency table results. The correction consists of reducing the absolute size of the deviation, $(O - E)$, by $\frac{1}{2}$ before squaring. This correction is undoubtedly correct for the special but rare case of a 2 × 2 contingency table for which both sets of marginal totals are predetermined. However, for the form of 2 × 2 contingency table which we have considered where only the row totals are determined before the data are collected, the important criterion is that under the null hypothesis the probability of obtaining a X^2 value greater than 3.84 should be as close as possible to 0.05, and this is achieved by ignoring the Yates' correction and completing the calculations as shown above.

13.3 MORE THAN TWO SITUATIONS OR MORE THAN TWO OUTCOMES

So far all the examples of contingency tables which we have considered have been 2 × 2 contingency tables in which we have examined a number of results in each of two situations where each result can have one of only two outcomes.

In our initial example the two regions of forest provided the two situations and the two outcomes allowed were 'sparse' or 'abundant' populations of ants. There is, however, no need to restrict ourselves either to two situations or to two outcomes. We might wish to compare several regions; we might want to classify the outcomes in more detail; either (a) 'abundant', 'intermediate', 'sparse', or 'absent' or (b) 'predominantly species A', 'predominantly species B', or 'equally A and B'.

The simplest extension is to the case when we are comparing several situations, but for each sample there are still only two possible outcomes. To return briefly to the analogy at the beginning of this chapter, the 2×2 contingency table is analogous to the use of the t-test to compare the means of two samples from two different situations. When we considered the comparison of means of more than two samples we used the F-test to assess the importance of the variation between means, though we also used t-tests to compare particular pairs of means. With data consisting of frequencies we can produce a test to compare the proportions of the two possible outcomes for a number of situations. However, whereas when comparing means we had to replace the t-test for two means by the F-test for several means, for frequency data we can continue to use the χ^2 test for comparing several proportions just as we did for two proportions.

Suppose we have k situations and that, of n_1 observations in situation 1, r_1 produce outcome 1 and $(n_1 - r_1)$ outcome 2 and similarly for n_2 observations in situation 2, and so on. We can represent the results as follows:

	Outcome 1	Outcome 2	Total
Situation 1	r_1	$(n_1 - r_1)$	n_1
Situation 2	r_2	$(n_2 - r_2)$	n_2
$\vdots$	$\vdots$	$\vdots$	$\vdots$
Situation k	r_k	$(n_k - r_k)$	n_k
Total	Σr	$\Sigma n - \Sigma r$	Σn

If the true proportions of outcome 1 are the same for all situations then we would estimate this proportions by the overall proportion of outcome 1 results, $\Sigma r / \Sigma n$. Hence we can work out the frequencies we would expect for outcome 1 in situation 1, outcome 2 in situation 1, outcome 1 in situation 2 and so on. The expected frequency for outcome 1 in situation 1 is the overall proportion of outcome 1 results multiplied by the total number of observations for situation 1, i.e.

$$(\Sigma r / \Sigma n)n_1$$

Hence we can construct a table of expected frequencies to match the observed frequencies given earlier:

	Outcome 1	Outcome 2
Situation 1	$\dfrac{(\Sigma r)}{\Sigma n}n_1$	$\dfrac{(\Sigma n - \Sigma r)}{\Sigma n}n_1$
Situation 2	$\dfrac{(\Sigma r)}{\Sigma n}n_2$	$\dfrac{(\Sigma n - \Sigma r)}{\Sigma n}n_2$
$\vdots$	$\vdots$	$\vdots$
Situation k	$\dfrac{(\Sigma r)}{\Sigma n}n_k$	$\dfrac{(\Sigma n - \Sigma r)}{\Sigma n}n_k$

To test the deviations of the observed frequencies from the expected frequencies we calculate

$$X^2 = \Sigma \frac{(\text{observed} - \text{expected})^2}{\text{expected}}$$

just as before and compare it with the χ^2 distribution. We are now dealing with a $2 \times k$ contingency table and the degrees of freedom for the χ^2 distribution are $(k-1)$. This is because we could arbitrarily fill in $(k-1)$ of the cells of the table while keeping the same row totals and column totals. Since the X^2 statistic is based on more terms than for the 2×2 table, a larger value of X^2 is required before the same level of significance is achieved and this is reflected in the 5% points of the chi-squared distributions in Table A.6.

Example 13.1 Analysis of a $2 \times r$ contingency table

These data are taken from a much larger set of data from a large factorial experiment. One object of the experiment was to examine the effects of different fertiliser treatments on the incidence of blackleg (*Bacterium phytotherum*) for potato seedlings. A number of seedlings were examined for each treatment and classified as contaminated by blackleg or free. The results for four treatments were as follows:

Observed frequencies	Blackleg	No blackleg	Total
No fertilizer	16	85	101
Nitrogen only	10	85	95
Dung only	4	109	113
Nitrogen and dung	14	127	141
Total	44	406	450

The overall proportion of seedlings with blackleg is $44/450 = 0.0978$. The expected frequency of seedlings from the no fertilizer plot with blackleg is therefore $101 \times 0.0978 = 9.9$ and the expected frequency of clean seedlings from the no fertiliser plot is 101×0.9022 (or equivalently $101 - 9.9) = 91.1$. The full table of expected frequencies is

Expected frequencies	Blackleg	No blackleg
No fertilizer	9.9	91.1
Nitrogen only	9.3	85.7
Dung only	11.0	102.0
Nitrogen and dung	13.8	127.2

Note that, although observed frequencies can only be whole numbers, expected frequencies must be calculated to one decimal to achieve the necessary accuracy in the test statistic.

It is sometimes useful to calculate the terms

$$(\text{observed} - \text{expected})^2/\text{expected}$$

for each separately to see which are contributing most to the values of X^2.

(Observed − expected)²/expected	Blackleg	No blackleg
No fertilizer	3.76	0.41
Nitogen only	0.05	0.01
Dung only	4.45	0.48
Nitrogen and dung	0.00	0.00

Hence

$$X^2 = \Sigma \frac{(\text{observed} - \text{expected})^2}{\text{expected}} = 9.16$$

The relevant χ^2 distribution is that with 3 d.f. for which the 5% point is 7.82. The calculated value of X^2 is greater than this. Hence we must conclude that the incidence of blackleg is affected by the fertilizer applied to the plot. The highest incidence is with no fertilizer, the lowest with dung only, the two plots with nitrogen giving intermediate levels. This pattern is recognizable as what we have called interaction when discussing factorial experiments: dung appears to have an effect on the incidence of blackleg when nitrogen is not applied but has no effect when nitrogen has been applied. We could go further and compare pairs of plots instead of all four together though in this case we would probably not learn much more. The most interesting comparison would be that between no fertilizer and nitrogen only and these results are given below, the expected frequencies being in brackets.

Observed (expected)	Blackleg	No blackleg	Total
No fertilizer	16 (13.4)	85 (87.6)	101
Nitrogen only	10 (12.6)	85 (82.4)	95
Total	26	170	196
Deviation $= \pm 2.60$			

$$X^2 = (2.6)^2 \left(\frac{1}{13.4} + \frac{1}{12.6} + \frac{1}{87.6} + \frac{1}{82.4} \right) = 1.20$$

Comparing the calculated value of X^2 with the tabulated 5% point for the χ^2 distribution of 1 d.f. (3.84) we see that there is no strong evidence that nitrogen causes a reduction in the incidence of blackleg.

Conclusion

Our conclusion must be that dung applied by itself without nitrogen reduces the incidence of blackleg but that the effects of nitrogen or nitrogen plus dung have not been clearly established.

Note that it is important to include all the data in calculating a X^2 statistic to compare proportions. Sometimes the results are expressed only in terms of the successful observations, for example 'three germination treatments applied to 100 seeds each, produced 88, 74 and 78 germinated seeds respectively'. To test whether the three treatments appear to alter the germination rate of the seeds we must consider not only the seeds which germinated but also those which did not (12, 26 and 22 respectively). The expected frequencies are 80 and 20 for germinated and non-germinated seeds respectively and the contributions to the X^2 statistic are

Germination	Non-germinated
0.80	3.20
0.45	1.80
0.05	0.20

giving a total X^2 statistic of 6.50 which is clearly mainly due to the relatively large differences in the numbers of non-germinated seeds.

13.4 GENERAL CONTINGENCY TABLES

Although it has been convenient to talk about contingency tables in terms of situations and outcomes, the idea of contingency tables is more general than this implies. A simple example of a 2×2 contingency table in which neither classification can be regarded as a set of situations occurs in genetics when the possible association of two characteristics is being investigated. Individuals can be classified into one of four groups by whether they do or do not possess each of two characteristics. The results can be tabulated in the following form:

| | | Characteristic A | | |
		Yes	No	Total
Characteristic B	Yes	n_1	n_2	$n_1 + n_2$
	No	n_3	n_4	$n_3 + n_4$
Total		$n_1 + n_2$	$n_2 + n_4$	$N = n_1 + n_2 + n_3 + n_4$

To test the hypothesis that the two characteristics occur independently of each other we require equivalently to test whether the proportions A and not A are the same for individuals with B and without B. This is exactly the same formulation as our 2×2 contingency table for the size of ant populations in two forest regions.

In general this form of contingency table may contain r rows by c columns, the row and column classifications being of any type. The principle of the analysis remains the same. Under the null hypothesis we assume that the proportions in the different columns are the same in the different rows. Hence we calculate the expected frequency for a particular row and column as

$$\frac{\text{column total}}{\text{overall total}} \times \text{total in row}$$

The expression is clearly symmetrical with regard to rows and columns as can be seen by rewriting it

$$\frac{\text{column total} \times \text{row total}}{\text{overall total}}$$

Having obtained the expected frequencies, the extent of the deviation from the null hypothesis is measured by the usual statistic

$$X^2 = \Sigma \frac{(\text{observed} - \text{expected})^2}{\text{expected}}$$

The X^2 statistic is compared with the χ^2 distribution, using the distribution with $(r-1) \times (c-1)$ degrees of freedom: by trial and error you will find that this is always the number of values which can be arbitrarily written into an $r \times c$ table while retaining the same row and column totals. The previous results about the degrees of freedom for 2×2 and $2 \times k$ tables are special cases of this general formula. The ideas of contingency tables can be extended to more than two classifications though clear thinking is required to identify the hypothesis being tested. We shall return to this subject using a more formal modelling approach in section 14.6.

Example 13.2 Analysis of a 3 × 4 contingency table

The frequencies of different combinations of hair and eye colours were recorded for 6800 males in Baden and are given below. The object of the exercise was to investigate the associations between hair and eye colours.

| | | Hair colour | | | | |
		Fair	Brown	Black	Red	Total
Eye colour	Blue	1768	807	189	47	2811
	Grey/green	946	1387	746	53	3132
	Brown	115	438	288	16	857
	Total	2829	2632	1223	116	6800

The expected frequencies assuming complete independence of the two classifications are as follows:

	Fair	Brown	Black	Red
Blue	1169	1088	506	48
Grey/green	1303	1212	563	53
Brown	347	332	154	15

Decimals are not required here because the numbers are so large. It is useful to tabulate both the deviations (observed − expected) and the contributions to X^2 (deviation^2/expected) *en route* to the test statistic as they both may aid the interpretation of the result.

Deviation	Fair	Brown	Black	Red
Blue	+599	−281	−316	−1
Grey/green	−357	+175	+183	0
Brown	−242	+106	+134	+1

(Deviation)^2/Expected	Fair	Brown	Black	Red
Blue	307	73	198	0
Grey/green	98	25	59	0
Brown	164	34	117	0

It is hardly necessary to calculate the value of $\chi^2 = 1075$ and to compare it with the χ^2 on 6 d.f. to see that the hypothesis that these characteristics are independent is quite untenable. Looking in detail at the table of deviations we see that the few individuals with red hair seem to be divided between the three eye colours in the average proportions over all other hair colours and we can therefore ignore red-haired individuals in seeking associations between hair and eye colour. The rest of the table shows a large surplus of blue-eyed fair-haired individuals with corresponding deficits of blue-eyed individuals with brown or black hair and of fair haired individuals with grey/green or brown eyes. It is therefore clear that blue eyes and fair hair are associated. The two most interesting further questions are whether the ratio of grey/green to brown eyes changes with hair colour and whether the ratio of brown to black hair changes with eye colour. We shall examine each of these in turn. The observed frequencies for each calculation are tabulated with the expected numbers in brackets.

	Fair	Brown	Black	Red	Total
Grey/green	946 (833)	1387 (1433)	746 (812)	53 (54)	3132
Brown	115 (228)	438 (392)	288 (222)	16 (15)	857
Total	1061	1825	1034	69	3989

$X^2 = 103.3$ which is very significant on 3 d.f.

	Brown	Black	Total
Blue	807 (680)	189 (316)	996
Grey/green	1387 (1456)	746 (677)	2133
Brown	438 (496)	288 (230)	726
Total	2632	1223	3855

$X^2 = 106.5$ which again is very significant on 2 d.f.

In spite of leaving out the blue eyes/fair hair combination there is clearly some dependence, and, again looking at the individual deviations, it is clear that in the first table the observed proportions for fair-haired individuals are discrepant, and in the second table the observed proportions for blue-eyed individuals are discrepant. Hence our last attempt is to omit all individuals who are either blue-eyed or fair-haired and to see whether the remaining colours occur independently.

	Brown	Black	Red	Total
Grey/green	1387 (1363)	746 (772)	53 (51)	2186
Brown	438 (462)	288 (262)	16 (18)	742
Total	1825	1034	69	2928

$X^2 = 5.43$ with 2 d.f., which is not quite significant at the 5% level

Conclusions

Blue eyes and fair hair occur together much more often than one would expect from the overall proportions of blue eyes and of fair hair. In addition the ratio of grey/green to brown eyes for fair-haired individuals is clearly higher than for other hair colours and similarly the ratio of brown to black hair colours for blue-eyed individuals is distinctly higher than for other eye colours. However, when individuals with blue eyes or fair hair or both are excluded from the analysis the remaining colours appear to be distributed reasonably independently. In other words the ratio of grey/green to brown eyes is essentially the same for individuals with brown, black or red hair.

EXERCISE 13.1

(1) Two batches of aphids are sprayed with different concentrations of sodium oleate. At a concentration of 1.1%, 3 out of 63 survive: at a concentration of 0.65% 12 out of 72 survive. Do the survival rates differ significantly?

(2) The number of summers (April–September) with periods of over 20 days without rainfall at Rothamsted between (a) 1852 and 1900 and (b) 1901 and 1956 are given below. Are the summers of the present century significantly more subject to drought than those of the last?

	Drought	No drought
(a) 1852–1900	4	45
(b) 1901–1956	14	42

(3) Four adjacent areas of heathland have been used in an experiment on the effects of burning. Area A was last burned 4 years ago, area B 3 years ago, area C 2 years ago and D last year. One hundred quadrat samples were taken at random from each of these areas and the presence or absence of *Agrostis tenuis* in each quadrat recorded. The numbers of quadrats, out of 100, in each area containing *Agrostis tenuis* were as follows:

Area	No. of quadrats containing *Agrostis tenuis*
A	44
B	36
C	40
D	28

Does the frequency of occurrence of *Agrostis tenuis* vary significantly from area to area? Summarize the conclusions to be drawn from the data.

(4) In an investigation into the growth characteristics of rose trees a number of cuttings for each of several combinations of scions and rootstocks were collected and transplanted. Records were taken of the state of each cutting at a stage when most cuttings were showing some growth, each cutting being classified as (a) both stems showing strong shoots, (b) one stem showing strong shoots, (c) shoots just initiated, or (d) dead. The observed frequencies for the four classes are given below for four scion-rootstock combinations. Do these show significantly different patterns of growth?

Scion	Rootstock	Growth			
		(a)	(b)	(c)	(d)
A	1	12	6	8	4
	2	6	6	12	6
B	1	5	4	14	7
	2	3	4	8	15

(5) Records of the number of lambs born to the ewes of various flocks are summarized below. The records are given for three breeds of ewe on each of three farms. Do the proportions of ewes giving birth to 0, 1, 2 or more lambs vary significantly for different farms and different breeds? Summarize your conclusions.

Farm	Breed	Number of lambs per birth				Total
		0	1	2	3 or more	
1	A	10	21	96	23	150
	B	4	6	28	8	46
	C	6	7	58	7	78
2	A	8	19	44	1	72
	B	5	17	56	1	79
	C	1	5	20	2	28
3	A	22	95	103	4	224
	B	18	49	62	0	129
	C	4	12	16	2	34

13.5 ESTIMATION OF PROPORTIONS

In the first four sections of this chapter we have been concerned with testing ideas about proportions. Now previously in this book we have emphasized that hypothesis testing by itself gives answers to only some of the questions which should be in the experimenter's mind, and that to complement hypothesis testing we should also be interested in estimation. So where does estimation come into the analysis of proportions? Let us again use the analogy between the analysis of proportions and the analysis of sample means. Testing the equality of two population proportions by comparing the sample proportions in a 2×2 contingency table is analogous to testing the equality of two population means by comparing the sample means using a t-test. When dealing with sample means we also considered estimates of the population means, both point estimates and confidence intervals. We further considered point estimates and confidence intervals for the difference between two population means.

The simplest example of the importance of estimation from proportions is the need for a seedsman to estimate the germination percentage of his seeds. Apart from legal restrictions on germination percentage the seedsman will obviously want to know what germination rate to advertise on the outside of his packets. This situation is the same as that of many industrial concerns who need to control the proportion of defective items they are offering for sale. Other areas with similar problems are in the control of pests or the viability of stored products. It is necessary to estimate from a sample the proportion of defective individuals in the whole store, or population. An extension of the problem is in comparing different storage methods; the research worker wants not only to test whether the proportion of defectives has changed, which he may be able to do by contingency tables, but also to estimate the possible reduction in the proportion of defectives to assess whether an improved storage system is economically viable.

In all cases there are 'obvious' estimates. Thus if from n seeds r germinate then the obvious estimate of the proportion of seeds in the population which will germinate under the same conditions is r/n. In general if in n trials we observe r successes our best estimate of the probability of a success is r/n. As with estimation of a population mean, where we wanted not just a point estimate but also an interval estimate, so we would like to calculate a confidence for a proportion.

As in our discussion of the construction of a confidence interval for a population mean (section 3.1), we have to consider the sampling distribution of r for given values of n and p. Later, in Chapter 14, we shall consider in detail the theoretical sampling distribution of r, called the binomial distribution. However, to calculate an interval estimate for p we shall use the general result that the distribution of a sample mean is approximately normal if the sample size is sufficiently large (section 2.5). The proportion r/n is a sample mean, being the mean number of 1s in a sample of n observations each of which is either 0 or 1. In Fig. 13.1 the sampling distributions of r for $n = 10$, 30 and 50 and for $p = 0.1$ and 0.5 are shown and, except for $n = 10$, $p = 0.1$, the distributions have shapes roughly like that of the normal distribution.

If we assume that the sampling distribution of r is approximately normal then we need only the mean and variance of r to be able to construct a confidence

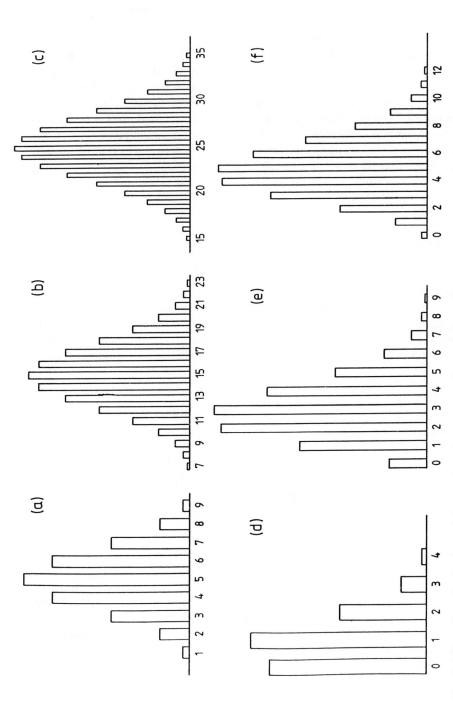

Fig. 13.1 Sampling distributions of the number of successes, r, from n independent trials with probability of success $= p$ for each trial; (a) $n = 10$, $p = 0.5$, (b) $n = 30$, $p = 0.5$, (c) $n = 50$, $p = 0.5$, (d) $n = 10$, $p = 0.1$, (e) $n = 30$, $p = 0.1$, (f) $n = 50$, $p = 0.1$.

interval. Using the theoretical results for the binomial distribution, the mean of the sampling distribution of r is np and the variance is $np(1-p)$. The result for the mean is the 'obvious' one which we used earlier to justify using r/n as an estimate of p. The result for the variance is generally sensible in that the most variable results for r will be obtained when p is about $\frac{1}{2}$; as p gets either larger, towards 1, or smaller, towards 0, the variance becomes smaller, because with very large or very small p values the scope for variation of r is limited. Thus if 98% of a population of seeds will germinate if tested, then from samples of 50 an average of 49 should germinate and the maximum possible of 50 not only makes deviation above the average small but also makes it most unlikely that fewer than 46 or 45 of a sample of 50 will germinate.

If the mean value for r is np then the mean value of r/n will be p. Similarly the variance of r/n will be $p(1-p)/n$, reduced by a factor $1/n^2$ from the variance of r. When the variance, σ^2, of the normal distribution is known then the 95% confidence interval for the population mean, based on an observation x, is $x \pm 1.96\sigma$. Therefore our 95% confidence interval for the true proportion, p, is

$$r/n \pm 1.96\sqrt{[p(1-p)/n]}$$

There is one final difficulty. We do not know p. If n is very small this causes considerable problems and we have to use a more complex argument to obtain a confidence interval. However, provided n is quite large then replacing the true value of p in the variance by the estimated value, r/n, has very little effect on the width of the confidence interval and we can calculate the confidence interval as

$$r/n \pm 1.96\sqrt{[(r/n)(1-r/n)/n]}$$

As an example consider a seed germination trial in which 80 seeds out of 100 germinate: $n = 100$, $r = 80$ and our best estimate of the germination percentage is 80%, or, if we work in proportions, 0.8.

The estimate of the variance is

$$(r/n)(1-r/n)/n = \frac{0.8 \times 0.2}{100}$$

$$= 0.0016$$

Hence the standard deviation $= \sqrt{\text{variance}} = 0.04$, and a 95% confidence interval is $0.80 \pm 1.96 \times 1.96 \times 0.04$, or approximately (0.72, 0.88).

The extension of this method from estimating one proportion to estimating the difference between two proportions is simple. Suppose from two populations we test n_1 and n_2 individuals respectively and we observe r_1 and r_2 successes, then the estimates of the true proportions of successes in the two populations are r_1/n_1 and r_2/n_2 and the estimate of the difference between the true proportions is

$$r_1/n_1 - r_2/n_2$$

The variance of this estimate is

$$\frac{p_1(1-p_1)}{n_1} + \frac{p_2(1-p_2)}{n_2}$$

since the variance of a difference between two quantities is the sum of their variances. Again, since we do not know p_1 or p_2 we use r_1/n_1 for p_1 and r_2/n_2 for p_2 and we can calculate our confidence interval as

$$\left(\frac{r_1}{n_1} - \frac{r_2}{n_2}\right) \pm 1.96 \sqrt{\left\{\left[\left(\frac{r_1}{n_1}\right)\left(1-\frac{r_1}{n_1}\right)\Big/ n_1\right] + \left[\left(\frac{r_2}{n_2}\right)\left(1-\frac{r_2}{n_2}\right)\Big/ n_2\right]\right\}}$$

Suppose that, in addition to the previous results on seed germination where 80 seeds out of 100 germinated, a trial of 200 seeds treated by a new process gave 175 germinating. We want to estimate the improvement in the germination proportion due to the new treatment. For the new treatment $n_2 = 200$, $r_2 = 175$ and the estimated germination proportion is $r_2/n_2 = 0.875$. The estimate of the variance of this estimation is

$$\frac{(0.875) \times (0.125)}{200} = 0.00055$$

We can now calculate the estimate of the improvement in the germination proportions as $0.875 - 0.8 = 0.075$ or 7.5%, and the 95% confidence interval for the improvement as

$$0.075 \pm 1.96 \sqrt{(0.0016 + 0.00055)}$$

$$\text{i.e.} \ 0.075 \pm 1.96 \times 0.046$$

$$\text{i.e.} \ (-0.015 \text{ to } 0.165)$$

Thus our 95% confidence interval for the improvement in the germination percentage is -1.5% (a decrease in germination) to $+16.5\%$.

As previously, there is a connection between confidence interval estimation and significance testing in that if the confidence interval includes zero as it does in this example then a significance test of the hypothesis that the two populations of seeds have the same germination proportion will not give a significant result at the corresponding significance level, in this case the 5% level. To verify this consider the contingency table comparison of the two proportions:

	Germinated	Failed to germinate	Total
First trial	80(85)	20(15)	100
Trial for new treatment	175(170)	25(30)	200
Total	225	45	300

Expected frequencies are given in brackets after the observed frequencies. Using the chi-squared test

$$X^2 = (5)^2\left(\frac{1}{85} + \frac{1}{15} + \frac{1}{170} + \frac{1}{30}\right) = 2.94$$

and the 5% point in the χ^2 distribution is 3.84. It may appear that there is no simple relation between the confidence interval for the difference between two proportions and the contingency table test of the hypothesis of no difference between the proportions. However, it can be shown that the X^2 test statistic is equivalent to the square of the ratio of the difference between the two observed proportions, to the standard deviation of that difference. For our example this is

$$\left(\frac{0.075}{0.046}\right)^2 = 2.66$$

the reason for the difference from $X^2 = 2.94$ being that for the latter the variances of the two proportions have been pooled whereas in calculating the standard deviation, 0.046, they were not pooled.

13.6 SAMPLE SIZES FOR ESTIMATING PROPORTIONS

Once we know the form of the sampling distribution of r/n in terms of p, there are a range of other questions concerning sampling which we can consider. We shall deal with these fairly rapidly using examples about sampling the germination rates of seeds though the arguments are applicable to any situation where the response can take one of only two values. The first question is how many seeds do we need to sample to estimate the germination rate with a prescribed accuracy? Since the variance of r/n depends on p we must obviously have an approximate value for p to answer this question. Suppose we know p to be about 0.85 or 85% then the variance of r/n is

$$\frac{p(1-p)}{n} = \frac{(0.85)(0.15)}{n} = 0.1275/n$$

If we require that the standard error of our estimate of p shall be not greater than 0.01 or 1%, then we could determine n by trying various values and determining the variance and hence the standard error r/n:

$n = 100$	variance $= 0.001275$	standard error $= 0.0357$
$n = 400$	variance $= 0.000319$	standard error $= 0.0178$
$n = 1000$	variance $= 0.000128$	standard error $= 0.0113$
$n = 2000$	variance $= 0.000064$	standard error $= 0.0080$

Our conclusion would be to use between 1000 and 2000 seeds. We could get a more exact figure by turning the problem round and asking how big a value of n is needed to make the standard error $\sqrt{(0.1275/n)}$ equal to 0.01? Clearly larger values of n give smaller standard errors. We require

$$\frac{0.1275}{n} = (0.01)^2$$

i.e. $\frac{0.1275}{(0.01)^2} = n$

i.e. $n = 1275$

confirming the results of our trial and error approach.

An alternative approach might be to pose the question in terms of confidence intervals. If the true probability, p, is about 0.3, what is the minimum number of seeds necessary to give a 95% confidence interval for p whose length is not greater than 0.1? The width of the 95% confidence interval is $2 \times 1.96\sqrt{[p(1-p)/n]}$, which for $p = 0.3$ is

$$3.92\sqrt{[(0.3)(0.7)/n]} = 3.92\sqrt{(0.21/n)}$$

This will be 0.1 if

$$3.92\sqrt{(0.21/n)} = 0.1$$

i.e. if $(3.92)^2 \times \dfrac{0.21}{n} = (0.1)^2$

i.e. if $\dfrac{(3.92)^2 \times 0.21}{(0.1)^2} = n$

i.e. if n is greater than 323

We can extend these considerations of sample size to the situation where we wish to compare two percentages. In most situations it will be reasonable to assume initially that the two percentages will be the same, the common proportion being p. The variance of the difference between the two estimated proportions will therefore be $(2p(1-p)/n]$ if the two sample sizes are to be the same, or $[p(1-p)/n_1] + [p(1-p)/n_2]$ if they can be different. For a given total number of observations the difference is estimated most precisely when the sample sizes are equal so we will confine our attention to this case. Having a good idea of the value of p we can calculate the approximate value of the standard error as $\sqrt{[2p(1-p)/n]}$ and hence calculate the length of the confidence interval for any n or calculate how large n needs to be for a given length of confidence interval. Thus suppose $p = 0.6$ and $n = 250$, the standard error of the difference between the two proportions is

$$\sqrt{[2(0.6)(0.4)/250]} = 0.044$$

and hence the expected length of the 95% confidence interval for the true difference in proportion would be $2 \times 1.96 \times 0.044 = 0.0172$ or 17%. Thus with

250 observations for each treatment we could expect to produce a 95% confidence interval of length about 17%.

We can also ask how many observations we need to be fairly confident of detecting a difference of a given size. The obvious argument is to say that we should detect the difference at the 5% significance level, if the observed difference is at least twice its standard error. However, if we choose n to make the standard error half the size of the difference which it is important to us to detect, then we shall only get a significant result if the observed difference is greater than the true difference. This would occur only 50% of the time, and so we would have only a 50% chance of detecting an important difference as significant at the 5% level. We should plan for a better chance than 50% and to achieve this we must reduce the standard error. A reasonable principle is to choose the standard error of a difference to be less than one-third of the size of the important difference when there would be a 4 to 1 chance of detecting the difference at the 5% significance level. For example suppose we expect a percentage germination of about 0.6 then the standard error of the difference between two percentages will be

$$\sqrt{[2(0.6)(0.4)/n]} = \sqrt{(0.48/n)}.$$

Suppose a difference of 10% is important then this standard error must be less than $(10/3)\%$ or in terms of proportions 0.033. Therefore we need n to be at least such that

$$\sqrt{(0.48/n)} = 0.0333$$

$$\text{i.e.} \quad \frac{0.48}{n} = (0.0333)^2$$

$$\text{i.e.} \quad \frac{0.48}{(0.333)^2} = n$$

which gives $n = 432$. Hence about 450 observations per treatment will give us a good chance of detecting a difference in percentage germination of 10% or more when the average percentage germination is around 60%. Note that this shows that small sample sizes are not very effective in comparing proportions. We shall discuss this problem of the amount of replication again in Chapter 16.

Another question is concerned not with the estimation of the germination rate but with control of the technological processes involved in producing batches of seed for marketing. A typical example of this form of question is if a seedsman is marketing seed where the normal germination rate is 85% and he tests 300 seeds per batch, what is the probability that he will get less than 80% of these 300 seeds germinating? It is assumed that the seedsman will reject the batch of seeds if less than 80% of his sample from the batch germinate. Again we go back to our results for the sampling distribution of r/n. Approximately r/n is normally distributed with mean p and variance $p(1 - p)/n$. In this case the mean is 0.85 and the variance $(0.85 \times 0.15/300) = 0.000425$ which gives standard deviation of 0.0206. We therefore require the probability of getting a value 0.05 less than the mean from a normal distribution of which the standard deviation is 0.0206. That is we require the probability of a value $(0.05/0.0206) = 2.43$ standard deviation less than the mean. From Table A.1, of the normal distribution, this is 0.0075 or rather less than 1%. Hence the probability of a 'false alarm' value of less than 80% when the germination rate of the whole batch is at the normal level of 85% is less than 1 in 100.

One final question might be how many seeds to test per batch if we want a probability of less than 1 in 1000 of a false alarm. An equivalent question concerns the marketing of small packets of seed. If there are n seeds per pack how large must n be to make the probability that the germination rate of the n seeds in a packet is less than 80% to be less than 1 in 1000? We can try various values of n, calculate the variance of distribution of r/n, calculate $z = ((0.80 - 0.85)/\sqrt{\text{variance}})$ and hence find the probability:

$$n = 300 \quad \text{variance} = 0.000425 \quad z = 2.43 \quad \text{probability } 0.00755$$
$$n = 1000 \quad \text{variance} = 0.0001275 \quad z = 4.43 \quad \text{probability } < 0.00001$$
$$n = 500 \quad \text{variance} = 0.0000255 \quad z = 3.13 \quad \text{probability } 0.00089$$
$$n = 400 \quad \text{variance} = 0.000319 \quad z = 2.80 \quad \text{probability } 0.00256$$

We would conclude that n needs to be nearly 500 for the probability of a false alarm to be acceptably low. As in previous questions about the necessary size of a sample we can go directly to the answer. We require the probability of a value less than $z = (-0.05/\sqrt{\text{variance}})$ to be less than 0.001. The value of z for which the probability is exactly 0.001 is -3.09. We must therefore choose n such that $(0.05/\sqrt{\text{variance}})$ is greater than 3.09. This variance is $(0.1275/n)$. Thus we require n to be at least large enough for

$$\frac{0.05}{\sqrt{(0.1275/n)}} = 3.09$$

$$\text{i.e. } \frac{n}{0.1275} = \left(\frac{3.09}{0.05}\right)^2$$

$$\text{i.e. } n = 0.1275\left(\frac{3.09}{0.05}\right)^2 = 487$$

The questions we have been dealing with are part of the much wider subject of quality control and the interested reader will find many books from which he may expand his knowledge of the subject, an appropriate book being Wetherill (1977).

EXERCISE 13.2

(1) In Exercise 13.1 we compared the frequencies of occurrence of *Agrostis tenuis* in four areas of heathland burnt 4, 3, 2 and 1 year ago respectively:

Area	No. of quadrats out of 100 containing *Agrostis tenuis*
A (burnt 4 years ago)	44
B (burnt 3 years ago)	30
C (burnt 2 years ago)	40
D (burnt last year)	28

Calculate an approximate 95% confidence interval for the true frequencies of occurrence of *Agrostis tenuis* for each area and also calculate approximate 95% confidence intervals for the increased frequency for each of the three areas, A, B, and C compared with the most recently burnt area D.

(2) In an experiment on methods of preserving onions, the experimenter proposes to compare two methods of storage by storing a large number of onions by each method and after 6 months to assess each onion as 'good' or 'defective'. He expects to find about 90% of the onions 'good'

(a) Approximately how many onions does he need to sample for each method to estimate the percentage of good onions with a standard error no bigger than 1%?

(b) Approximately how many onions does he need to sample for each method to obtain a 95% confidence interval for the difference between percentages of good onions for the two treatments whose length is not more than 5%?

(c) Approximately how many onions does he need to give a good chance of detecting a true difference of 3% between the percentages of good onions for the two treatments?

(3) A series of laboratory experiments on the effects of different concentrations of chemical on the larval stages of insects is being planned. Over the range of concentrations to be assessed it is expected that the mortality rate will vary from less than 5% to more than 95%.

(a) Calculate the standard error of the estimated mortality rate for samples of 20, 50, 100 and 200 for mortality rates of 0.05, 0.10, o.20, 0.30, 0.40 and 0.50. If a sample of 200 is used when the concentration is expected to be about 0.5, how big a sample is required to achieve the same standard error for each of the other mortality rates?

(b) If two concentrations giving mortality rates of about 0.2 and 0.8 are tried with samples of 25, what will be the standard error of the estimated difference between the two mortality rates? How large would the sample sizes have to be for the standard error of the estimated difference to be 0.3?

14

Models and distributions for frequency data

14.1 MODELS FOR FREQUENCY DATA

This is the third chapter in which we consider models for the process from which the data is produced. The models we are considering are of an empirical nature, attempting to describe the anticipated structural patterns in the data rather than building a mechanical model to describe the components of the biological and physical processes involved. Such empirical models are frequently able to summarize the observed patterns and to pinpoint important aspects requiring more information.

In both of the previous chapters on models (Chapters 11 and 12) the measurement variable has been assumed to be a continuous quantity for which a Normal distribution may be an acceptably approximate description. Our modelling there has been of the causes of variation in the yield or output and an analysis of variance has provided the summary of the importance of various causal factors and variables.

When our data is in the form of frequencies we have already seen, in Chapter 13, that the statistical methods are of an apparently very different form. We have been assessing the evidence from the data of whether some quite simple hypotheses are credible. The X^2 goodness of fit statistic can be used to test the deviations of observed frequencies from theoretical frequencies for any model proposed for frequency data. In the next four sections we shall consider various models from very simple models implying that alternative outcomes are equally likely (based on ideas of symmetry), through more structured models, often used in population genetics, to formal distributions of counts, the Binomial and Poisson distributions. In each case testing whether the data are compatible with the model provides us with knowledge about the assumptions underpinning the model.

When we were discussing the assumptions required for the analysis of variance in Chapter 8 we noted that a more general form of model was required for the analysis of experimental data when the measured variable was a frequency count. The models appropriate to this situation in which we wish to investigate the strength of different causes of variation, while assuming a particular distribution for the random variation of counts are called generalized linear models. In the last three sections of this chapter we look at three particular forms of generalized linear models, for data from an experimental design structure, for complex contingency tables and for regression relationships when the observed response is a proportion based on counts. Such generalized models are a major development of modern statistical methodology and should become increasingly used in the analysis of data.

14.2 TESTING THE AGREEMENT OF FREQUENCY DATA WITH SIMPLE MODELS

There are many situations where we may expect a particular pattern of frequencies under one simple model but departure from this pattern for alternative, more complex models. The simplest form of hypothesis is, sometimes, that different outcomes are equally probable. Examples with two possible outcomes are the tossing of a coin for which heads and tails should be equally probable if the coin is 'fair' but not if the coin is biased; the sex of individual births for which the model of equal probability for male and female births is the simplest model which it is reasonable to accept until disproved; or a simple tasting test in which each tester is asked to indicate their preference for one of two identically presented samples. The tasting test example can obviously be extended to a choice from several samples, all of which may be assumed equally likely to be selected. A similar situation exists in many psychological experiments where individuals have to select one of a number of objects.

The analysis of this form of data is very simple. For example, suppose that we wish to test the model that male and female births are equally likely for a particular strain of a species and that a random sample of births gives 32 males and 41 females. We would expect half of the total of 73 births to be male and half female. Hence we would expect 36.5 of each. In testing this model of equal probability we will determine the significance of the deviation of the observed number of male births, 32, from the expected number, 36.5. To determine the significance we use the continuous χ^2 distribution, as if the number of male births was a continuous scale, between 31.5 and 32.5. In measuring the deviation of 32 from 36.5 we apply the continuity correction and consider the probability of a deviation at least as big as $36.5 - 32.5$. Hence our goodness of-fit-test statistics is

$$X^2 = \Sigma \frac{(|\text{observed} - \text{expected}| - \tfrac{1}{2})^2}{\text{expected}} = \frac{(32.5 - 36.5)^2}{36.5} + \frac{(40.5 - 36.5)^2}{36.5}$$

$$= \frac{2(4.0)^2}{36.5} = 0.88$$

Since we have two observed frequencies and one restriction, that the two frequencies total 73, we have 1 degree of freedom, the second frequency being determined once the first is chosen. Comparing 0.88 with the 5% point of the χ^2 distribution on 1 d.f. (3.84) we find that the data do not provide statistically significant evidence against the model of equally probable male and female births.

As another example suppose that 40 testers were asked to compare four different cheeses produced by different procedures and identified only by the letters A, B, C and D. If each tester makes one choice and the preferences were

A	5
B	7
C	18
D	10
Total	40

we might suspect that this shows an overall preference for C. To test the simple model that testers are equally likely to prefer A, B, C or D we would calculate the expected frequency for each cheese to be preferred as the total number of testers divided by $4 = 40/4 = 10$.

Then we calculate:

$$X^2 = \frac{(5-10)^2}{10} + \frac{(7-10)^2}{10} + \frac{(18-10)^2}{10} + \frac{(10-10)^2}{10} = 9.80$$

There is no continuity correction in this case because with four alternatives these results cannot be represented on a one-dimensional scale. This time we have four frequencies with one overall restriction that they total 40, and so there are 3 d.f. The 5% point of the χ^2 distribution on 3 d.f. is 7.82 so that the unevenness of the preferences is significant. The evidence suggests that the model of equally likely choices is incorrect.

To assess the extent to which it is the preference for cheese C which contradicts the model we might decide to do a further test to compare this preference for C, 18, with those for some other cheese, 22. Again on the model of equally likely choices, we would expect 10 and 30. Because there is only one deviation we again use the continuity correction and calculate

$$X^2 = \frac{(17.5-10)^2}{10} + \frac{(22.5-30)^2}{30} = 7.50$$

This is significant at the 1% level of the χ^2 distribution on 1 d.f. confirming that there is a strong preference for cheese C. The last test could well be omitted since it is 'obvious' that it is causing the discrepancy. However, more specific tests after an initial general test are often valuable.

The method of analysis for models that specify unequal proportions is essentially the same. Many such models arise in genetics where under the assumptions of Mendelian inheritance the expected ratios of numbers of individuals with different characteristics can be calculated for various genetic models. There are other fields of application, however, such as multiple choice examination questions, and another form of tasting test in which the taster is presented with three samples, two of which are the same. The requirement for the taster is to identify the odd one out, and for the experimenter to test whether the taster is having any success. If the taster cannot distinguish between the two substances then he has one chance in three of guessing the odd one correctly. Therefore on the null hypothesis of no discrimination we would expect one-third of the trials to be successful and two-thirds to be failures. A significantly larger proportion of successes indicates a detectable difference between the two substances. A significantly smaller proportion of successes indicates that the taster is behaving in a curious fashion!

As an example of the χ^2 test for models with unequal proportions we shall consider some classical data on primula plants. A total of 560 plants were classified by the type of leaf (flat or crimped) and the type of eye (normal or Primrose Queen), with the following results:

Leaf	Flat	450
	Crimped	110
Eye	Normal	405
	Primrose Queen	155

On the hypothesis of a Mendelian 3:1 ratio we would expect, for each characteristic, three-quarters of the total 560 observations in the first class of the character and the remaining quarter in the second class. Thus the expected frequencies are 420 and 140 for each character. Again because we are considering only a single deviation we use the continuity correction.
For the leaves,

$$X^2 = \frac{(449.5 - 420)^2}{420} + \frac{(110.5 - 140)^2}{140} = 8.29$$

For the eyes,

$$X^2 = \frac{(405.5 - 420)^2}{420} + \frac{(154.5 - 140)^2}{140} = 2.00$$

In each case we have one degree of freedom for which the 5% point of the χ^2 distribution is 3.84. Hence we would conclude that although it is reasonable to accept a 3:1 ratio for the normal:Primrose Queen eyes, the data give strong evidence against the ratio of flat:crimped leaves being 3:1, the observed ratio being close to 4:1.

14.3 INVESTIGATING MORE COMPLEX MODELS

The figures for the primula plants do not of course summarize the situation fully, just as in section 6.3 the average differences between two levels of phosphate and between two levels of potash did not fully summarize the result of a 2×2 factorial experiment. The figures actually obtained for the primula plants were:

	Normal eye	Primrose Queen eye	Total
Flat leaves	328	122	450
Crimped leaves	77	33	110
Total	405	155	560

An extension of the previous Mendelian model would be that three-quarters of the plants should have flat leaves and, quite independently, three-quarters of the plants should have normal eyes. Consequently this model predicts that three-quarters of the flat-leaved plants should have normal eyes, giving $3/4 \times 3/4$ of all the plants or 9/16 with flat leaves and normal eyes; the remaining one-quarter of the flat-leaved plants, which is $1/4 \times 3/4$ or 3/16, should have Primrose Queen eyes. Similarly 3/16 of the plants should have crimped leaves and normal eyes and 1/16 crimped leaves and Primrose Queen eyes. Hence the hypothesis predicts ratios 9:3:3:1 for the four classes (flat normal:flat Primrose Queen:crimped normal:crimped Primrose Queen). The expected frequencies are calculated as 9/16, 3/16, 3/16 and 1/16 of 560 giving 315, 105, 105, 35 and to test this overall model we calculate:

$$X^2 = \frac{(328-315)^2}{315} + \frac{(122-105)^2}{105} + \frac{(77-105)^2}{105} + \frac{(35-35)^2}{35}$$
$$= 0.54 + 2.75 + 7.47 + 0.11 = 10.77$$

We have four class frequencies and only one restriction, the overall total, and hence have to compare 10.77 with the 5% point of the χ^2 distribution on 3 d.f. (7.82). Not surprisingly, in view of our previous conclusions about the proportions of the two types of leaves, we conclude that the 9:3:3:1 model is not acceptable.

Consider now the situation after rejecting this model. We know that something is wrong with the model for this data but since the model is a composite one, one or more of three simple hypotheses could be faulty. The three simple hypotheses are:

(a) The two types of leaves occur in a 3:1 ratio
(b) The two types of eyes occur in a 3:1 ratio
(c) The two classifications are independent

These hypotheses correspond to the two main effects and an interaction in a factorial experiment. In the same way that the full treatment sum of squares in a factorial set of treatments can be split into three independent sums of squares to test the two main effects and the interaction independently, so the X^2 value can be split into three components, each of 1 d.f. to test the three hypotheses (a), (b) and (c). However, there are some complications in testing (c). We already know that hypotheses (a) and (b) can each be tested using X^2 compared with the χ^2 distribution on 1 d.f. If we subtract the two values of X^2 calculated for testing (a) and (b) from the overall X^2 on 3 d.f. then the remainder is approximately distributed as χ^2 on 1 d.f. on the hypothesis that the two classifications are independent **and** each of the classification independently has a 3:1 ratio. Thus if the two X^2 for testing hypotheses (a) and (b) are small enough for us to continue to accept that each ratio is 3:1 then we can use the remainder of X^2, called in the genetic examples the linkage component, to test for independence. If, on the other hand, one or both of the X^2 for (a) and (b) are significantly large then we cannot use the linkage component to test for independence because we believe the hypotheses necessary for this remainder component to have a χ^2 distribution on 1 d.f. to be untrue. In other words a significant value of the remainder X^2 would not be evidence of linkage because it could be due to the deviation of one classification from 3:1. Conversely a non-significant value could arise from simultaneous breakdowns of the 3:1 ratio hypothesis and the independent hypotheses. If, therefore one of the classifications deviates from a 3:1 ratio the only way to test the independence of the two classifications is to use a 2×2 contingency table making no assumptions about the proportions in the different rows and columns.

We can illustrate these arguments by further reference to the primula data. The calculated values of X^2 can be set out as follows:

Variations	X^2	d.f.
Total X^2 for testing 9:3:3:1	10.77	3
X^2 for testing leaves 3:1	8.29	1
X^2 for testing eyes 3:1	2.00	1
Difference X^2 (linkage)	0.48	1

If the two X^2 for testing the 3:1 ratios had both been small the difference X^2 of 0.48 would have been used to test for independence. As the leaves X^2 is significant this is not valid, though a value as low as 0.48 makes it very probable that the two classifications appear to be independent. We therefore calculate the 2×2 contingency table:

	Normal	Primrose Queen	Total
Flat	328 (325.4)	122 (124.61)	450
Crimped	77 (79.6)	33 (30.4)	110
Total	405	155	560

$$X^2 = (2.6)^2 \left(\frac{1}{325.4} + \frac{1}{124.6} + \frac{1}{79.6} + \frac{1}{30.4} \right) = 0.38$$

which confirms our earlier conclusions that the two classifications appear to behave independently.

We have seen how X^2 can sometimes be split into components and we shall now consider the combining of X^2. For this we take advantage of a most useful mathematical property of the χ^2 distribution, namely that if you have two independent observations which should have χ^2 distributions with n_1 and n_2 degrees of freedom respectively, the sum of the two observations should have a χ^2 distribution with $(n_1 + n_2)$ degrees of freedom. The particular use of this property of additivity is in the situation where we have a number of small sets of data from similar sources. For each set of data we can calculate the goodness of fit X^2 for the agreement of the observed and theoretical frequencies. In such cases it is not uncommon to find that some of the sets of data show significant disagreement with theory while some do not. We wish to distinguish three different situations:

(i) The assumed model does hold for all the different sources and the significant results are the result of the sampling procedure from which we should expect one X^2 value in twenty to be significant.

(ii) A model different from that assumed holds for all the different sources, the apparent variation in the results being again due to the sampling procedure, the deviation from the assumed model being for some samples insufficient to achieve significance.

(iii) The sources are heterogeneous with different models being appropriate for different sources.

One way of differentiating between (i) and (ii) is to apply the χ^2 test to the observed frequencies totalled over the samples from the various sources. Suppose that we have k sets of data each with m classes. Then we can calculate the individual values of X^2, each on $(m-1)$ degrees of freedom; we also have the single overall X^2 on $(m-1)$ d.f. using the totals for each class. If we sum the individual X^2 to give ΣX^2 this should, if **all** the null hypotheses are correct, be distributed as χ^2 with $k(m-1)$ d.f. and will be bigger than the single X^2 on $(m-1)$ d.f. Since the overall X^2 measures the average disagreement of the data with the theoretical hypotheses the difference between ΣX^2 and the single overall X^2 must measure the divergence between the different data sets, generally called heterogeneity. Note that if we were interested only in testing heterogeneity, and not in testing a model which specifies the porportions, then the frequencies for the different data sources would be compared using a contingency table test. The full calculations can be summarized in the table given on the next page.

The analysis may be interpreted in several ways. We can look at the individual X^2 values, interpret them individually and try to summarize the various conclusions. Alternatively, we can use only the overall test of the hypothesis provided by X_T^2 together with the test for heterogeneity. A third possibility is to look at ΣX^2 first and then split it into individual values or into the overall

Source	X^2	d.f.	Testing
First data set	X_1^2	$m - 1$	divergence of set 1 from hypothesis
Second data set	X_2^2	$m - 1$	divergence of set 2 from hypothesis
⋮	⋮	⋮	⋮
kth data set	X_k^2	$m - 1$	divergence of set k from hypothesis
	ΣX^2	$k(m - 1)$	disagreement of data sets with all null hypotheses
Combined data sets	X_T^2	$m - 1$	divergence of pooled data from the hypothesis
Difference	$\Sigma X^2 - X_T^2$	$(k - 1)(m - 1)$	heterogeneity of data sets

X_T^2 and the heterogeneity component, or both. There is an interesting parallel between the various patterns of this analysis and the analysis of a two-factor factorial experiment. In the factorial analysis we first calculate the sum of squares for variation of all treatment combinations. This can then be split either into variation between levels of one factor and then variation between the levels of the second factor for each level of the first factor in turn, or we can use the more standard division into sums of squares for each main effect and for the interaction.

Example 14.1   Analysis of goodness-of-fit X^2 values

In genetical research a number of cells from a hybrid plant obtained by crossing **Secale** and **Aegilops** plants were classified as either AA, AS or SS. Under assumptions that the chromosomes form associations independently it can be shown that the three types of cell should occur in the ratio 3:7:3. The data for five hybrid plants are given below:

| Hybrid | Type of association | | | |
	AA	AS	SS	Total
(1) *Ap. caudata* × *S. cereale*	35	76	23	134
(2) *Ap. caudata* × *S. cereale*	27	57	27	111
(3) *Ap. caudata* × *S. vavilovii*	27	76	42	145
(4) *Ap. comosa* × *S. cereale*	23	32	25	80
(5) *Ap. umbellulata* × *S. cereale*	40	99	58	197

For each hybrid we can calculate the expected frequencies on the 3:7:3 hypothesis and hence the value of X^2 on 2 d.f. to test the model for that hybrid.

Hybrid	Observed (expected) frequencies			X^2
(1)	35 (30.9)	76 (72.2)	23 (30.9)	2.76 (2 d.f.)
(2)	27 (25.6)	57 (59.8)	27 (25.6)	0.28 (2 d.f.)
(3)	27 (33.5)	76 (78.0)	42 (33.5)	3.47 (2 d.f.)
(4)	23 (18.5)	32 (43.1)	25 (18.5)	6.19 (2 d.f.)
(5)	40 (45.5)	99 (106.1)	58 (45.5)	4.57 (2 d.f.)

Of these five X^2 values only one is significant at the 5% level, the others being non-significant. If we consider the totals over the five hybrids for the three types of association and calculate the overall X^2 we have the following:

Observed frequencies	152	340	175	Total 667
Expected frequencies	153.9	359.2	153.9	
X^2	0.02	1.03	2.89	Total 3.94 (2 d.f.)

This value is some way below the 5% significance point of 5.99 so that overall there is not a great deal of evidence against the 3:7:3 model. To look at the possible heterogeneity among the hybrids in the degree of agreement with the model we total the X^2 for the five separate sets of data giving 17.27 on 10 d.f., again not significant at the 5% level though quite close to the significance point of 18.31. We now split the 17.27 into the overall deviation component of 3.94 and the heterogeneity component of $17.27 - 3.94 = 13.33$ on 8 d.f., once again not significant. We may summarize the results as follows:

Test	X^2	d.f.	5% point of χ^2 distribution
Hybrid (1)	2.76	2	5.99
(2)	0.28	2	5.99
(3)	3.47	2	5.99
(4)	6.19	2	5.99
(5)	4.57	2	5.99
Sum of deviations from 3:7:3	17.27	10	18.31
Overall deviation (on data)	3.94	2	5.99
Heterogeneity of data sets	13.33	8	15.51

Conclusion

The evidence is not strong enough to reject, at a 5% significance level, the hypothesis that the chromosomes form associations independently to give 3:7:3 proportions of the three types of associations. The single significant X^2 value for data set (4) can be viewed as one of the 5% of values that by chance are

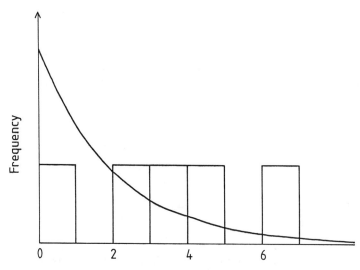

Fig. 14.1 Histogram representing the X^2 values for five data sets of Example 14.1, with the χ^2 distribution on 2 d.f. for comparison.

more extreme than the 5% significance point. On the other hand while the departures from the 3:7:3 model are certainly not proven, several of the X^2 values are close enough to the 5% significance level to provoke some doubt about the hypothesis, and perhaps further experimentation. In considering data from many sources it is often useful to present the results in a histogram of X^2 values for comparison with the χ^2 distribution and this is done in Fig. 14.1, though five data sets are too few for this representation to be useful.

EXERCISE 14.1

(a) In a television panel game the three members of each team give three definitions of the meaning of an unusual word, one definition being correct and the other two false. The opposite team has to identify the correct definition. Over a number of contests the frequencies of correct and incorrect identifications were:

Correct 25   False 40

Test the hypothesis that the identifications are pure guesses giving a 1 in 3 probability of picking the true definition.

(b) The frequencies of presence and absence of two characters in poppies are given below for two independent sets of data. A, B denote the presence of the characters; a, b absence.

Characters	Data set (1) frequencies	Data set (2) frequencies
AB	72	123
Ab	29	30
aB	40	27
ab	12	21

Treating each set of data independently,

(i) Test the hypothesis that the true frequencies are 9:3:3:1.
(ii) Partition the X^2 on 3 d.f. obtained from (i) into the components for testing A:a = 3:1, B:b = 3:1; independence (or linkage).
(iii) If a further test of independence of the two characters is necessary, calculate the approximate X^2.

Summarize your conclusions

(c) In an experiment to investigate methods of estimating population size and other characteristics of an insect population, insects were captured and released on each of five days. On the first day 57 were captured and marked with a green spot. On the second day 52 were captured of which 25 had green spots showing they had been captured the day before, the remaining 27 being new captures. All 52 were marked with a white spot. Thus at the end of the second day there were available for catching on subsequent days 32 insects with green spots, 27 with white spots and 25 with both. If these groups of insects behave in the same way then the proportions caught on subsequent days should be in the ratios 32:27:25. If differential effects of the 'spotting' or of mortality exist these will produce deviations from the expected ratio. On the next three days the numbers caught were as follows:

	Day 3	Day 4	Day 5
Green spot	8	7	10
White spot	9	4	7
Green and white spots	11	2	5

Test the 32:27:25 hypothesis for each day in turn and for the totals over the three days. Discuss your conclusions.

14.4 THE BINOMIAL DISTRIBUTION

In the last chapter and in the earlier sections of this chapter, we have considered various form of inference about proportions; these have included the comparison of two proportions, the comparison of a single proportion with a theoretical expectation, and the calculation of a confidence interval for a population proportion. In many of the examples to which we have applied these methods—the germination of seeds, the response of insects to chemicals, the quality of stored fruit—the basic units have been tested in groups. Typically in a germination test a batch of 50 or 100 seeds are placed together, under standard conditions, in a dish and those which have germinated after a set number of days are counted. All the inferences we have discussed have assumed that the probability of response is the same for each unit and that the responses of individual units are independent of each other; that is, the germination of one seed in the dish does not alter the environment of the other seeds so as to make it either more or less likely that any other particular seed will germinate. The validity of this assumption of independence will affect the conclusions we draw and we should therefore consider how to test the assumption. More generally by investigating the implications of a model which assumes that individuals in a group behave independently, we may hope to understand more about the system for which such a model is proposed.

The model for the sampling distribution of the number of successes from a set of independent trials is the binomial distribution and, after the normal distribution, the binomial is probably the most frequently used standard distribution. We shall develop the formulation of the distribution through the rather trivial example of the number of heads from tossing coins. Suppose we toss 3 coins and record the number of heads, how often should we expect to get 3 heads, 2 heads, 1 head or no heads? The possible outcomes are:

First coin	Second coin	Third coin	Number of heads
H	H	H	3
H	H	T	2
H	T	H	2
H	T	T	1
T	H	H	2
T	H	T	1
T	T	H	1
T	T	T	0

Assuming that the probability of a head is 1/2 for each toss, these eight outcomes are equally likely, so that the proportions of 3, 2, 1 and 0 heads should be 1/8, 3/8, 3/8 and 1/8 respectively. If we got more sets of 3 heads or 3 tails than this suggests we might suspect either biased coins or some non-inedependence of the results for different coins and might inquire into how the tossing was done—for example, which side was face up before the toss?

Suppose, instead of tossing coins, we were looking at the frequencies of 3 boys, 2 boys and a girl, 1 boy and two girls, and 3 girls in complete families of three children. Let us assume that the proportion of boys is 0.515 in the

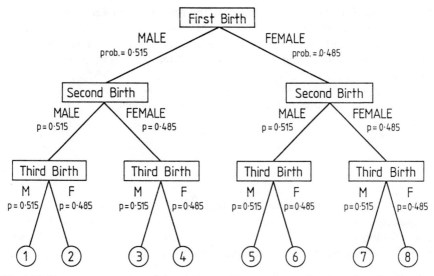

Fig. 14.2 Tree representation of the eight possible sequences of male or female offspring from three births.

population. Then, if we equate the results for the first, second and third children to the results for the first, second and third coins it is clear that the number of combinations giving 3, 2, 1 and 0 boys out of 3 is again 1, 3, 3 and 1 respectively. However, because of the greater probability of each birth being male the eight possible outcomes are not equally likely. To calculate the relative probabilities consider the 'tree' of events, shown in Fig. 14.2. Thus the probability of three successive male births along route I, is

$$0.515 \text{ (for the first birth)} \times 0.515 \text{ (for the second birth)}$$
$$\times 0.515 \text{ (for the third birth)} = (0.515)^3$$

Three routes, 2, 3 and 5 give two out of three male births and the total probability is

$$0.515 \times 0.515 \times 0.485 + 0.515 \times 0.485 \times 0.515$$
$$+ 0.485 \times 0.515 \times 0.515 = 3 \times (0.515)^2 \times (0.485)$$

We can calculate the probabilities of 1 or 0 males similarly and summarize the results as follows:

Number of males	Expected proportion (probability)	
3	$1 \times (0.515)^3$	$= 0.1366$
2	$3 \times (0.515)^2 (0.485)$	$= 0.3859$
1	$3 \times (0.515) (0.485)^2$	$= 0.3634$
0	$1 \times (0.485)^3$	$= 0.1141$

Note the pattern. The initial number is the number of different routes to the particular results, the power to which (0.515) is raised is the number of males, and the power of (0.485) is the number of females. Suppose we had information on the number of boys and girls in 200 completed families of three children as follows:

Number of males	Frequency of families
3	24
2	84
1	82
0	10

We could test the agreement of these frequencies with the theoretically expected frequencies based on the proportions given above. Thus we would expect $200 \times 0.1366 = 27.3$ families with 3 boys, 77.2 with 2, 72.7 with 1 and 22.8 with none. To compare observed frequencies with expected frequencies we use the χ^2 test.

Observed	Expected	$(O - E)^2/E$
24	27.3	0.40
84	77.2	0.60
82	72.7	1.19
10	22.8	7.19

$$X^2 = \Sigma \frac{(O - E)^2}{E} = 9.38$$

X^2 is significant at 5% (the 5% point of the χ^2 distribution on 3 d.f. is 7.82) so that the binomial model and the data are not compatible. The implication of such a result might not be that successive births are not independent. Looking at the proportion of males in the 600 births giving rise to the data we see that 322 births are male, a proportion of 0.537, rather higher than the assumed proportion of 0.515 based on a national average. Quite apart from any effect of this deviation it is possible that patterns in families of three children are not determined randomly. For example parents with three daughters but no sons may be more likely to have further children than those with two sons and a daughter so that complete families with three female (or male) children might well be under-represented not because of the non-independence of the sex of successive births but because of decisions on the size of family.

   The general formula for probabilities or proportions in a binomial distribution is:

$$\text{probability } (r \text{ successes out of } n \text{ trials}) = \frac{n!}{r!(n - r)!} p^r (1 - p)^{n-r}$$

where p is the probability of success with a single trial—equivalently the long term proportion of successes and $n!$ is $n \times (n - 1) \times (n - 2) \times \ldots \times 3 \times 2 \times 1$. $n!$

is a convenient piece of shorthand and is said 'n factorial' though there is no connection with factorial experiments; 1! is 1, $2 \times 1 = 2$, 3! is 6, 4! is 24, etc. Zero factorial (0!) is 1 because when you multiply it by 1 it gives $1! = 1$.

In the cases we have been looking at, n is 3 and the four probabilities of 3, 2, 1, 0 heads, or males are:

	Coins	Births
$p_3 = \dfrac{3!}{3!\,0!}p^3 = p^3$	$\frac{1}{8}$	$(0.515)^3$
$p_2 = \dfrac{3!}{2!\,1!}p^2(1-p) = 3p^2(1-p)$	$\frac{3}{8}$	$3(0.515)^2\,(0.485)$
$p_1 = \dfrac{3!}{1!\,2!}p(1-p)^2 = 3p(1-p)^2$	$\frac{3}{8}$	$3(0.515)\,(0.485)^2$
$p_0 = \dfrac{3!}{0!\,3!}(1-p)^3 = (1-p)^3$	$\frac{1}{8}$	$(0.485)^3$

If we write the general binomial probabilities in their algebraic form we can see a quick way of calculating them sequentially:

p_n, the probability of successes in all trials $\dfrac{n!}{n!\,0!}p^n \quad = p^n$

$$p_{n-1} = np^{n-1}(1-p) \qquad = n\left(\frac{1-p}{p}\right)p_n$$

$$p_{n-2} = \frac{n(n-1)}{2}p^{n-2}(1-p)^2 \qquad = \left(\frac{n-1}{2}\right)\left(\frac{1-p}{p}\right)p_{n-1}$$

$$p_{n-3} = \frac{n(n-1)(n-2)}{2 \times 3}p^{n-3}(1-p)^3 \qquad = \left(\frac{n-2}{3}\right)\left(\frac{1-p}{p}\right)p_{n-2}$$

So, if we know n and p we can easily calculate the binomial probabilities and hence the expected frequencies to compare with the observed frequencies. The only remaining problem is the value of p. Since the primary object of the investigation will frequently be to estimate p we cannot assume that we will always know p. Hence we must use the obvious estimate for p, which is the overall proportion of successes, r/n (as in section 13.5).

Finally, before we go on to consider an example of fitting and testing the binomial distribution, there are two new points concerning the use of the χ^2 test to compare observed and expected frequencies. Usually in one or other tail of the binomial distribution the expected frequencies become very small. This can lead to spurious large values of X^2 because a deviation of only 1 or 2 between observed and expected frequencies is compared with a very small expected frequency. When small expected frequencies occur it is necessary to combine frequencies for two or more neighbouring classes. There is no hard and fast rule as to how small 'small frequencies' are. It is desirable to avoid expected frequencies less than 1 and to have at most two expected frequencies less than 5. Note that small observed frequencies cause no problems—it is because, in the calculation of X^2, we divide by each expected frequency that small expected frequencies have to be treated with care.

The other point concerns the degrees of freedom of the χ^2 distribution with which the X^2 statistic is to be compared. In previous uses of the χ^2 test to compare observed and expected frequencies the model has defined the proportions in the different classes exactly. In testing the fit of the binomial we choose the estimate of p to agree with the observed overall proportions of successes. Thus we make the agreement of observed and expected frequencies more likely. We adjust the test by subtracting one from the degrees of freedom for the χ^2 distribution. Hence the degrees of freedom are the number of classes for which frequencies are compared (after pooling classes if necessary) minus 2.

Example 14.2 Fitting the binomial distribution and testing the goodness-of-fit

Eight hundred seeds were tested for germination in batches of 10. The number germinating in each batch was recorded and the results for the 80 batches were as follows:

> 6 batches with 0 seeds germinating
> 20 batches with 1 seed germinating
> 28 batches with 2 seeds germinating
> 12 batches with 3 seeds germinating
> 8 batches with 4 seeds germinating
> 6 batches with 5 seeds germinating

These results can be summarized:

Number germinating per batch of 10	Number of batches	Total seeds germinating
0	6	0
1	20	20
2	28	56
3	12	36
4	8	32
5	6	30
Over 5	0	0
Total	80	174

To test the hypothesis that seeds in each batch germinate independently we calculate the expected frequencies of batches with 0, 1, 2, 3, 4, 5 and over 5 seeds germinating from the binomial distribution. To do this we estimate p from the overall proportion of seeds germinating.

$$p = \frac{\text{total number of seeds germinating}}{\text{total number of seeds}} = \frac{174}{80 \times 10} = 0.2175$$

The calculation of the expected frequencies is most easily set out as follows:

Number of seeds germinating	Probability	Expected frequency = 80 × probability
0 $p_0 = (1-p)^{10}$	$= (0.7825)^{10}$	6.89
1 $p_1 = n\left(\dfrac{p}{1-p}\right)p_0$	$= 10(0.278)p_0 = 0.2392$	19.14
2 $p_2 = \dfrac{n-1}{2}\left(\dfrac{p}{1-p}\right)p_1$	$= \frac{9}{2}(0.278)p_1 = 0.2992$	23.94
3 $p_3 = \dfrac{n-2}{3}\left(\dfrac{p}{1-p}\right)p_2$	$= \frac{8}{3}(0.278)p_2 = 0.2218$	17.75
4 $p_4 = \dfrac{n-3}{4}\left(\dfrac{p}{1-p}\right)p_3$	$= \frac{7}{4}(0.278)p_3 = 0.1079$	8.63
5 $p_5 = \dfrac{n-4}{5}\left(\dfrac{p}{1-p}\right)p_4$	$= \frac{6}{5}(0.278)p_4 = 0.0360$	2.88
6 $p_6 = \dfrac{n-5}{6}\left(\dfrac{p}{1-p}\right)p_5$	$= \frac{5}{6}(0.278)p_5 = 0.0083$	0.67
7 $p_7 = \dfrac{n-6}{7}\left(\dfrac{p}{1-p}\right)p_6$	$= \frac{4}{7}(0.278)p_6 = 0.0013$	0.11
8 $p_8 = \dfrac{n-7}{8}\left(\dfrac{p}{1-p}\right)p_7$	$= \frac{3}{8}(0.278)p_7 = 0.0001$	0.01

We now compare the observed and expected frequencies

Number germinating	Observed frequency	Expected frequency	$\dfrac{(O-E)^2}{E}$
0	6	6.9	0.12
1	20	19.1	0.04
2	28	23.9	0.70
3	12	17.8	1.89
4	8	8.6	0.04
5 and over	6	3.7	1.43
Total	80	80.0	$X^2 = 4.22$

Note that, to avoid small expected frequencies, we have combined the classes for 5 or more seeds germinating. The 5% point of the χ^2 distribution with 4 d.f. is 9.49, and the value of 4.22 for the X^2 statistic gives us no reason to doubt that the binomial model fits these data. To interpret this result, we consider again the assumption of the binomial model for these data. It has been assumed that the germination of seeds within each group of 10 is independent, and the probability of germination, p, is the same for the whole population from which the 80 samples were taken. The estimate of p is 0.2175 and, since the binomial models fits the data, we can treat this estimate as coming from a large sample of 800 independent seeds. Consequently, using the methods of section 13.5, we can calculate the variance of the estimate;

$$\text{variance} = \frac{(0.2175)(0.7825)}{800} = 0.000213$$

and an approximate 95% confidence interval for the germination probability:

$$0.2175 \pm 1.96\sqrt{0.000213} \text{ or } (0.189, 0.246)$$

14.5 THE POISSON DISTRIBUTION

The second theoretical distribution for counts which is important in statistics is the Poisson distribution. This arises in two distinct ways. The first, and more important, is as a representation of randomness. If events in either time or space occur independently and randomly, then if the number of events in periods of time or in areas or volumes of space are counted the distribution of such counts should be Poisson. In time, simple examples are the emission of particles from a radioactive source, or the movement of cars past an observer on a road with relatively little traffic. The emission of a particle at a particular point in time makes it neither more nor less likely that a further emission will occur in the immediately following period of time. With cars, the assumption of randomness assumes a low level of traffic, such that the passing of cars is not affected by, i.e. predictable from, the presence of other cars. The Poisson model is closely related to the exponential decay models of section 12.4. For series of random events in time, the counts of events in intervals of time have a Poisson distribution; the intervals between successive events have an exponential distribution.

An example in two-dimensional space is the spatial distribution over the ground of plants of a given species. If the plants are spread evenly over the area under consideration, so that there is no trend of plant density over the whole area, and the siting of individual plants is such that the position of one plant is unaffected by the positions of other plants, then the distribution of plants is said to be homogeneous and random. If we count the numbers of plants in randomly placed quadrats (or marked areas of ground) of a constant size, then the number of plants per quadrat should have a Poisson distribution.

The other situation giving a Poisson distribution arises from a particular form of the binomial distribution. We have found that it is possible, though laborious, to calculate the probabilities of r successes out of n for a binomial distribution with n up to 10. As n gets larger and p gets smaller, the average number of successes remains much the same but the calculations get increasingly difficult. However, it can be shown mathematically that, if n is large and p is small, then the binomial distribution is very like the Poisson distribution, and the similarity becomes closer as n increases, provided that np, the expected number of successes, does not change. To show just how rapid this tendency

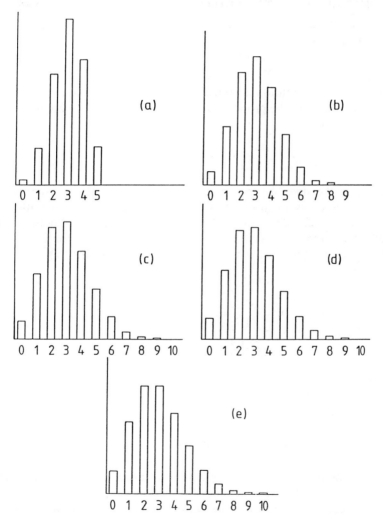

Fig. 14.3 Binomial distributions for (a) $n = 5$, $p = 0.6$, (b) $n = 10$, $p = 0.3$, (c) $n = 20$, $p = 0.15$, (d) $n = 40$, $p = 0.075$, and (e) the limiting Poisson distribution for $\lambda = np = 3$.

is, Fig. 14.3 shows the binomial distributions for $n = 5$, $p = 0.6$; $n = 10$, $p = 0.3$; $n = 20$, $p = 0.15$; $n = 40$, $p = 0.075$; and the Poisson distribution with $np = 3$. So the Poisson is a useful limiting case of the binomial distribution and provided n is large and np is not large we can use the Poisson distribution to test for independence of events. The classical example, which is too good to omit, concerns the frequency of death by horse-kick in Prussian corps. Bortkieicz recorded the number of deaths by horse-kick for each of 14 corps for each of 20 years, giving a total of 280 observations. Obviously the probability for each man of a corps is tiny but, with so many men per corps, the average number (np) of deaths per corps/year is not negligible. Bortkiewicz's data are as follows:

Death per corps per year	0	1	2	3	4
Frequency of occurrence	14	91	32	11	2

The Poisson distribution is completely defined by one parameter, λ, which is the average count. The probabilities of counts of 0, 1, 2, ..., are given by the general formula:

$$\text{prob (count} = r) = \frac{e^{-\lambda}\lambda^r}{r!}$$

where $r!$ is the quantity 'r factorial' defined previously for the binomial distribution and e is a constant, which is the base for natural logarithms, and whose value is 2.7182818285.... Values of $e^{-\lambda}$ are easily obtained using a calculator or computer. As with the binomial distribution, there is a simple method of calculating the Poisson probabilities sequentially:

$$p_0 = e^{-\lambda}$$

$$p_1 = \frac{\lambda e^{-\lambda}}{1} = \lambda p_0$$

$$p_2 = \frac{\lambda^2}{2}e^{-\lambda} = \frac{\lambda}{2}p_1$$

$$p_3 = \frac{\lambda^3}{2 \times 3}e^{-\lambda} = \frac{\lambda}{3}p_2$$

$$p_4 = \frac{\lambda^4}{2 \times 3 \times 4}e^{-\lambda} = \frac{\lambda}{4}p_3, \text{ and so on}$$

Only rarely is there a theoretical reason for expecting a particular value of λ. More usually we have to estimate λ from the average count from the sample. When we come to comparing the observed frequencies with the expected frequencies, this estimation of λ gives a reduction of one degree of freedom for the χ^2 distribution again as with the binomial distribution.

Example 14.3 Fitting the Poisson distribution and testing the goodness-of-fit

In an investigation into the spatial distribution of trees in a red pine plantation outside Ontario, counts of the numbers of pine trees in 60 randomly placed quadrats, each 13 ft 4 in square, were made as follows:

Count of trees	No. of quadrats (frequency)
0	6
1	18
2	7
3	13
4	2
5	3
6	2
7	5
8	3
9	1

To test the randomness of the spatial pattern it was proposed to fit the Poisson distribution to these data and test the fit. The average count is

$$\lambda = \frac{1(18) + 2(7) + 3(13) + 4(2) + 5(3) + 6(2) + 7(5) + 8(3) + 9(1)}{60}$$

$$= \frac{174}{60} = 2.9$$

$$e^{-\lambda} = e^{-2.9} = 0.0550$$

Hence the expected frequencies are calculated as follows:

Count	Poisson probability	Expected frequency (60 × probability)
0	0.0550	3.30
1	$2.9 \times 0.0550 = 0.1595$	9.57
2	$\frac{2.9}{2} \times 0.1595 = 0.2313$	13.88
3	$\frac{2.9}{3} \times 0.2313 = 0.2236$	13.42
4	$\frac{2.9}{4} \times 0.2236 = 0.1621$	9.73
5	$\frac{2.9}{5} \times 0.1621 = 0.0940$	5.64

(Continued)

Count	Poisson probability	Expected frequency (60 × probability)
6	$\dfrac{2.9}{6} \times 0.0940 = 0.0454$	2.72
7	$\dfrac{2.9}{7} \times 0.0454 = 0.0188$	1.13
8	$\dfrac{2.9}{8} \times 0.0188 = 0.0068$	0.41
9	$\dfrac{2.9}{9} \times 0.0068 = 0.0022$	0.13

We calculate the expected frequencies, for the Poisson distribution, of counts greater than 9 by subtracting the sum of the frequencies up to and including 9 from the total 60. This gives an expected frequency of 0.07 for counts greater than 9. To compare observed and expected frequencies by the χ^2 test we must group the counts to avoid very small expected frequencies.

Count	Observed	Expected	$\dfrac{(O - E)^2}{E}$
0	6	3.30	2.2
1	18	9.57	7.4
2	7	13.88	3.4
3	13	13.42	0.0
4	2	9.73	6.1
5	3	5.64	1.2
6 and over	11	4.46	9.6
Total	60	60.00	$29.9 = X^2$

The 0.1% point of the χ^2 distribution on 5 d.f. is 20.5. Thus the data provide extremely strong evidence that the trees are not randomly distributed.

In most examples of temporal or spatial distributions of counts the Poisson distribution is found to be incompatible with the data. This is not surprising. The Poisson model of randomness assumes that each individual in the population is quite independent of all other individuals. But, to consider just the example of plants, we know that, on the one hand, plants take up an amount of space and there is a limit to how close plants can get to each other; on the other hand, the propagation of plants through seed dispersal from parent plants or vegetation reproduction leads to clustering of plants, clearly not spatially random. Similar arguments could be advanced for other situations. The value of the Poisson model is that it provides a very simple model from which we can attempt to describe the pattern of deviations and go on to develop a more complex model.

Just to demonstrate that the Poisson distribution does sometimes fit data, we will carry out the same calculations for Bortkiewicz's data on deaths by horse-kick.

The mean death rate is

$$\lambda = \frac{1(91) + 2(32) + 3(11) + 4(2)}{280} = \frac{196}{280} = 0.7$$

$e^{-\lambda} = 0.4966$

Deaths	Probability	Expected	Observed	$\frac{(O-E)^2}{E}$
0	0.4966	139.0	144	0.18
1	0.3476	97.3	91	0.41
2	0.1217	34.1	32	0.13
3	0.0284	8.0	11	1.12
4 or more	0.0057	1.6	2	0.10

$X^2 = 1.94$ compared with the 5% point of the χ^2 distribution on three degrees of freedom which is 7.82. Thus the data provide no reason to doubt the assumptions leading to the Poisson distribution—that the probability of death by horse-kicks is independent for each individual in a corps, and the probability is the same for each corps, and for each year.

EXERCISE 14.2   *Fitting the binomial and Poisson distributions*

(i) It is claimed that housewives cannot tell product A from product B. To test this assertion, 100 housewives are each given eight pairs of samples, each pair containing one sample of product A and one of product B, and are asked to identify the A sample. The number of times (out of 8) that each housewife is correct is recorded as follows:

Time correct	Number of housewives
0	2
1	0
2	5
3	13
4	36
5	13
6	18
7	10
8	3

(a) Calculate the probabilities for a binomial distribution with $n = 8$ and $p = \frac{1}{2}$ and hence obtain the expected frequencies of 0, 1, 2, ..., 8 correct identifications. Compare the expected frequencies with the observed frequencies and hence test the hypothesis that the housewives cannot distinguish the two products.

(b) Accepting that there is some discriminating ability it is required to test whether the discriminating powers are the same for different housewives and whether the results within each set of 8 are independent. In other words, we now want to test whether the data are compatible with a binomial distribution with p, the probability of a correct identification, greater than $\frac{1}{2}$. Estimate p from the overall proportion of correct identifications and calculate the expected frequencies for a binomial distribution with this value of p and $n = 8$. Hence test the hypothesis of independence of results and equality of discriminating powers.

(ii) Pieces of eggshell are dipped in concentrated nitric acid for 25 seconds so that pores in the shell are large enough to let light through. The number of pores in a $1\,mm^2$ area are counted for 320 pieces of shell, and the results are summarized below. Fit the Poisson distribution to the data and test whether the agreement of observed and expected frequencies is acceptable.

Number of pores	0	1	2	3	4	5
Frequency	18	74	139	70	17	2

(iii) One hundred squares of a haemocytometer were examined and the number of yeast cells per square counted. The results are given below. Fit the Poisson distribution to the data and test the agreement between the observed frequencies and the expected frequencies.

Number of cells per square	Observed frequency of squares
0	7
1	11
2	20
3	25
4	14
5	13
6	7
7	3
More than 7	0

14.6 GENERALIZED MODELS FOR ANALYSING EXPERIMENTAL DATA

Suppose we have data from an experiment with a design structure of one of the various forms discussed in Chapters 5, 6 and 7, but for which one of the response variables is a proportion or count, and clearly not normally distributed. The subdivision of variation which we would obtain from the analysis of variance corresponds to the questions for which it is hoped the statistical analysis will provide answers. The model explaining the variation caused by the experimental factors will be in a form such as,

$$y_{ij} = b_i + t_j + e_{ij}$$

or

$$y_{ijk} = b_i + f_j + v_k + (fv)_{jk} + e_{ijk}$$

as in Chapter 11.

Because the y_{ij} are proportions or counts and not normally distributed, we have not only to model y_{ij} in terms of b_i, t_j, f_j, v_k, $(f_v)_{jk}$, etc, but also have to postulate a model for the e_{ij}. That is we have a two-component model, modelling the influence of the causal factors on the mean response and modelling the variation in terms of a binomial or Poisson distribution. Although we can think about the two modelling processes separately, when we come to the actual calculation in a statistical analysis the combination of the two models has implications for the manner in which the two models are specified.

When the measurement variable is a proportion then if we simply write the model

$$y_{ij} = p_{ij} = b_i + t_j + e_{ij} \tag{14.1}$$

then for particular block-treatment combinations it is possible to obtain predictions for p_{ij} which are not between 0 and 1. For example suppose the proportions of infected seedlings from two control treatments in three sites are

	\multicolumn{4}{c}{Site}			
	1	2	3	Mean
Treatment 1	0.05	0.15	0.55	0.25
Treatment 2	0.15	0.55	0.95	0.55
Mean	0.10	0.35	0.75	

The average difference between the proportions for the two treatments is $0.55 - 0.25 = 0.30$. If we assume a simple additive model, implying that the treatment difference should be expected to be the same at each site then the predicted proportions for each site–treatment combination should be 0.15 above the site mean (for treatment 2) or 0.15 below the site mean (for treatment 1). The resulting predicted values,

	Site 1	Site 2	Site 3
Treatment 1	−0.05	0.20	0.60
Treatment 2	0.25	0.50	0.90

include an impossible, negative, proportion for the site 1–treatment 1 combination.

To avoid such unacceptable outcomes we modify our modelling process in a manner analogous to the use of transformations in Chapter 8. For generalized linear models we use a link function so that the model (14.1) is replaced by

$$\text{logit } (p_{ij}) = \log (p_{ij}/(1 - p_{ij})) = b_i + t_j + e_{ij} \tag{14.2}$$

In this model the logit (p_{ij}) is the link function and it will usually be appropriate to use the logit function as a link for binomially data distributed data. An alternative link function used for binomially distributed data is the probit function used extensively when modelling dose–response data. The probit is defined in terms of a normal distribution and its use is illustrated in section 14.8.

When counts are recorded as the response variable but they do not occur as proportions from a fixed total, then the most common alternative to the Binomial distribution to describe the random variation is the Poisson distribution. An example where the Poisson distribution might be a reasonable assumption is the poppy count data first discussed in section 8.3. The most commonly used link function with a Poisson error distribution is the log link function.

Once we have identified a reasonable assumption for the error distribution and a suitable link function, we can fit sets of terms for treatment and block factors, and interpret the effects of various terms through their contribution to an analysis of variation and through the estimated effects and their standard errors. There are, however, two respects in which the analysis differs from that for experiments discussed in Chapters 5, 6 and 7.

Because the variance of a binomial or Poisson distribution is related directly to the mean we can predict what the expected error sum of squares should be if the model we are fitting is correct. Reflecting this changed theoretical situation the residual sum of squares is called the deviance. If the model fits correctly the deviance should be about equal to the residual degrees of freedom.

The second difference is that because we have gone beyond the simple linear model structure the change in deviance as we add an additional term to our model depends on which other terms are already included in the model. That is, the interpretation of an analysis of deviance resembles that for the analysis of a multiple regression (section 10.3). We may fit terms in various orders to examine the importance of particular terms in the context of differing assumptions about other terms to be included in the model.

Example 14.4 Binomial error structure for germination data

As part of a larger study on the effects of various chemicals on the germination of seeds under various temperature regimes, four different concentrations of a chemical were used for treating seeds stored in four temperature regimes. For each of the 16 combinations, four replicate dishes of 50 seeds were stored and the germination of all 64 sets of seeds tested under standard conditions. The numbers of seeds (out of 50) germinating in each dish are shown. If seeds are assumed to germinate independently within each set of 50, and with no differences between germination rate between dishes treated identically, then a binomial distribution plus a logit link might be expected to be an appropriate model. A sequence of models is fitted to assess the relative importance of the concentration and temperature regime effects and their interaction. Note that there is no blocking of the four replicates.

Temperature Regime		Water	Chemical concentration		
			0.1	1.0	10
T_1	Rep 1	9	13	21	40
	2	9	12	23	32
	3	3	14	24	43
	4	7	15	27	34
T_2	Rep 1	19	33	43	48
	2	30	32	40	48
	3	21	30	37	49
	4	29	26	41	48
T_3	Rep 1	7	1	8	3
	2	7	2	10	4
	3	2	4	6	8
	4	5	4	7	5
T_4	Rep 1	4	13	16	13
	2	9	6	13	18
	3	3	15	18	11
	4	7	7	19	16

The deviance for each of several fitted models is now listed with the terms included in the model.

Fitting	Deviance	
Overall Mean	1193.8	(63 d.f.)
Mean + Temp	430.1	(60 d.f.)
Mean + Conc	980.1	(60 d.f.)
Mean + Conc + Temp	148.1	(57 d.f.)
Mean + Conc + Temp + Conc × Temp	55.6	(48 d.f)

From these results we can construct an analysis of deviance.

Source (in order of fitting)	Deviance Change	d.f.	Mean Square	F
Temperatures	763.7	3	254.6	220
Concentration	282.0	3	90.7	81
T × C	92.5	9	10.3	8.9
Residual	55.6	48	1.16	

We can see that there are large main effects for both factors and clear evidence also of an interaction between the factors. We have called the (deviance change/d.f.) mean squares by analogy with the normal-distribution-based analysis of variance; the ratios of mean squares are only approximately F-distributed but the results in this case are very strong and would be highly significant. Because the error distribution is assumed to take a specific, binomial, form we can also test the residual deviance against a χ^2 distribution on 48 d.f. to assess whether our error distribution assumption is reasonable. The observed residual deviance is very similar to its d.f. and is not significant compared with the χ^2 distribution on 48 d.f. Hence the binomial distribution is credible for this data. We may note that this is unusual in laboratory germination tests which more usually show non-independence and greater than binomial variation.

We can calculate values on the logit scale or transformed back to the probability scale for germination. For illustration we will demonstrate both, though with effects of this strength the conclusions from the analysis of deviance can be adequately interpreted in terms of the fitted values for the germination probabilities

Mean logit

Temperature	Water	Concentration 0.1	1.0	10	Mean
T_1	−1.82	−0.99	−1.10	+1.07	−0.46
T_2	−0.02	+0.43	+1.42	+3.32	+1.29
T_3	−2.14	−2.84	−1.70	−2.20	−2.22
T_4	−2.04	−1.36	−0.71	−0.90	−1.25
Mean	−1.50	−1.19	−0.27	+0.32	

Germination probability fitted values

		Concentration		
Temperature	Water	0.1	1.0	10
T_1	0.14	0.27	0.48	0.74
T_2	0.50	0.60	0.80	0.96
T_3	0.10	0.06	0.16	0.10
T_4	0.12	0.20	0.33	0.29

The mean source of the interaction is seen in the absence of a response to increasing concentration of the chemical for temperature regime T_3. For all the other temperature regimes germination increases with increasing concentration, the larger increases being for T_1 and T_2, for which the logit means show very similar responses. It is also clear that the germination is highest under T_2 and lowest for T_4 and T_3.

Example 14.5

We will analyse the poppy data, used in Chapter 8, to illustrate the use of a Poisson distribution plus a log link. The data counts of poppies per plots are

Treatment	Block 1	Block 2	Block 3	Block 4
A	538	422	377	315
B	438	442	319	380
C	77	61	157	52
D	115	57	100	45
E	17	31	87	16
F	18	26	71	20

The deviances for various models are as follows

Model	Deviance	d.f.
General mean	3878.8	23
Blocks	3803.3	20
Treatments	331.7	18
B + T	256.3	15

The analysis of deviance is

Source (in order)	Deviance Change	d.f.	Mean Square
Blocks	75.5	3	25.2
Treatments	3547.1	5	709.4
Residual	256.3	15	17.1

There are two main points to notice from this analysis. First, the treatments do, of course, account for a very large proportion of the total variation and we can produce fitted values for the six treatments and corresponding fitted counts.

Treatment	Fitted log value	Fitted count
A	6.014	409
B	5.969	391
C	4.454	86
D	4.363	78
E	3.622	37
F	3.505	33

The standard errors for differences of the fitted log values (based on the residual mean square) are

	B	C	D	E	F
A	0.145	0.244	0.253	0.351	0.370
B		0.245	0.254	0.352	0.320
C			0.321	0.403	0.419
D				0.409	0.425
E					0.489

These standard errors confirm that the treatments split clearly into three pairs (A, B), (C, D), (E, F) each group having a different poppy count, though the differences between C or D and E or F are only marginally significant.

The other notable aspect from the analysis is that the residual deviance of 256 is very strongly significant showing that the assumption of Poisson distribution for the poppy counts is not tenable. Presumably the inflated residual variation is caused by clustering of poppies within plots so that wherever there is a poppy there are almost always several. When the residual deviance is significantly large it would usually be appropriate to use the residual deviance mean square for the calculation of standard errors.

14.7 LOG-LINEAR MODELS

In Chapter 13 we saw how X^2 values, calculated by comparing observed and expected values in contingency tables, can be used to assess the statistical significance of differences between proportions based on samples from different populations. This approach can be extended to test hypotheses about the proportions of individuals occurring in different categories when several classification factors are involved in the definition of the categories. For example, consider the following results for mating pairs of two species of hawks:

	Successful pairs	Failed pairs	Total pairs
Species A	45	60	105
Species B	75	20	95
	120	80	200

The question of whether the success rate is independent of species is of course the same as the question whether the proportion of successful pairs is the same for the two species. The independence of success and species would imply that we should expect the proportion of the total 200 pairs that are both successful and from species A to be the product of the overall proportion of successful pairs, 120/200, and the proportion of pairs belonging to species A, 105/200. Hence the expected number of successful species A pairs would be $200 \times 105/200 \times 120/200 = 63$. In more formal terms, the expected number of individuals E_{ij} in row i and column j of a contingency table will be:

$$E_{ij} = N \times \frac{R_i}{N} \times \frac{C_j}{N}$$

where R_i, C_j and N are the total of row i, column j and the whole table respectively. Thus

$$E_{ij} = \frac{R_i C_j}{N}$$

exactly as when we were comparing proportions in Chapter 13. Hence in testing the model of independence the calculations leading to the X^2 statistic are exactly the same as those in section 13.2. For the hawk data,

$$X^2 = (45 - 63)^2 \left(\frac{1}{63} + \frac{1}{42} + \frac{1}{57} + \frac{1}{38} \right) = 27.07$$

which is very large compared with the 5% point of the χ^2 distribution on 1 degree of freedom, 3.84, showing that the two categories of species and success/failure are clearly not independent. We can describe the form of dependence either by contrasting the different success rate for species A (43%) and species B (79%) or by contrasting the proportions of species A in successful pairs (38%) and unsuccessful pairs (75%). Note that when testing independence we do not need to assume that either the row or column totals represent 'sample size' in the sense of being predetermined in any way. The total N is, of course, usually predetermined by the experimenter.

With a table having only two factors, species and result (success or failure), there is nothing more to be said about testing independence. Assume now that the data on species and result had a further classification because the data were collected in two different valleys.

Species		Valley			Result	
A		1			S	25
A		1			F	10
A		2			S	20
A		2			F	50
B		1			S	60
B		1			F	5
B		2			S	15
B		2			F	15
						200
Totals	A	105	1	100	S	120
	B	95	2	100	F	80
		200		200		200

If all three categories (species, valley and result) are independent then the expected number of successful pairs of species A in valley 1 will be

$$200 \times \frac{105}{200} \times \frac{100}{200} \times \frac{120}{200} = 31.5$$

or, in formal terms,

$$E_{A1S} = N \cdot \frac{S_A}{N} \cdot \frac{V_i}{N} \cdot \frac{R_S}{N} \qquad (14.1)$$

Using expected values of this kind we obtain the X^2 value,

$$X^2 = 103.72$$

There are four degrees of freedom for the X^2 value because there are eight categories, three proportions are 'fitted' and the total number is fixed. The X^2 value is obviously highly significant indicating that there is at least one pattern of dependence among the three proportions corresponding to species, valley and result. Simple examination of the frequencies in the eight classes suggest some strong patterns. The success failure ratio varies from 5:2 for species A in valley 1, to 2:5 for the same species in valley 2, and to 12:1 for species B in valley 1. To investigate further the nature of the dependence we consider some possible models to represent that dependence.

On taking logarithms, (14.1) becomes

$$\log E_{A_1S} = \log S_A + \log V_1 + \log R_S - 2 \log N \qquad (14.2)$$

which is linear in the terms $\log S_A$, $\log V_1$ and $\log R_S$. For this reason, models of this kind are called log-linear models. Dependencies among the proportions can be modelled by adding terms to model 14.2 which are essentially equivalent to the interaction terms in the models that we have already used to analyse the results from factorial experiments (section 6.4). Thus to model the possibility that the success ratio and the valley are not independent we must allow the four combination of success/failure and valley to have their own individual probabilities, with a constant ratio between species. Thus the model becomes

$$\log E_{A_1S} = \log S_A + \log V_1 + \log R_S + \log (VR)_{1S} - 3 \log N$$

where the $\log (VR)_{1S}$ term represents the interaction between valley and result, and $(VR)_{1S}$ is not to be confused with the product $V_1 R_S$. The expected values needed to calculate the X^2 values for most of the models incorporating dependencies among the proportions can be easily derived by using the model or, often more simply, by considering the logic of the hypothesis on which the model is based.

Before analysing the hawk data, one difficulty which we have already met in the context of multiple regression must be mentioned. The exclusion of a factor from a model and the resultant pooling of the numbers over the levels of that factor can lead to logical difficulties if the omitted factor is not independent of all the factors that are still retained in the model.

We can now investigate possible hypotheses about the dependencies that we have found in the hawk data. For example, we could hypothesize any of the following explanations:

(i) The two species have the same distribution over the four categories defined by the two valleys and success or failure. Data which exactly satisfy this hypothesis (within the limits of observing counts) are as follows:

	V_1 success	V_1 failure	V_2 success	V_2 failure
Species A	45	8	18	34
Species B	40	7	17	31

(ii) Success or failure is independent of valleys for each species separately. The following data would satisfy this hypothesis:

	Species A				Species B	
	Success	Failure			Success	Failure
Valley 1	15	20	Valley 1		51	14
Valley 2	30	40	Valley 2		24	6

(iii) There are dependencies between pairs of factors but not between all three factors simultaneously. This is the most complex situation and will be discussed later.

A model allowing for all possible dependencies, including that between all three factors simultaneously, will be a complete model with all four possible interaction terms, and will fit the data exactly. In this complete model each category has its own individual probability which is estimated by the proportion in that category.

In the log-linear model the emphasis is on the symmetry of the factors. None is necessarily considered to be a response to the others. We are investigating the joint occurrences of different 'levels' of the various factors. The only constraint we have in choosing which factors to include in a log-linear model

Table 14.1

Model	Dependent pairs of categories	d.f.	X^2
1	None	4	103.72
2	(Result, species)	3	59.22
3	(Species, valley)	3	60.23
4	(Result, valley)	3	35.43
5	(Species, results) & (species, valley)	2	39.71
6	(Species, result) & (result, valley)	2	8.82
7	(Result, valley) & (species, valley)	2	12.21
8	All three pairs of categories	1	0.77

		Observed values	Expected values Model							
			1	2	3	4	5	6	7	8
A 1	S	25	31.5	22.5	21.0	44.6	15.0	31.9	29.8	26.2
	F	10	21.0	30.0	14.0	7.9	20.0	11.2	5.2	8.8
A 2	S	20	31.5	22.5	42.0	18.2	30.0	13.1	24.5	18.8
	F	50	21.0	30.0	28.0	34.1	40.0	13.1	24.5	18.8
B 1	S	60	28.5	37.5	39.0	40.4	51.3	48.8	45.5	51.2
	F	5	19.0	10.0	26.0	7.1	13.7	3.8	9.8	6.2
B 2	S	15	28.5	37.5	18.0	16.6	23.7	21.9	10.5	16.2
	F	15	19.0	10.0	12.0	30.9	6.3	16.2	19.5	13.8

is that any factors with predetermined proportions must be included and generally so must the dependencies among them. Predetermined proportions are clearly not estimates of any meaningful population parameters. In the hawk data, only the total numbers of pairs in each valley were predetermined.

The X^2 values and the expected values that are used in their calculation are shown in Table 14.1 for eight different models. The usual warnings are needed about small expected values but there is clearly no problem with the data of our example. Note that models 4 and 5 are those discussed earlier as (i) and (ii). Clearly there are statistically significant differences between the observed values and the expected values derived from all the models except model 8, which allows for dependencies among all three pairs of categories. The pattern in model 8 is not immediately apparent, but there is a pattern. If we consider any pair of categories and the four combinations derived from the two categories then the 'ratio of the ratios' is the same for both levels of the third category.

| | Species A | | | Species B | |
	Success	Failure		Success	Failure
Valley 1	26.2	8.8	Valley 1	58.8	6.2
Valley 2	18.8	51.2	Valley 2	16.2	13.8
Ratio 1/2	1.4	0.17	Ratio 1/2	3.6	0.45
	Ratio 0.8			Ratio 0.8	

| | Success | | | Failure | |
	Valley 1	Valley 2		Valley 1	Valley 2
Species A	26.2	18.8	Species A	8.8	51.2
Species B	58.8	16.2	Species B	6.2	13.8
A/B	0.44	1.2	A/B	1.4	3.7
	Ratio 0.4			Ratio 0.4	

| | Valley 1 | | | Valley 2 | |
	Species A	Species B		Species A	Species B
Success	26.2	58.8	Success	18.8	16.2
Failure	8.8	6.2	Failure	51.2	13.8
Ratio S/F	3.0	9.5	S/F	0.36	1.2
	Ratio 0.3			Ratio 0.3	

The fitting of log-linear models has become much easier with the availability of computer packages for fitting generalized linear models. More powerful methods of maximum likelihood are used in the computer packages but the user is required to state only the appropriate link function and error distribution plus the sets of model terms to be fitted. As will be plain from the discussion of the hawk data the models are based on a log link function, corresponding to proportional changes in frequencies. The Poisson distribution is the obvious one to model counts.

In choosing terms to be included in the fitted model we must distinguish between explanatory factors with predetermined proportions and response factors. Explanatory factors are fixed by the definition of the scope of the investigations. Thus 100 pairs of hawks were chosen in each valley. The 100:100 proportion is fixed and the 'valley' must be included in any model. The species and result factors are not controlled in the investigation and are response factors. We are interested in the dependence of the response factors and their inter-dependence and we compare models with and without various combinations of the response factors. Further information on the use of log-linear models is given in Krzanowski (1988) or McCullagh and Nelder (1990).

For an example of using a computer package we return to a set of data from the exercises at the end of Chapter 13.

Example 14.6

The data are the frequencies of occurrence of different numbers of lambs per birth for ewes from three breeds at each of three farms,

Farm	Breed	Number of live lambs per birth				Total
		0	1	2	3 or more	
1	A	10	21	96	23	150
	B	4	6	28	8	46
	C	6	7	58	7	78
2	A	8	19	44	1	72
	B	5	17	56	1	79
	C	1	5	20	2	28
3	A	22	95	103	4	244
	B	18	49	62	0	129
	C	4	12	16	2	34

Here we have two explanatory factors, Farms (F) and Breeds (B), and the numbers of sheep giving birth for each of the nine farm/breed combinations is fixed. Hence all models fitted must include the terms F + B + F B. The response factor is number of lambs (L) and our interest is in the interdependence of L with F or B or both. To provide answers the following models were fitted and deviances calculated.

Model	Deviance	d.f.
F + B + FB	683.7	27
F + B + FB + L	131.2	24
F + B + FB + L + BL	116.3	18
F + B + FB + L + FL	18.8	18
F + B + FB + L + BL + FL	14.6	12

Looking at these deviances directly we can see that for each of the last two models the deviance is approximately equal to the d.f. showing that each of these two models provides an acceptable fit to the data. That is the departure from the predictions of either of those two models are of the size that should be expected because of random variation. If we summarize the changes in deviance by an analysis of deviance as for the examples in the previous section we have the following

Source (in order fitted)	Deviance Change	d.f.	m.s.
L	552.5	3	184.2
FL	112.4	6	18.7
BL	4.2	6	0.7
Residuals	14.6	12	1.2

Clearly the BL 'interaction' is insignificant and the model we should use is F + B + FB + L + FL. The implications from this selection of model are that the proportions of births giving 0, 1, 2 or 3 + live lambs are independent of breeds within a farm but the proportions do vary between farms. The results should therefore be summarized by the two-way table for Farms × Number of live lambs, in the form of proportions.

	Number of live lambs per birth			
	0	1	2	3+
Farm 1	0.073	0.124	0.664	0.139
Farm 2	0.078	2.229	0.670	0.022
Farm 3	0.111	0.418	0.456	0.015

The pattern of the interaction can be detected from the fitted interaction terms or directly from this table. The fitted interaction terms (log scale are)

$$
\begin{array}{cccc}
-0.34 & -0.77 & -0.06 & +1.17 \\
+0.01 & +0.13 & +0.24 & -0.38 \\
+0.33 & +0.64 & -0.18 & -0.79
\end{array}
$$

The largest interaction terms are for (farm 1, 3 + lambs) which has a positive interaction effect and for (farm 1, 1 lamb) and (farm 3, 3 + lambs) which have negative interaction effects. The major pattern is that farm 1 produces more multiple births and farm 3, fewer multiple births.

14.8 PROBIT ANALYSIS

Our final example of the use of a model to interpret data on proportions arises when groups of experimental units are treated with different levels of some stimulus. In each group the number of units responding to the stimulus is counted and the proportion of units which respond is related to the level of stimulus. The model used to express this relationship is a non-linear model, and so the ideas discussed in Chapter 12 are important for the method of probit analysis to be discussed in this section.

Typically the stimuli applied to the different groups are different concentrations of some chemical. We shall discuss the philosophy and details of the method of probit analysis in terms of the application of a series of concentrations of an insecticide to groups of larvae, the recorded response being the number of deaths per group. The objective is usually the estimation of the concentration of insecticide which will achieve a 'kill' of a specified proportion of the population. Most commonly it is the concentration producing a 50% kill which is estimated, though other percentage kills may also be appropriate.

At a low concentration all larvae survive; as the concentration increases the proportion succumbing increases and at high concentrations all larvae are killed. A typical set of data with groups of 20 larvae tested for each concentration is given below

Insecticide concentration	0.375%	0.75%	1.5%	3%	6%	12%	24%
Number of larvae killed	0	1	8	11	16	18	20
Proportion killed	0	0.05	0.40	0.55	0.80	0.90	1.00

These data are plotted in Fig. 14.4.

The model on which probit analysis is based is that in the population of units to be used in the experiment there is a normal distribution of the concentration which each unit will tolerate. Experience shows that the assumption of a normal distribution of tolerance levels is often most reasonable when the concentrations are transformed to a log scale. One consequence of this is that sets of concentrations are often chosen so that consecutive concentrations differ by a factor of 2, as in the example. If the concentration which an individual larva receives is greater than the tolerable level for that larva then it dies. If the applied concentration is below the tolerated level then the larva survives.

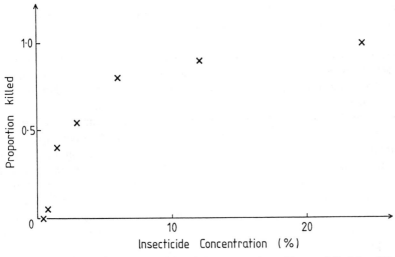

Fig. 14.4 Data showing the relationship of the proportion of larvae killed by different insecticide concentrations.

Of course we do not know the tolerance levels of the 20 different larvae for each concentration. Nor do we know the mean and standard deviation of the normal distribution of tolerance. However, we can estimate them from the observed data. Even without a formal method of analysis we could make rough guesses from the raw data. If we code the concentrations 0 (for 0.375%) to 6 (for 24%) then the mean tolerance level is obviously about 3(3%) or 2(1.5%) because the percentage killed at these levels is about 50%. The standard deviation can be guessed by noting that 100% survival occurs at coded levels 0 but only 95% at 1; and that 100% death occurs at level 6 but only 90% at 5. A standard deviation between 1 and $1\frac{1}{2}$ would fit these figures. If we assume that the mean tolerance level is 2.8 and the standard deviation is 1.2 then, by calculating the proportion of the normal distribution greater than each coded level, we have the 'fitted' proportions.

Level (x)	$z = \dfrac{x - 2.8}{1.2}$	$\Phi(z)$
0	-2.33	0.0099
1	-1.50	0.0668
2	-0.67	0.2525
3	0.17	0.5662
4	1.00	0.8413
5	1.83	0.9666
6	2.67	0.9962

The 'z' values are the probits, and the relationship between the proportion (of the normal distribution) Φ and the normal deviate, or probit, z defines the link function.

Clearly, by trying various values for the mean and standard deviation of the tolerance distribution, we could find values giving the best fit to the observed data. The procedure would be essentially the same as for the other non-linear models in Chapter 12. There is, however, a complication. As we have argued in Chapter 13, on the estimation of proportions, the sample estimates of a proportion of about 0.5 will be relatively more variable than the corresponding estimates of a proportion of either 0.05 or 0.95. In fact, the standard deviation of sample estimates of a true proportion p is $\sqrt{[p(1-p)/n]}$ where n is the sample size. In assessing the agreement between observed and fitted proportions, we must allow for the inherently greater variation when p is about 0.5. However, we do not know the true value of p for the different concentrations, we have only estimates of their values, either from the observed proportion or from the fitted proportions.

The method of probit analysis, consists of several stages or iterations. Initially, we estimate the variances of each fitted value by assuming that the observed proportions are the true proportions. We then obtain estimates of the mean and standard deviation of the tolerance distribution by linear regression of the normal deviate corresponding to each proportion killed on the dose level, x, weighting each observation inversely by its estimated variance. From the estimates of the mean and standard deviation, we calculate fitted proportions and for the next iteration we use the variances appropriate to these fitted proportions. This leads to different weightings in the regression and to different estimates of the mean and standard deviation, though often the estimates will be similar to those from the first iteration. These new estimates of mean and standard deviation give new fitted values, which in turn give new variances, and hence different weightings and new estimates of mean and standard deviations. The system of estimation usually settles down, or converges, quite quickly.

Formally, the method of probit analysis is defined as follows: we wish to fit a linear regression of the normal deviate corresponding to each proportion p killed, probit (p), on the dose level, x, and the regression relationship is written

$$\text{probit } (p) = \alpha + \beta x$$

The estimates of α and β may be obtained manually using the iterative procedure outlined above but are more readily found using a statistical package with facilities for fitting generalised linear models. The probit function provides the link; the distribution of the numbers killed is assumed to be Binomial.

Example 14.7 Probit analysis

We shall analyse the data discussed earlier in the section displayed in Fig. 14.4.

Concentration(%)	Number killed	Proportion killed (p)
0.375	0	0
0.75	1	0.05
1.5	8	0.40
3	11	0.55
6	16	0.80
12	18	0.90
24	20	1.00

The model to be fitted is:

$$\text{Probit } (p) = \alpha + \beta \times \log_{10} (\text{Concentration})$$

When this is fitted as a generalized linear model with a probit link and binomial error distribution the following estimates are obtained:

	Estimate	s.e.
α	-1.022	0.202
β	2.415	0.331

The adequacy of the fit of the model can be assessed by comparing the residual deviance (which is analogous to the residual sum of squares in a standard analysis of variance) with the appropriate χ^2 distribution, or as a quick rule of thumb looking for a mean deviance of about one. For this model the residual deviance is 3.777 with 5 degrees of freedom (mean deviance $= 0.7554$), since this deviance is less than the 5% point for a χ^2 distribution with 5 degrees of freedom the model gives an adequate fit.

The estimate of the log concentration which would kill 50% of larvae, the LD50, is given by probit $(0.5) = 0$. The resulting equation is

$$0 = -1.022 + 2.415 \times \text{LD50},$$

giving $\qquad\qquad\qquad 0.423 = \text{LD50}$

This is equivalent to a concentration of $10^{0.423} = 2.65\%$

An approximate standard error for the estimate of the log dose 50 can be obtained using the formula

$$\text{SE(LD50)} = \sqrt{\left(\frac{\text{Var(a)} - 2.\text{LD50.Cov(a, b)} + \text{LD50}^2 \, \text{Var(b)}}{b^2} \right)} \qquad (14.3)$$

where a and b are the estimates of α and β and the variances and covariances are obtained from the variance – covariance matrix of the parameter estimates. The approximation is reasonable if the standard error of the slope is relatively small.

For this data the variance covariance matrix values are

$$Var(a) = 0.04072$$
$$Var(b) = 0.10968$$
$$Cov(a, b) = -0.04715$$

substituting these in (14.3) gives a standard error for LD50 of 0.131. The mean, which is the LD50, and the standard deviation of the distribution of $\log_{10}$ tolerance concentration are calculated from a and b by $m = -a/b = 0.423$ and $s = 1/b = 0.414$, respectively.

Another transformation of proportions which can be used instead of the probit is the logistic transformation, discussed in section 14.6

$$logit (p) = \log \{p/(1 - p)\}$$

It is widely used in epidemiological studies since it can be directly interpreted as the log of the odds ratio in favour of success. There is a very large literature on the fitting of probit and logistic models and we have described the simplest possible case here. Any experimenter who wishes to analyse data on proportions by fitting probit or logistic models will find the books by Finney (1971) and Collett (1991) useful and would be well advised to discuss both design and data analysis with a statistician before embarking on the analysis.

15

Making and analysing many experimental measurements

15.1 DIFFERENT MEASUREMENTS ON THE SAME UNITS

In our discussions of the analysis of data from designed experiments we have considered a single set of data from each experiment. In practice it would be extremely rare for only a single measurement to be recorded for each unit. Much more commonly several measurements characterizing different aspects of the response of the units to the various treatments would be recorded and used together to summarize the effects of the different treatments.

For example, to compare different varieties and management systems for carrots the measurements could include germination rates per plot at several dates, total harvested weights of roots and of stem plus foliage, possibly at several different harvest times, numbers and weights of roots in several size grades, incidence of disease, colour and taste of the roots.

In laboratory experiments with plants, detailed measurements of plant components (height, leaf number, leaf area, leaf shape) and timings of physiological stages may be recorded. For animal experiments records of dimensions of parts of the animal, chemical analyses of blood samples, timing of developmental phases and information about behavioural patterns may be recorded.

Later in this chapter we shall consider the analysis of two sets of multiple variable data in detail. The two data sets are listed in full in Examples 15.1 and 15.2. The first data set consists of the weights of 27 rats on each of five occasions, each rate being treated with one of three chemicals (Control for 10 rats, Thyroxin for 7 rats and Thiouracil for 10 rats). The second data set consists of yields of Cowpea and Maize for 72 plots from an intercropping experiment to examine the effects of different varieties and nitrogen levels on the two crops grown together.

In general, the total set of initial information from an experiment will consist of the design information on the structure of experimental units and the treatments applied to the units, together with values for each unit of a number of observed variables (measurements). In the computer this information is usually stored as a matrix array in which each row contains all the information about a particular unit and each column contains the information from all units on a particular variable.

The information on the response of units to treatments is not restricted to those variables which are directly measured. Other variables can be calculated, for each unit, as functions of the recorded variables. For example if plant height has been recorded on each of two occasions we can calculate, for every experimental unit, the increase in height or, by dividing the difference in height by the length of the time interval between the occasions, the rate of increase in height.

If we are recording plant height on several occasions then we will measure the same plants on each occasion. Similarly the weights of animals can be recorded, for the same animals, on different occasions. However, if we wish to record plant weights of parts of plants, or weights or other measurements on internal parts of animals, the measurement process will usually require the destruction of the plant or animal. Hence, for each experimental unit, we will have to use different sub-samples of plants or animals for the different measurements times. This means not only that we have to have sufficient plants or animals available in each experimental unit for sampling on the various occasions, but also that our estimates of changes in weight over a time interval will be less precise because they are based on two different sets of plants or animals.

Calculations of new variables derived from the original variables are very simply achieved on the computer. If the mean weights of samples of plants from each unit on the first sampling occasion are stored as wt_1 and the corresponding means weights from the second occasion are stored as wt_2 then an instruction in the form

$$wt_{12} = wt_2 - wt_1$$

would produce the set of mean weight increases (wt_{12}) from occasion 1 to occasion 2, for all experimental units. A slightly more complex calculation

$$rgr_{12} = (\log(wt_2) - \log(wt_1))/t_{12}$$

where t_{12} is the time gap between the two occasions, produces, the set of relative growth rates (rgr_{12}) for all experimental units.

Typically a computer package will provide facilities for calculating derived variables using basic arithmetical functions, powers, logs, square roots, sines and cosines and often more complex functions. The form of the analysis of variance and comparison of treatments means will be exactly the same for each variable, observed or derived. For all variables the pattern of the analysis is defined by the structure of the experimental design.

15.2 INTERDEPENDENCE OF DIFFERENT VARIABLES

The analysis for each observed or derived variable provides new and different information. However, the analyses for different variables may have some similarities of pattern because each set of values of a variable is based on exactly the same set of experimental units. Consider the results from a simple experiment on bees. The experiment included nine experimental hives, three hives for each of three diets. The brood area in each hive was measured on five occasions, giving the results shown. The results of an analysis of variance for each time are also given.

Hive	Diet	Brood Area				
		Time 1	Time 2	Time 3	Time 4	Time 5
1	1	368	920	944	1084	1108
2	1	308	980	1014	1008	894
3	1	284	500	754	908	518
4	2	684	974	808	490	208
5	2	288	624	570	280	334
6	2	468	778	638	310	178
7	3	444	1330	1010	748	448
8	3	280	290	248	118	80
9	3	380	828	830	598	432

Analysis of Variance Mean Squares

Between diets (2 d.f.)	23284	448	52732	344064	318400
Within diets (6 d.f.)	16125	128227	63919	42997	46443

We can see that, although there are changes in the pattern of the nine results at the different times, the patterns also show some strong consistencies. Thus, for diet 3, the areas from hive 7 are always greater than those for hive 9, with the areas for hive 8 much smaller than either. Similar, but less striking patterns occur for both other diets. When we analyse and interpret the patterns of treatment differences at each time, part of each pattern will reflect differences between the units which are consistent over the times simply because we repeatedly measure the same units. In our analysis we have to be able to allow for this inevitable consistency while seeking to interpret the change, or lack of change, of the treatment effects.

One way of characterizing the extent to which the patterns of variation amongst the area for the nine units are similar at different times is by examining the relationship between the area measurements on different occasions. The relationship between two measurements is best displayed by a graph plotting one against the other. In Fig. 15.1 the relationships between some pairs of times are shown. The relationship between the areas at times 2 and 3 is clearly strong, that between times 2 and 4 rather less strong, and the relationships between times 2 and 5, or between times 2 and 1 are hardly discernible. This is not surprising; the treatments began immediately after time 1, producing some large initial changes, with more gradual changes through subsequent times, the changes being less between more clearly adjacent times.

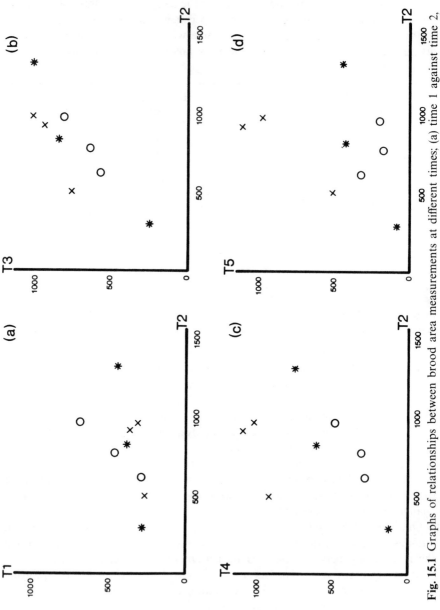

Fig. 15.1 Graphs of relationships between brood area measurements at different times; (a) time 1 against time 2, (b) t_3 against t_2, (c) t_4 against t_2, (d) t_5 against t_2.

Notice that in the plots of pairs of variables in Fig. 15.1 the overall relationship of all nine values does not appear so clearly as the separate relationships for each different treatment diet (distinguished by different symbols). In experimental data we have seen that the total variation of a single yield variable can be split into components, representing different sources of variation, the splitting being displayed in an analysis of variance. In the same way, the relationship between variables may well vary in different components of variation, since different causes of variation apply in these different components.

Consider the measurement of two components of plants (roots and leaves) or of two crop yields when the crops are grown in a mixture. For plants with the same treatment on plots of differing fertility root weight and leaf weight may tend to be large together on the better plots and small together on the poorer plots. But the treatments may change the balance between the roots and leaves so that when comparing different treatments root weight and leaf weight may be negatively correlated. Similarly crops competing in a mixture may display negative correlation across replicate plots of the same treatment, but over a series of increasing soil fertility treatments the weights of both crops may increase together as fertility improves. We shall consider further examples where different patterns of relationship occur in different components of an experimental structure in sections 15.4 and 15.5.

The results from the analyses of different variables will usually not be independent, but neither will the results for one variable be totally predictable from another; each additional variable provides some additional information. In such situations we have to be careful in interpreting the combined significance from the analysis of several different variables. If we did have two completely unrelated variables, each providing evidence of a genuine difference between treatment effects at the 5% significance level, then we would be justified in treating the combined evidence as very strong indeed. The effective significance level from the two analyses, for testing the hypothesis that there are no treatment differences for either variable, would be $(1/20) \times (1/20)$, or a 0.25% significance level.

With interdependent information, as in the data on brood areas at different times, the overall significance level is not increased to anything like the same degree. The complete set of results from all five occasions clearly provides more information than the results from any single occasion but the interdependence between the results means that the combined information is not equivalent to a five-fold increase from that for a single time. In general, we should obviously not expect to be able to increase our information about any situation simply by taking more and more records at increasingly short intervals. Evenually we would be recording virtually identical results at successive occasions! The second measurement could not then provide any information additional to that from the first.

Apart from exercising caution in the interpretation of significance from analyses of multiple measurements, how can we improve our statistical analyses of such situations? In the following sections of this chapter we consider, first, the particular case of a sequence of repeated measurements of the same variable on the same set of units. Then we consider the joint analysis of two different variables. Finally we look at the situation where the experimenter is particularly interested in the relationship(s) between measurements, either in addition to, or even instead of, the direct treatment comparisons.

15.3 REPEATED MEASUREMENTS

Probably the most common form of multiple measurements is that of repeated measurements of a single variable on the same experimental unit at a number of times. Particularly with the increasing availability of sophisticated technology it is now easy to take large numbers of measurements in a time sequence, and it is natural to assume that this provides more information. An extreme case is encountered when measurements such as temperature, light intensity or blood pressure, are recorded 'continuously'. The analysis of repeated measurements has acquired a special status and much has been written on alternative approaches. A good book on the subject is that by Crowder and Hand (1990). Here we shall consider only the basic concepts for the analysis.

We shall assume that the experimental design structure for the experimental units, on which multiple measurements are recorded, is clearly defined and understood. The design structure may be a simple completely randomized design with three or more replicates of each of several treatments, or it may be quite complex involving blocking, Latin squares, incomplete blocks or split plots. Whatever the design structure is, it defines the basic form of analysis and all proposed analyses must reflect this underlying design structure.

One method of analysis that has been suggested, and often used, for repeated measurements is a split-plot analysis. The argument to justify this is that the set of measurements at different times for each unit is analogous to the set of measurements on split plots which have been treated or managed in different ways. Is it reasonable for time to be regarded as a split-plot factor?

The general assumptions in a split-plot analysis is that observations on the same unit at different times will tend to be more similar than observations on different units at different times, which is generally reasonable for repeated measurements. However the particular form of the assumption when we consider time as a split-plot factor is that the similarity of observations at different times is independent of the particular times. In other words, observations at the first and last times should have the same degree of similarity as observations at two consecutive times. This is not usually credible.

A further drawback for the assumptions required for a split-plot analysis is that the assumption of uniform similarity of split-plot yields within a main plot is supported by the randomization of treatment allocation to individual split plots. Time, however, cannot be randomized!

The split-plot analysis oversimplifies the true situation, in its assumption of the homogeneity of similarities between times, and fails to use the ordering of time. Time is treated as a factor rather than as a repetition of measurements. An alternative approach, in which no assumptions are made about the patterns of similarity of observations at different times is multivariate analysis of variance in which the relationship between times are estimated wholly from the data, ignoring the order of times (see Crowder and Hand (*loc cit*)).

The approach to the analysis of repeated measurements which reflects the ordering of time and emphasizes the concern with the response over time, has two stages. The first is to examine and summarize, in one or more parameters the response pattern of change through time for each experimental unit. Any of the models developed in earlier chapters for examining relationships may be appropriate for this summary. The model should be chosen to match the biological situation and the pattern in the data. The parameters of the model, which are particular derived variables can then be analysed.

Suppose we are examining the growth of fungal cultures and the diameters of each culture are measured each day. Typically the diameter increases by very similar amounts each day and it is probably entirely adequate to represent the growth rate by

$$[(\text{diameter day 6}) - (\text{diameter day 1})]/5$$

An alternative would be to calculate the linear regression of diameter on days, which will give a very similar figure because the first and last values in such a regression provide most of the information about the slope of the line.

Whichever way we estimate the growth rate, we obtain an estimate for each experimental unit and we can then analyse the set of growth rate values to separate variation between treatments, blocks and error. We use the data set of growth rates just as we would use any other set of measurements for each unit. The difference in mean growth rates for different treatments is assessed relative to the variation of growth rates across units receiving the same treatment.

For the honeybee data it is clear that the response pattern has an initial rise from occasion 1 to occasion 2, followed by trends, either down or, less strongly, up through occasions 2 to 5. A simple linear trend would not be an appropriate summary. Possibly the change in area from occasion 1 to occasion 2 defined as one derived variable representing the initial response to the treatments, and the linear regression from occasions 2 to 5 as a second derived variable, representing subsequent changes, would encapsulate most of the pattern of variation of area over the five times. The derived data and analyses of variance are

Occasion 2-Occasion 1

Diet 1	Diet 2	Diet 3
552	290	886
672	336	10
216	310	448

	S.S.	d.f.	m.s.
Between diets	47744	2	23872
Within diets	544240	6	90707

Diet	Mean	
1	2480	
2	312	S.E. (difference) = 246
3	448	
Mean	413	S.E. (mean) = 100

Regression Coefficients (2–5)

Diet 1	Diet 2	Diet 3
+ 70.4	− 261.6	− 290.8
− 29.4	− 116.0	− 76.0
+ 20.8	− 212.8	− 142.0

	S.S.	d.f.	m.s.
Between diets	84180	2	42090
Within diets	40176	6	6696

Diet	Mean	
1	+ 20.6	
2	− 196.8	S.E. (difference) 67
3	− 169.6	

As anticipated there is an increase in brood area from occasion 1 to occasion 2, which is consistent across the diets. There are clear differences between diets after occasion 2 with similar strong rates of decline for diets 2 and 3, and a slight (non-significant) increase for diet 1. Each case has to be considered individually using information on the biological situation together with the apparent patterns in the data to determine a proper choice of derived variables.

Example 15.1

A much used set of data for the growth of rats provides an example, again with simple design structure. Three treatments (a control and two chemicals added to drinking water) were applied to 10, 7 and 10 rats respectively, each rat being treated independently. The design was completely randomized. The weights of the rats were recorded at the start of the experiment and after one, two, three and four weeks of treatment. The full data are given on the opposite page:

Control	0	1	Week 2	3	4
Rat 1	57	86	114	139	172
Rat 2	60	93	123	146	177
Rat 3	52	77	111	144	185
Rat 4	49	67	100	129	164
Rat 5	56	81	104	121	151
Rat 6	46	70	102	131	153
Rat 7	51	71	94	110	141
Rat 8	63	91	112	130	154
Rat 9	49	67	90	112	140
Rat 10	57	82	110	139	169
Thyroxin					
Rat 1	59	85	121	156	191
Rat 2	54	71	90	110	138
Rat 3	56	75	108	151	189
Rat 4	59	85	116	148	177
Rat 5	57	72	97	120	144
Rat 6	52	73	97	116	140
Rat 7	52	70	105	138	171
Thiouracil					
Rat 1	61	86	109	120	129
Rat 2	59	80	101	111	126
Rat 3	53	79	100	106	133
Rat 4	59	88	100	111	122
Rat 5	51	75	101	123	140
Rat 6	51	75	92	100	119
Rat 7	56	78	95	103	108
Rat 8	58	69	93	114	138
Rat 9	46	61	78	90	107
Rat 10	53	72	89	104	122

As mentioned previously we can analyse the results for each measurement time and for any difference between times. The results from analyses of weights at week 1, at week 2, at week 4 and the change in weight (week 4 − week 0) are given below as examples.

Week 1	s.s.	d.f.	m.s.	F
Between treatments	36	2	18	0.3
Within treatments	1650	24	69	
Treatments means	(1) Control	78.5		
	(2) Thyroxin	75.9		
	(3) Thiouracil	76.3		
	S.E. ((1)–(3))			3.7
	S.E. (other differences)			4.1

Week 2	s.s.	d.f.	m.s.	F
Between treatments	601	2	300	2.5
Within treatments	2876	24	120	
Treatment means	(1) Control	106.0		
	(2) Thyroxin	104.9		
	(3) Thiouracil	95.8		
	S.E. ((1)–(3))			4.9
	S.E. (other differences)			5.4

Week 4	s.s.	d.f.	m.s.	F
Between treatments	9204	2	4602	17.1
Within treatments	6442	24	268	
Treatment means	(1) Control	160.6		
	(2) Thyroxin	164.3		
	(3) Thiouracil	124.4		
	S.E. ((1)–(3))			7.3
	S.E. (other differences)			8.1

Week 4–Week 0	s.s.	d.f.	m.s.	F
Between treatments	10304	2	5152	22.6
Within treatments	5476	24	228	
Treatment means	(1) Control	108.2		
	(2) Thyroxin	108.7		
	(3) Thiouracil	69.7		
	S.E. ((1)–(3))			6.8
	S.E. (other differences)			7.4

These analyses show clearly that growth is reduced by Thiouracil, particularly when total growth over the complete four-week period is considered. However, to identify the patterns of changed growth we will learn more by examining the form of the relationship between size and time. A first step is to plot mean size against time for each treatment and this is shown in Fig. 15.2. This suggests the the rats' growth for the control and thyroxin treatments is essentially linear but that Thiouracil, as well as producing a lower overall growth, may have a curved growth pattern with gradually diminishing growth.

Since we have a sample of rats for each treatment we should compare the growth relationships for different treatments by assessing the relationship for each rat and analysing the variation of the relationships between treatments

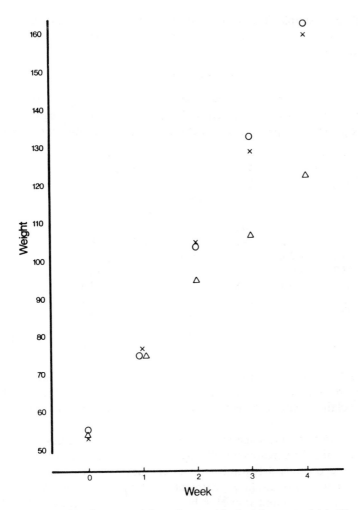

Fig. 15.2 Relationship of mean weights of rats with time for control (x), Thyroxin (o) and Thiouracil (Δ).

relative to the variation within treatments. Comparison of relationships is achieved through comparing quantitative characteristics of the relationships.

Since the growth relationships appear to be fairly linear, with only a small amount of curvature, we first fit a linear regression contrast:

$$b = [2(\text{week } 4) + (\text{week } 3) - (\text{week } 1) - 2(\text{week } 0)]/10$$

and also calculate a curvature contrast:

$$c = [2(\text{week } 4) - (\text{week } 3) - 2(\text{week } 2) - (\text{week } 1) + 2(\text{week } 0)]/14$$

The particular, simple, form of these coefficients, or contrasts, depends on the equal spacing of the five times. These two contrasts are the same as the regression coefficients in the model

$$y = a + b(x - \bar{x}) + c(x - \bar{x})^2$$

An advantage of defining the two contrasts in the above form is that they are orthogonal, in the sense discussed in Chapter 6, which ensures that the two contrasts provide independent information. The resulting values for b and c are:

	Control		Thyroxin		Thiouracil	
Rat	b	c	b	c	b	c
1	28.3	+ 0.4	33.5	+ 1.2	17.0	− 3.1
2	28.7	− 0.8	20.7	+ 1.6	16.5	− 1.6
3	33.3	+ 2.2	34.2	18.7	18.7	− 0.9
4	29.2	+ 2.1	29.9	+ 0.5	14.9	− 2.6
5	23.0	+ 0.3	22.2	+ 1.1	22.6	− 1.3
6	27.5	− 0.5	21.9	+ 0.1	16.1	− 1.4
7	21.9	+ 1.1	30.6	+ 2.0	12.9	− 3.1
8	22.1	− 0.8			20.5	+ 1.6
9	22.7	+ 1.4			15.1	− 0.1
10	28.1	+ 0.8			17.0	0.3

For each rat almost all the variation between the set of five weights can be shown to be accounted for by regression contrasts. Hence analysis and interpretation of the variation of these two contrasts should provide all the information available.

These figures provide some immediate suggestions of differences in the relationships for the three treatments. Plainly the linear effect is smaller for Thiouracil than for the other two treatments. The curvature effect seems to suggest a possible separation of all three treatments, with negative values for Thiouracil, positive for Thyroxin and values on both sides of zero for the control. We can see these patterns more clearly if we display the linear slopes and curvature effects graphically as in Figs. 15.3 and 15.4.

We can also analyse the values of the contrasts in analyses of variance.

Slope (b)	s.s.	d.f.	m.s.	F
Between treatments	610.8	2	305.4	17.9
Within treatments	409.5	24	17.1	
Treatments means	(1) Control	26.5		
	(2) Thyroxin	27.6		
	(3) Thiouracil	17.1		
	S.E. ((1)–(3))			= 1.83
	S.E. (other differences)			= 2.01

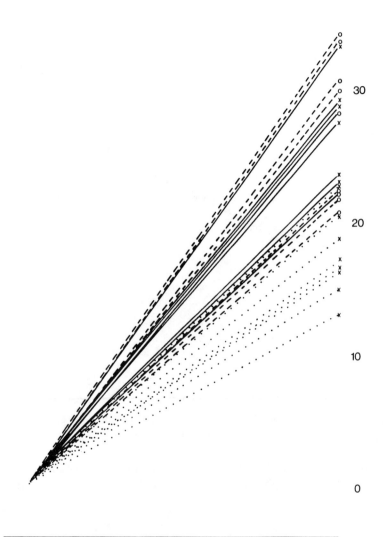

Fig. 15.3 The distribution of slope values for the 27 rats; Control (x ————), Thyroxin
(o————) and Thiouracil (x ....).

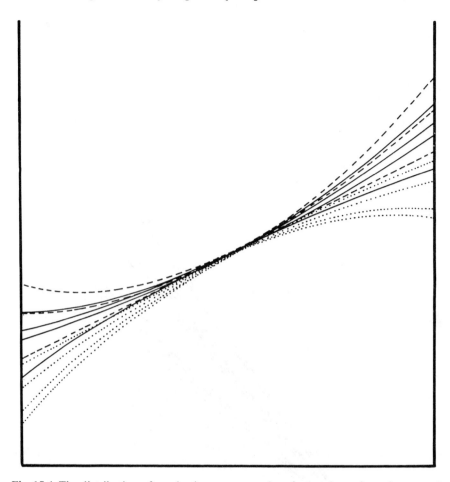

Fig. 15.4 The distribution of quadratic curvature values for a subset of rats for control
(————), Thyroxin (-----) and Thiouracil (.....)

The analysis confirms that the values for Thiouracil are very significantly lower
and also shows negligible difference between Thyroxin and the control.

Curvature (c)	s.s.	d.f.	m.s.	F
Between treatments	33.19	2	16.60	10.7
Within treatments	37.30	24	1.55	
Treatments means	(1) Control	+ 0.62		
	(2) Thyroxin	+ 1.41		
	(3) Thiouracil	− 1.28		
	S.E. ((1)–(3))			0.56
	S.E. (other differences)			0.62

As anticipated, this analysis also shows that the response to Thiouracil is quite different from the other two treatments, the negative curvature indicating a gradually declining growth rate. The mean curvature effects for the control and Thyroxin treatments are both positive and the difference between the two values is not significant at the 5% level. Nevertheless there is a mild suggestion that whereas the growth rate for control rats has very little curvature, the growth rate for the Thyroxin treated rates may be increasing with time (1.41 is significantly larger than zero). Note that there is a possible outlier for the curvature for Thiouracil from rat 8. This observation inflates the error mean square considerably and omitting it reduces the standard errors by about 10%.

We can summarize the information provided by the analyses of the observed and derived variables.

Variable		Treatment means		
	F ratio	Control	Thyroxin	Thiouracil
Week 1	0.3	78	76	76
Week 2	2.5	106	105	96
Week 3	9.8	130	134	108
Week 4	17.1	161	164	124
Week 4–Week 0	22.6	108	109	70
Slope	17.9	26.5	27.6	17.1
Curvature	10.7	+ 0.62	+ 1.41	− 1.28

The slope and curvature contrasts together provide a very clear description of the contrasts the treatments. The fitted mean response equations are:

$$\text{Control} \quad \text{Weight} = 104.7 + 26.5(t - 2) + 0.62(t - 2)^2$$

$$\text{Thyroxin} \quad \text{Weight} = 104.2 + 27.6(t - 2) + 1.41(t - 2)^2$$

$$\text{Thiouracil} \quad \text{Weight} = 94.4 + 17.1(t - 2) - 1.28(t - 2)^2$$

and the fitted values are

Week	Control	Thyroxin	Thiouracil
0	54.0	54.6	55.1
1	78.1	78.0	76.1
2	104.7	104.2	94.4
3	131.1	133.2	110.2
4	159.5	165.0	123.5

This approach to repeated measurements data requires thought from the user. In the first stage it is important to choose a sensible form of the relationship of measurement with time and to look at the random variation of that relationship between replicate individuals or experimental units. Of course, it is always important to investigate a relationship graphically before summarizing it, whatever form of summary is ultimately chosen (parameter values, indices or simple averages and differences).

For the second stage of comparing relationships for different treatments we can only assess the differences between the mean relationships for the different treatments if we have a measure of the variation of the relationship between individuals or units treated identically. Simply fitting the relationship to the mean values for the five times for each treatment would give the same values as the means of the values calculated for each rat, but we would not be able to assess the precision of the differences between the relationships for different treatments.

In our calculation of linear and quadratic regression contrasts we have simply used the observations at the different times as if they were independent. There are more complex ways of calculating contrasts allowing for varying degrees of interdependence but the methodology is beyond the scope of this book (again see Crowder and Hand (1990)). A final point to remember is that the form of the summary of the response pattern over time for each unit is dictated only by what is sensible for the particular biological situation. Non-linear models can be used for asymptotic response patterns; in a periodic, oscillating, form of response a sine curve or a series of parallel straight lines could be appropriate. But the form of model chosen must always be biologically reasonable within the restriction of simplicity.

15.4 JOINT (BIVARIATE) ANALYSIS

Analysing each variable from an experiment separately can provide misleading information or fail to display the real situation. A very simple example of this is provided by some data on the chemical contents of different leaf samples. Concentrations of two chemicals were measured for the same two sample pieces from each of three leaflets from each of four leaf clusters from four different plants. For one plant the results were as follows:

Leaflet	1		2		3	
Sample	1	2	1	2	1	2
Chemical 1						
Leaf						
1	2.43	1.88	1.80	1.80	2.66	2.88
2	2.36	3.25	3.25	3.08	3.45	3.45
3	2.92	3.92	3.38	3.81	3.58	3.52
4	3.66	4.17	3.83	3.44	3.81	3.19

Chemical 2

Leaf						
1	4.21	5.47	5.97	5.77	3.91	3.71
2	4.48	3.34	3.34	3.52	3.16	3.16
3	3.74	2.82	3.22	2.92	3.11	3.13
4	3.05	2.68	2.88	3.23	2.92	3.44

The objective was to examine the variability of the two chemical concentrations at different levels of the plant structure. The separate analyses of variance, for observations from the single plant are

	s.s	d.f.	m.s.	F
Chemical 1				
Between leaves	7.475	3	2.492	12.5 (3, 8 d.f.)
Between leaflets within leaves	1.598	8	0.200	1.5 (8, 12 d.f.)
Between samples within leaflets	1.589	12	0.132	
Chemical 2				
Between leaves	12.373	3	4.124	6.6 (3, 8 d.f.)
Between leaflets within leaves	4.974	8	0.622	3.4 (8, 12 d.f.)
Between samples within leaflets	2.216	12	0.185	

At first sight the analyses provide similar but distinguishably different patterns of information. For chemical 1 differences between leaves are much stronger than those between leaflets within leaves and variation within leaflets is almost as large as that between leaflets. For chemical 2 there is more variation between leaflets than within leaflets and also more between leaves than between leaflets for the same leaf. Certainly the two analyses provide information which would encourage us to draw contrasting conclusion about the two chemicals. It is therefore slightly surprising to find that if we plot the relationship between the two chemical concentrations we get Fig. 15.5.

The graph provides a very clear display of the strong degree of interdependence of the two variables but the tables of original data, do not suggest this. It is also very easy to fail to detect such patterns when data are analysed using a computer package. Computer analysis packages usually provide summary information about each separate variable analysed but not about relationships between variables. We return to this set of data later in the chapter.

To understand just how relationships between two variables can lead to misleading conclusions from separate analyses we consider different patterns of

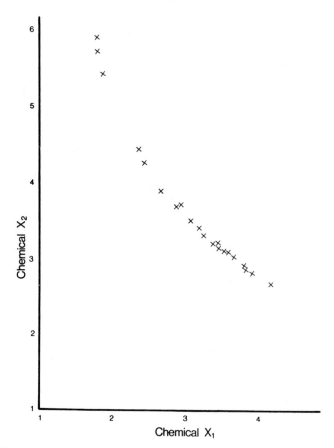

Fig. 15.5 Relationship between recorded values for chemical X_1 and chemical X_2.

joint variation of two variables for two treatments. In Fig. 15.6(a), (b) and (c) three different patterns of joint variation of two variables (on the horizontal and vertical axes) for two treatments (represented by circles and crosses) are shown.

The two sets of values for the two treatments on the horizontal axis are identical for each of the three diagrams; similarly the information on the vertical axis about the two treatments is the same in each diagram. The conclusions from the separate analyses would therefore be identical for the three situations. However, the joint information is clearly different in the three diagrams. In Fig. 15.6(a) the two sets of points representing the pairs of yields for the two treatments are almost totally distinct. The distinctness is the result of the negative correlation between the two yields for each treatment, which contrasts with the difference between the treatments in which both mean yields are higher for the circles.

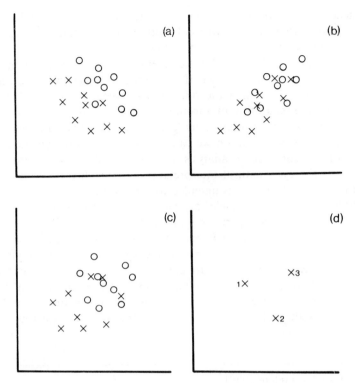

Fig. 15.6 Patterns of joint variation of two variables: (a) negative correlation within each treatment group; (b) positive within-group correlation; (c) no within-group correlation; (d) three treatment pairs of means.

In Fig. 15.6(b), in contrast, there is little separation between the two groups of points because the difference between the treatments is very similar to the positive correlation between the two yields within each treatment. The difference between the treatments could well be merely a reflection of the general trend for the two yields to increase or decrease together. The situation in Fig. 15.6(c) is intermediate, in that there is no correlation pattern between the two yields within each treatment. The distinction between the two treatment mean pairs is rather clearer for the joint variation than for each variable separately, because the independence of the two variables each showing some separation produces stronger overall evidence of a difference than does either separate variable.

The interpretation of differences of yield for each of two variables must take account of the underlying relationship between the two variables. If we consider the interpretation of the differences between the three treatment mean pairs of yields in Fig. 15.6(d) then the visual similarity of the three distances between each pair of treatment points should not be translated into an assessment that the evidence for differences between each treatment pair is similar. If the underlying correlation of the two yields within each treatment is negative, as in Fig. 15.6(a) then the difference, in Fig. 15.6(d), between treatments 1 and 2 is not large compared with the within treatment variation, but treatment 3 does appear rather different from either. If the within treatment variation is more like that of Fig. 15.6(b), then the difference between treatments 1 and 2 is impressive but the separation of 3 from either 1 or 2 is not convincing.

Without information about the interdependence of the two variables our eyes may be misled. In this situation we can choose either to re-educate our eyes or, alternatively, consider an analysis and method of graphical presentation which makes the eye's natural interpretation correct. Bivariate analysis and plotting means on skew axes is intended to achieve this.

A situation where two variables are measured and where we should expect them to be related is in experiments on mixtures of crops. There is a lot of interest in experimentation on intercropping (two agricultural crops grown together) and agroforestry (a perennial tree species grown with an annual agricultural species) and the experimenter will want to assess the effects of treatments on both crop yields simultaneously. We consider, first, a small set of data from an experiment on the joint yields from a mixture of millet and sorghum grown alternate rows. The yields for the two crops for four different sorghum varieties, in four blocks are:

Millet Yield (kg per plot)

Sorghum Variety	Block				Total
	1	2	3	4	
1	2.14	1.74	2.38	1.64	7.90
2	2.65	1.56	1.68	1.93	7.80
3	2.04	3.88	1.94	1.47	9.33
4	1.07	2.02	1.95	1.23	6.27
Total	7.88	9.20	7.95	6.27	31.30

Sorghum Yield (kg per plot)

Sorghum Variety	Block				Total
	1	2	3	4	
1	1.12	0.69	1.57	1.48	4.86
2	0.84	2.24	1.14	1.41	5.63
3	2.68	2.08	2.00	2.36	9.12
4	4.90	2.98	2.98	3.24	15.10
Total	9.54	7.99	8.64	8.49	34.71

The plot of millet yield against sorghum yield for all 16 observations is shown

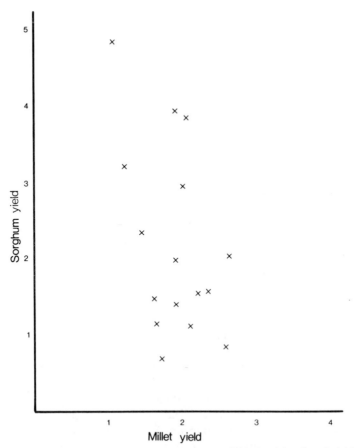

Fig. 15.7 Relationship of sorghum and millet yields for 16 pairs of yields.

in Fig. 15.7 and displays a slight negative correlation, mainly due to the low millet and high sorghum yields for sorghum variety 4.

Bivariate analysis and plotting on skewness are described by Pearce and Gilliver (1979) and Dear and Mead (1983). An analysis of variance is calculated for each of the two variables in the usual manner, providing sums of squares for each component of the analysis. To include the information about the relationship between the two variables we also calculate the sums of products (SP) for each component of the analysis. For the millet/sorghum data the results (and some calculations) are:

$$\text{Total SP} = (2.14)(1.12) + (1.74)(0.69) + \cdots + (3.24)^2 - (34.71)^2/16$$
$$- (31.30)(34.71)/16 \qquad\qquad = -3.904$$

$$\text{Block SP} = [(7.88)(9.54) + \cdots + (6.27)(8.49)]/4 - (31.30)(34.71)/16$$
$$= 0.151$$

Source	d.f.	S.S.(Millet)	S.P.	S.S.(Sorghum)
Blocks	3	1.081	− 0.151	0.313
Treatments	3	1.172	− 2.383	16.326
Error	9	4.056	− 1.371	3.766
(Error mean squares)		(0.451)	(− 0.153)	(0.418)
Total	15	6.309	− 3.904	20.405

The sums of products terms are analogous to S_{xy} used in linear regression calculations.

If the pairs of treatment mean yields are plotted on the usual horizontal and vertical axes, the negative yield correlation shown in the error line of the analysis means that the standard errors for comparing two treatment mean pairs would be of different lengths in different directions. The negative correlation means that the standard error would be longer in the directions in which the increase of one variable is balanced by a decrease in the other (North–West and South–East); in contrast the standard error would be shorter in the directions in which both variables both increase or both decrease. The 'natural' pattern of negative correlation means that treatment differences in which the balance between yields changes are less statistically significant than those in which both decrease or increase.

If we imagine squeezing the graph by pushing the axes closer together, then the variation of the standard error in different directions can be reduced so that the standard error will be uniform in all directions. The principle is demonstrated for the four millet/sorghum mean yields in Fig. 15.8. The first graph uses rectangular axes with the standard errors of precision for each treatment mean pair represented by ellipses; the second, with squashed skew axes, drawn symmetrically about the vertical, has circular standard errors, and portrays exactly the same pattern of precision and of significance. The symmetrical arrangement of axes is appropriate when both yields are of roughly equal interest. The squeezing transformation changes the relative pattern of the four means slightly, but the main benefit is in basing the interpretation on equal significance of equal distances, in any direction.

The appropriate angle between the two axes to give circular S.E.s should be ϕ, where $\cos(\phi)$ is calculated from the error mean squares (V_1 and V_2) and the error mean product (V_{12});

$$\cos(\phi) = V_{12}/\sqrt{(V_1 \times V_2)}$$

For the millet/sorghum data this gives

$$- 0.153/\sqrt{[(0.451)(0.418)]} = 0.353$$

Details of the construction of the skew axes diagram are given in Dear and Mead (1983).

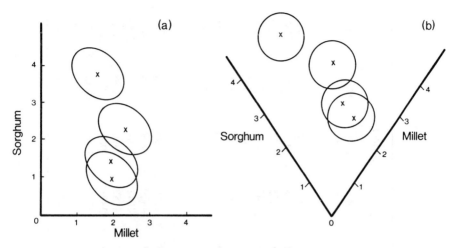

Fig. 15.8 Representation of joint precision for four treatment pairs of mean yields of sorghum and millet: (a) elliptical confidence regions for rectangular axes; (b) circular confidence regions for skew axes.

The differences between the treatment mean pairs in the symmetric skew axes diagram may be interpreted using 95% confidence regions shown in Fig. 15.8 in terms of four directions. A difference parallel to the x axis corresponds to a change in the x yield; a difference parallel to the y axis corresponds to a difference in the y yield. A difference in the vertical direction corresponds to a simultaneous increase for both yields. A horizontal difference corresponds to a change in the balance between the two yields. We shall use these interpretations in an example later.

Pearce and Gilliver also show how to calculate an F-test for joint variation of the two variables. The F-test for overall variation between treatments for the millet/sorghum data is calculated from the sums of squares and products for (Treatments + Error) and Error

$$L = [(5.228)(20.092) - (-3.754)^2]/[(4.056)(3.766) - (-1.371)^2] = 6.79$$

$$F = (\sqrt{2} - 1)(9/3)4.818 \text{ on 6 and 16 d.f.}$$

Since the 1% point of $F(6, 16)$ is 4.2 the treatment variation is significant at the 1% level. The skew axes graph shows that the dominant difference between the results for the sorghum varieties is, not surprisingly, in the sorghum yields. The difference between varieties 1 and 2 is negligible; variety 4 gives the highest sorghum yield but the comparison with variety 3 suggests that there is some compensatory loss of millet yield. Variety 4 is clearly superior to varieties 1 and 2, and the difference between varieties 4 and 3 is also significant at the 5% level.

Table 15.1 Yield data (kg/ha) for cowpea and maize intercropping in Nigeria

Cowpea variety	Nitrogen level	Maize variety	Block 1			Block 2			Block 3		
			I	II	III	I	II	III	I	II	III
						Cowpea					
A	N_0		259	645	470	523	540	380	585	455	484
A	N_1		614	470	753	408	321	448	427	305	387
A	N_2		355	570	435	311	457	435	361	586	208
A	N_3		609	837	671	459	483	447	416	357	324
B	N_0		601	707	879	403	308	715	590	490	676
B	N_1		627	470	657	351	469	602	527	321	447
B	N_2		608	590	765	425	262	612	259	263	526
B	N_3		369	499	506	272	421	280	304	295	357
						Maize					
A	N_0		2121	2675	3162	2254	3628	4069	2395	2975	4576
A	N_1		3055	3262	3749	3989	3989	4429	4429	4135	4429
A	N_2		3922	3955	4095	4642	4135	4642	5589	4429	5156
A	N_3		4129	4129	4022	3975	4789	4282	5990	5336	5663
B	N_0		2535	2535	2288	4209	3989	2321	2901	4429	3482
B	N_1		2675	3402	3122	4789	4936	3342	3555	4936	4135
B	N_2		3855	3815	3535	5083	4496	3702	6023	5296	4069
B	N_3		3815	4202	3749	5656	5576	5223	5516	5083	5369

Example 15.2

The yield data in kg/ha for an intercropping experiment on cowpea and maize in Nigeria are given in Table 15.1. There were three treatment factors, 2 cowpea varieties, 3 maize varieties and 4 nitrogen levels and the design consisted of three randomized complete blocks of the 24 treatment combinations.

The bivariate analysis of variance in Table 15.2 shows large F ratios for maize varieties and nitrogen levels with some indication that there is also an interaction effect for cowpea variety by nitrogen. The three skew axis diagrams for these three important effects provide considerable insight into the form of these significant effects, and they are shown in Fig. 15.9.

The differences between the mean pairs of yields for the three maize varieties are principally in the direction of changing the balance between the yields of the two crops; variety 1 produces less maize yield than varieties 2 and 3, the loss being compensated by more cowpea yield; the difference between varieties 2 and 3 is negligible.

The effect of increasing nitrogen level is, as might be expected, to increase the maize yield, though we can also notice a small degree of compensation in a reduction of cowpea yield as maize yield increases. Perhaps the most interesting

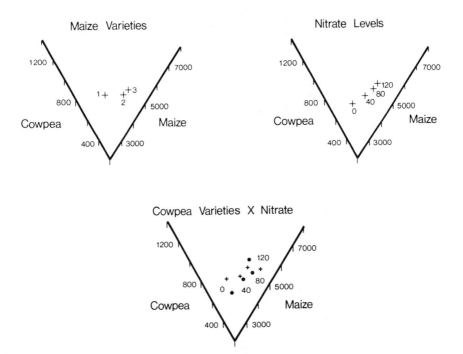

Fig. 15.9 Skew-axis representation of treatment effects on pairs of cowpea and maize yields; (a) Maize variety effects; (b) Nitrate effects; (c) combinations of Cowpea variety and Nitrate.

Table 15.2

Source	d.f.	Maize S.S. (X_1)	Cowpea S.S. (X_2)	S.S. for ($X_1 + X_2$)	Sum of products	F	Correlation
Blocks	2	0.29	0.0730	0.0247	−0.058	1.75	−0.40
M variety	2	17.52	0.4094	12.665	−2.632	11.90	−0.98
C variety	1	0.03	0.0060	0.062	0.013	0.44	1.00
Nitrogen	3	28.30	0.1131	25.081	−1.766	10.59	−0.98
M × C	2	1.11	0.0099	0.922	−0.099	0.82	−0.95
M × N	6	1.25	0.0676	0.920	−0.199	0.64	0.93
C × N	3	0.24	0.1724	0.152	−0.130	2.40	−0.64
M × C × N	6	1.28	0.1354	1.349	−0.03	1.40	−0.08
Error	46	15.90	0.5993	13.671	−1.414		−0.466
(MS)		(0.346)	(0.0130)		(−0.031)		
Total	71	66.13	1.5861	55.080	−6.318		

pattern comes when we interpret the form of the interaction between nitrogen and cowpea variety. For cowpea variety A(circles) the trend with increasing nitrogen is purely an increase in maize yield, whereas for cowpea variety B(+) the effect of increasing nitrogen is predominantly to change the balance between the two yields, the gain in maize yield being compensated by reduced cowpea yield.

15.5 INVESTIGATING RELATIONSHIPS WITH EXPERIMENTAL DATA

The bivariate analysis in the previous section enables us to assess the effects of treatments on two variables, allowing for the interrelationship between the variables. In that analysis we estimate the relationship between the variables at the random error level and use that relationship to examine the way different treatments produce patterns of variation in both variables, including the relationship between variables at the treatment level. Our primary interest is the effects of treatments rather than the relationships between the variables.

In other situations the relationship itself is of interest, either as a subsidiary concern to the treatment comparison, or even as the primary interest of the comparison, the treatments being utilized to provide a wide range of conditions. For example, the latter would be the case if the primary interest is in allometric relationships between the weight of the part of an animal and the whole animal weight. This could be investigated by measuring a random sample of animals but a more comprehensive assessment could be arranged by deliberately choosing animals from several different breeds, at several different ages, possibly with different dietary regimes. We could then assess whether the allometric relationship within breeds was the same as that between breeds, or between ages, or even whether it was invariant over age–breed combinations.

The analysis of variance for a single variable divides the total variation into components corresponding to different sources or causes of variation. We normally expect different amounts of variability in the different components of the analysis because we know that these components represent different forms of change of the experimental units. The same arguments should be relevant to the consideration of the analysis of variation of relationships. We have information about the relationship between variables at each level of the analysis of variance structure. For viable comparisons of relationships at different levels we need more observations than for comparing the variation of a single variable. A single treatment degree of freedom provides adequate information about a treatment difference, relative to the error variation (provided the error s.s. is based on at least 10 d.f.). A single treatment d.f. for a relationship between two variables provides no information because the relationship based on two points must fit exactly, and to ask how well an inevitably exact model fits is pointless. In fact we would not usually hope to get much information about a relationship from a random sample of less than about eight observations (seven d.f.), though if we are allowed to choose the values for the 'independent variable' fewer points could be adequate.

Questions of which components of an analysis of variance are appropriate for estimating a relationship are not trivial, and it is very important to take account of the structure when using data from a structured experiment to examinine relationships. Consider the following simple example using artificial data from an experiment to compare four nitrogen levels in four randomized blocks.

		Blocks			
Nitrogen	1	2	3	4	Mean
0	7	3	9	5	6.0
100	6	11	18	7	10.5
200	9	12	18	17	14.0
300	9	15	14	16	13.5

This data could be analysed to provide information about the relationship between yield and nitrogen level in two ways, and these are illustrated in Fig. 15.10. We could simply use the four treatment mean values or we could use the complete data set of 16 values. Many people's intuitive feeling is that the whole data must provide the better information; this is totally wrong.

The plots in Fig. 15.10 suggest that the means provide much clearer information, the whole data producing such a variable cloud of points that it might even be doubted whether there is any real relationship. A quadratic model for the yield response to nitrogen explains almost all of the treatment variation

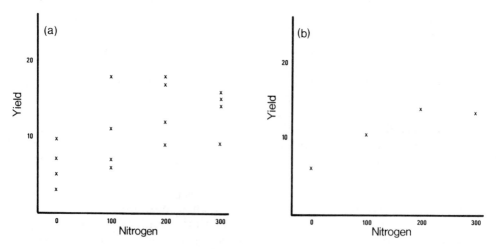

Fig. 15.10 Relationship between yield and nitrogen application; (a) individual plot values; (b) treatment mean values.

and the analysis of variance for the design including the regression can be written:

Source	Regression s.s.	Residual s.s.	Total s.s.	Ratio of m.s. (Reg./Res.)
Treatments	162(2 d.f.)	2(1 d.f.)	164(3 d.f.)	40.5
Blocks	0	88(3 d.f.)	88(3 d.f.)	not defined
Error	0	104(8 d.f.)	104(9 d.f.)	not defined
Total	162(2 d.f.)	192(13 d.f.)	354(15 d.f.)	5.5

The regression sums of squares for blocks and error are zero because neither component of variation provides any information about the yield–nitrogen relationship. This should be obvious for blocks, each of which includes all four nitrogen levels; for error, remember that error variation is that which remains after eliminating all treatment effects. It should now be clear that the regression from the total set of 16 observations fails to give good information about the regression because it includes additional (and irrelevant) residual variation, thereby inflating the residual variation and masking the real relationship, which is clearly visible in the subdivision of the treatment variation. In the plot of all 16 points in Fig. 15.10(a) the vertical variation for each treatment includes the substantial systematic differences between blocks. If the block differences were even larger the same differences between the nitrogen levels would appear less and less significant compared with the range of variation of yields for each nitrogen level, whereas in fact the evidence for the trend with nitrogen would remain unchanged.

This situation, where there is positively no information about the relationship in two components of the analysis, is an extreme case which will usually occur only in designed experiments where the levels of the explanatory variable are the applied treatments. More generally we will have different amounts of information about the relationship in the different components of the analysis of variance.

For a more substantial example of regression analysis within an analysis of variance for a designed experiment, we consider the cowpea/maize date from the intercropping experiment in example 15.2 and suppose that we are interested in the relationship between the two yields. The simple form of the bivariate analysis is:

Source	d.f.	Maize S.S.	S.P.	Cowpea S.S.
Blocks	2	0.29	− 0.058	0.0730
Treatments	23	49.93	− 4.849	0.9138
Error	46	15.90	− 1.414	0.5993
Total	71	66.13	− 6.318	1.5861

There is insufficient information (2 d.f.) about the variation between blocks to assess the variation of the regression relationship in that component of the analysis of variance, but we can calculate the regression of maize yield on cowpea yield (or *vice versa*) separately for the treatment and error components as well as for the total:

Treatments

$$\text{Regression coefficient} = -4.849/0.9138 = -5.30$$

$$\text{Regression s.s.} = (-4.849)^2/0.9138 = 25.73$$

$$\text{Residual s.s.} \quad = 39.93 - 25.73 \quad = 24.20$$

Error

$$\text{Regression coefficient} = 1.414/0.5993 = -2.36$$

$$\text{Regression s.s.} = (-1.414)^2/0.5993 = 3.34$$

$$\text{Residual s.s.} \quad = 15.90 - 3.34 \quad = 12.56$$

Total

$$\text{Regression coefficient} = 6.263/1.5131 = -4.14$$

$$\text{Regression s.s.} = (-6.263)^2/1.5131 = 25.92$$

$$\text{Residual s.s.} \quad = 65.86 - 25.92 \quad = 39.94$$

To compare the regression coefficients for the treatments and error components and to assess whether they are convincingly different we could use the methods of analysis of regressions from Chapter 11. The analysis of residual sums of squares is

	Residuals s.s.	d.f.	Mean square
Treatments	24.20	22	1.10
Error	12.56	45	0.28
Total	39.94	68	
Sum of (T + E)RSS	36.76	67	
Difference	3.18	1	3.18

We cannot pool the residual mean squares for the treatment and error regressions because we know, from our earlier analysis, that the treatment variation is clearly greater than the error variation. However, it is clear that the difference mean square is large compared with either residual mean square.

The formal conclusion from this analysis is therefore that the regression of maize yield on cowpea yield shows a larger regression coefficient for the variation between treatments than for the error variation. For the random, error, level of variation the plots with more cowpea yield produce less maize yield with an 'exchange rate' of about two and a half less units of maize yield for every extra unit of cowpea yield. The corresponding exchange rate estimated from variation between different treatment plots is more than five units of maize yield less for every extra unit of cowpea yield. We should try to find a biological explanation of why treatment variation shows a higher exchange rate.

For a final example of the comparison of regressions in different components of an analysis of variance we consider the relationship between the two leaf chemicals, discussed earlier in this chapter. The bivariate analysis of variance, splitting the variation within one plant into that between the four leaves per plant, the leaflets within each leaf cluster, and the samples within each leaflet is shown below with the regression summary.

	d.f.	S.S.(X_1)	S.P.	S.S(X_2)	b	Residual s.s.	m.s.
Between leaves	3	7.47	-9.56	12.37	-0.77	0.18 (2 d.f.)	0.09
Between leaflets	8	1.60	-2.67	4.97	-0.54	0.17 (7 d.f.)	0.02
Between samples	12	1.59	-1.74	2.22	-0.78	0.23 (11 d.f.)	0.02
Total	23	10.66	-13.97	19.56	-0.71	0.68 (22 d.f.)	0.03

The difference between the separate residual ss for the three components and that for the total variation is

$$0.68 - (0.18 + 0.17 + 0.23) = 0.10 \ (2 \ \text{d.f.})$$

with a mean square of 0.05, which is larger than the separate residual mean squares but, given the rather small d.f., not convincingly larger. Repeating the analyses for sets of data for the three other plants gives similar results. The regressions appear consistent between different levels of plant variation for each plant.

A combined analysis gives

Between leaves	12	24.36	-35.04	54.14	-0.65	1.68 (11 d.f.)	0.15
Between leaflets	32	16.82	-21.97	37.02	-0.59	3.78 (31 d.f.)	0.12
Between samples	48	15.42	-17.00	24.54	-0.69	3.64 (47 d.f.)	0.08
Total	92	56.60	-74.01	115.70	-0.64	9.26 (91 d.f.)	0.10

The difference between the sum of the separate residual ss and that for the total variation is

$$9.26 - (1.68 + 3.78 + 3.64) = 0.16 \ (2 \ \text{d.f.})$$

A summary of the residual sums of squares is

	Residual s.s.	d.f.	Residual m.s.
Between leaves	1.68	11	0.15
Between leaflets	3.78	31	0.12
Between samples	3.64	47	0.08
Total	9.26	91	
Difference	0.16	2	0.08

There is no evidence for variation in the relationships between the two chemicals at different levels of the plant sampling. The relationship seems to be very consistent.

There are other ways in which the data can be split and other comparisons of regressions which could be made. In the same way as in Chapter 11 we could compare the regressions for different plants, using all the sample information for each plant.

	d.f.	S.S.(X_1)	S.P.	S.S(X_2)	b	Res s.s.
Plant 1	23	10.66	− 13.97	19.56	− 0.71	0.68 (22 d.f.)
Plant 2	23	24.71	− 34.79	58.83	− 0.59	4.14 (22 d.f.)
Plant 3	23	14.12	− 17.12	24.72	− 0.69	0.83 (22 d.f.)
Plant 4	23	7.29	− 9.02	12.59	− 0.72	0.83 (22 d.f.)
Overall	92	56.60	− 74.01	115.70	− 0.64	9.26 (91 d.f.)

The analysis of residual s.s. is

	Residual s.s.	d.f.	Residual m.s.
Plant 1	0.68	22	0.03
Plant 2	4.14	22	0.19
Plant 3	2.26	22	0.10
Plant 4	0.83	22	0.04
Sum	7.91	88	0.09
Total	9.26	91	
Difference	1.35	3	0.45

Although the residual mean squares from the separate plant regressions vary quite widely, the size of the difference mean square is sufficiently large to make it difficult to believe that the regressions for the four plants are consistent. The conclusions therefore are that although the regression relationship appears to be consistent between the different levels within each plant there appear to be differences between the regressions for the four plants.

16

Choosing the most appropriate experimental design

16.1 THE COMPONENTS OF DESIGN; UNITS AND TREATMENTS

In the earlier chapters of this book we have discussed the analysis of data from various experimental designs. The merits and deficiencies of these designs were discussed in a piecemeal fashion without a consistent development of the underlying principles of experimental design. In this chapter we shall review the important overall concepts of experimental design, placing the various designs previously considered in context and indicating where more complex designs not previously discussed, are appropriate. Two concepts which we have already discussed at some length are control of variation by blocking (Chapter 5) and factorial structure for treatments (particularly in Chapter 7). We shall not add much to our earlier discussion of these topics but we must emphasize that these two concepts are extremely important in designing experiments.

It is essential to distinguish two components of an experimental design, the structure of the experimental units, and the structure of the treatments. Although there are areas of design philosophy where the two components must be considered together, a proper understanding of experimental design requires that each component is first understood separately. Understanding the structure of the experimental units requires an identification of the units and a description of the likely patterns of variation between them if a single treatment were applied to all the units. The important ideas about units are those of replication, blocking and randomization.

Treatment structure is concerned with the choice of different treatments to be included in the set of experimental treatments. This choice must be determined by the objectives of the experiment but, even when there are very precisely formulated objectives, there are important statistical considerations in choosing the treatments. Important matters to be discussed later in this chapter include the distinction between qualitative and quantitative factor levels, and screening or selection experiments that often include a very large number of treatments.

16.2 REPLICATION AND PRECISION

Let us go back to square one! The reason for doing an experiment is to examine differences between the effects of two or more different treatments. Here treatment will be used in its widest possible sense to include not only applied treatments such as different concentrations of an applied chemical, but also inherent treatments such as different strains or varieties of crop; sex or age of an animal, or different histories of the experimental units. In order to compare two treatments, and the same arguments apply for more than two treatments, we must obviously observe the effects of the treatments on different experimental units, plots of land, groups of plants or animals, or individual plants or animals. The need for a formal experiment aries from the knowledge that if we have two units treated in apparently identical fashion then there will be differences between the units in respect of any quantity which we care to consider. Hence to interpret the difference in any particular measurement between two units with different treatments we need to have some idea of the variation we would have expected to have found had the two units been treated identically. The logical solution is to have several units with the first treatment and several more with the second treatment. By contrasting the average difference between the two sets of units treated differently with the variation between similarly treated units we have the basis for a judgement on the extent to which the difference between the units with different treatments can be ascribed to differences between the effects of the treatments. Hence by having several observations, or replications, for each treatment we have a viable experiment.

How many units should we have per treatment? This question of replication per treatment is probably the one most frequently put to statisticians and as it stands there is no answer. Before we can decide how many units per tretment and thus what precision we require for a treatment mean it is necessary to have some idea of the size of the difference between treatments that the experiment is required to detect. An experiment which detects as significant differences of 20% in the means for different treatments is appropriate if this is the level of difference the experimenter is interested in; but if the experimenter is trying to detect differences of 2% then the experiment is worthless; and if his interest is in differences of 50% then the experiment has been wasteful of resources because a smaller experiment would have been sufficient.

So, in deciding the size of an experiment we need to compare the size of difference which is important to the scientist with the variability per unit which might be expected in the experiment. Suppose the experimenter wishes to detect any differences between the effects of two treatments as large as d units of yield. Suppose also that a reasonable guess based on previous experiments with similar experimental material suggests that the variance per unit should be about σ^2. Then, using n plots per treatment, the standard error of a difference between treatment means should be about $\sqrt{(2\sigma^2/n)}$. Now an estimated difference of twice the standard error will be significant at about the 5% level. But the estimate of the treatment difference, whose real value is d, may well be less than d. Therefore to allow for this possibility a reasonable rule to use is that an estimated difference, whose true value is three times the standard error, will 5 times out of 6 be significant at the 5% level.

For example, it is known that in experiments with cabbages using 25 plants per plot the standard deviation per plot is about 10% of the mean, $\sigma = m/10$.

With four blocks in a randomized block experiment the s.e. of a difference between means will be

$$\sqrt{\left[2\left(\frac{m}{10}\right)^2 / 4\right]} = \sqrt{\left[\left(\frac{m}{10}\right)^2 / 2\right]} = 0.07\,m$$

Thus a true difference of 21% of the mean value will be detected as significant at the 5% level, 5 times out of 6.

To detect a true difference of 15% what value of n do we need? The standard error is

$$\sqrt{\left[2\left(\frac{m}{10}\right)^2 / n\right]}$$

This must be less than one third of $15m/100$. Thus

$$3\sqrt{\left[2\left(\frac{m}{10}\right)^2 / n\right]} < 0.15\,m \qquad 9\left[2\left(\frac{m^2}{10}\right)/n\right] < (0.15)^2 w^2$$

Hence

$$18\left(\frac{m}{10}\right)^2 / n < (0.15)^2\,m^2 \qquad 18/$$

therefore

$$18\left(\frac{m^2}{10}\right) < n(0.15)^2 m^2$$
$$n > 18(0.1)^2 / (0.15)^2 = 8$$

Thus at least 8 blocks will be required to have a good chance of detecing a true difference of 15% of the mean as significant at the 5% level.

In this argument and in most of the experimental designs which we have considered previously it is assumed that all comparisons of treatments or of factor levels are of equal interest. However, this is not necessarily true and, in defining his experiment, an experimenter should think about the relative precision which he wants for different treatment comparisons. The most common situation when some comparisons should be more precise than other is when a control treatment is included in the experiment and part of the interest is in the comparisons among various new treatments, and part in the comparison of each new treatment with the control. We have already considered, in section 6.1, an example in which the control treatment was applied to two units in each block, the other treatments to one unit in each block.

For a given total number of units we may, by replicating one treatment several times in a block, while at the same time using fewer blocks, be able to get different standard errors according to our interest in the various different comparisons. To illustrate this possibility, suppose we are comparing five treatments with a control, and that 24 plots are available with no clearly defined blocks. Two possible designs are

(i) Four blocks of six plots each with each treatment and the control represented once in each block,

(ii) Three blocks of eight plots each with the control applied to three plots in each block.

The standard errors for (a) comparing two treatments, and (b) a treatment with the control are as follows:

(a) Two treatments

$\quad$ (i) $\sqrt{(2\sigma^2/4)} = 0.707\sigma$ $\qquad\qquad$ (ii) $\sqrt{(2\sigma^2/3)} = 0.816\sigma$

(b) Treatment and control

$\quad$ (i) $\sqrt{(2\sigma^2/4)} = 0.707\sigma$ $\qquad\qquad$ (ii) $\sqrt{[\sigma^2(\frac{1}{3} + \frac{1}{9})]} = 0.667\sigma$

If the comparison of treatment and control is sufficiently important then the second design should be chosen.

There is another reason why we may choose to have unequal replication of treatments and this is when there is a shortage of material for one or more treatments. Consider again the variance of a difference between two treatment means based on n_1 and n_2 observations respectively:

$$\text{Var} (\bar{x}_1 - \bar{x}_2) = \sigma^2(1/n_1 + 1/n_2)$$

If the total number of observations $n = n_1 + n_2$ is fixed then this variance is made as small as possible by choosing $n_1 = n_2$. But if n_1 is fixed then we can make the variance smaller by increasing n_2. This may occur for example in experiments to compare different breeding lines. Often for some lines there may be only enough seed material for a small number of plots, in extreme cases possibly only one plot. There is no reason not to include such lines in the experiment. Comparisons with control varieties will be less precise than for lines with more replicates, but may still be informative.

Two other objections which might be raised are the difficulty of the analysis if the seed material is insufficient for the inclusion of some varieties in all blocks, and the effects of this loss of replication for some varieties on the estimation of σ^2. On the first question the availability of computer packages for general linear models (section 11.2) means that there is no difficulty in obtaining a valid analysis and comparison of lines even if the blocks contain different sets of treatments. On the question of replication to estimate σ^2, we have assumed in all the analyses of experimental designs that we can pool estimates of variance because we assume that the variance of observations is the same for each treatment. If this assumption is true it does not matter which treatment replications provide the estimate of σ^2; if the assumption is not true then we must consider other methods of analysis, as discussed in Chapter 8.

16.3 DIFFERENT LEVELS OF VARIATION AND WITHIN-UNIT REPLICATION

A major difficulty which can arise in considering replication is to determine the correct level of replication to use as the basis of particular treatment comparisons. We have already met, in the split plot designs of section 7.6, a situation where some treatments are applied to large units and other treatments to small units, which then leads to different standard errors for different treatment comparisons. The general principle, of which the split plot analysis is a particular example, is that the appropriate level of replication, from which the variance to compare treatments should be derived, is the replication of the experimental units to which the treatments are applied.

For example, if an experimental treatment is applied to a plot of land on which a number of wheat plants are growing then treatments can be compared only in terms of the total yields from different plots. Measurement of weights or yields of individual plants provides information about the variation between plants within a plot, and it may be of interest to see how this variation is affected by different treatments, but this variation is not the relevant variation for the comparison of treatment mean yields. Within-plot variation of plant yields may give an overestimate of plot-to-plot variation, because of interplant competition, or it may give an underestimate of plot-to-plot variation, because of a very homogeneous within-plot environment. The crucial point is that we have no way of telling whether the within-plot variation is the same as between-plot variation and therefore we cannot use within-plot replication to compare treatments applied to different plots.

The same argument applies to measurements of individual animals when a treatment is applied to whole litters, or to measurements of individual insects in a cage when the treatment is applied to the whole cage, or to individual seeds when a treatment is applied to a batch of seeds. In each case there may be particular arguments that the within-unit replication must, on biological grounds, be equivalent to the between-unit replication, and in some cases an experimenter may decide to use within-unit replication as the basis for treatment comparisons. The experimenter must then recognize the risk that the results of such a procedure will be open to doubt and that an act of faith is required on the part of the experimenter which may not be shared by others evaluating the experimenter's results.

To illustrate quantitatively the difficulties of dealing with variation at more than one level, we shall consider a method of analysis that has been used occasionally in soil analysis. To compare different soils in respect of some

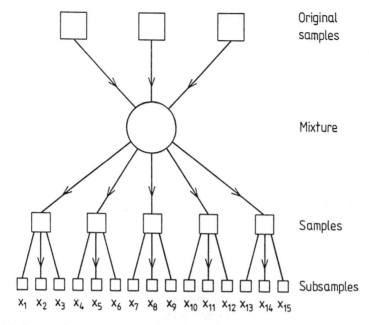

Fig. 16.1 A possible sampling procedure for estimating characteristics of soil, which produces spurious accuracy.

variable, or to estimate the population mean value for a single soil the following type of sampling procedure is proposed (illustrated in Fig. 16.1).

(a) Several samples (say three) of the soil are obtained, the object of using several samples being to get a representative estimate.
(b) The samples are mixed.
(c) A number of samples (say five) are taken from the mixture and for each sample a separate solution is produced.
(d) Subsamples (say three) are now taken from each sample solution and the variable of interest measured.
(e) The mean and variance of the 15 values are calculated and the standard error of the mean obtained from the variance.

At first sight the procedure may appear sensible. Certainly the estimate of the soil characteristic provided by the mean of the 15 values should be very precise. But how precise? The variance from which the standard error is calculated is a measure of the variation due to (i) the mixing technique and (ii) the technique of determination of the value of the variable. It gives no real information on the real variation of this variable in the soil of which the original three samples are representative and consequently we do not have a measure of the precision of our estimate.

Suppose the distribution of the variable in the original soil has mean 10 and standard deviation 2. This distribution can be thought of as the distribution of values for all possible samples of soil, each sample being entirely converted into a solution and the values of the variable averaged over all possible samples from the solution. Suppose further that, after mixing, the distribution of values in the samples taken from the mixture has a standard deviation of 0.2 and a mean equal to the mean of the values for the original three samples. Finally suppose the distribution of values between subsamples of the solution has standard deviation 0.1. Using random values from a normal distribution at each stage of sampling we can model the behaviour of the system.

Example 16.1   Simulation of a soil sampling system

The three random values of the variable in the origional samples are 8.46, 10.96 and 8.10 (the standard deviation being 2, these are not unusually inconsistent values). The mean of these three values is 9.17.

The values of the variables in the five samples (random values added to 9.17) are 9.25, 9.37, 9.26, 9.50 and 9.02. The values obtained from the subsamples (random values added to the five sample means) are

Sample 1:	9.39	9.21	9.14
Sample 2:	9.16	9.29	9.39
Sample 3:	9.26	9.29	9.13
Sample 4:	9.48	9.66	9.34
Sample 5:	9.04	8.95	9.11

The mean is 9.26, the variance 0.0329, and the standard error of the mean 0.047. This standard error is very small and implies that the mean of 9.26 is highly precise. However, since the true mean value is known to be 10 the apparent precision is misleading.

The results for two other simulations of the system are given in Table 16.1 and the three sets of results are represented diagrammatically in Fig. 16.2. In fact the three overall mean values derived from the initial sets of three samples of soil are rather closer together than would be expected. Nevertheless, the conclusions reached from comparing the three means of 9.26, 10.42 and 9.79 using their standard errors 0.047, 0.051 and 0.039 would of course be that the three means were highly significantly different and that the three sets of soil samples were clearly different whereas we know that the sets of three samples each were taken from the same distribution.

This example shows clearly the danger of a spurious accuracy and the corresponding need to ensure that, when we compare a difference with its standard error, the standard error is appropriate to the difference. In other words we must ensure that we are working at the correct level of variability.

Table 16.1 Simulation of soil sample analysis

Run 2:
Three initial sample values: 12.96   8.46   10.16.   Mean = 10.53
Five sample values from mixture: 10.60   10.47   10.20   10.27   10.60
Three subsample values from solution for each sample:

Sample 1 = 10.67   10.56   10.71
Sample 2 = 10.28   10.30   10.55
Sample 3 = 10.32   10.06   10.31
Sample 4 = 10.61   10.37   10.27
Sample 5 = 10.61   10.56   10.62

Mean (of 15 values) = 10.42; variance = 0.0398; s.e. of mean = 0.051

Run 3:
Three initial sample value: 11.36   9.22   8.78   Mean = 9.79
Five sample values from mixture: 9.71   9.95   9.81   9.86
Three subsample values from solution for each sample:

Sample 1 = 9.86   9.58   9.69
Sample 2 = 9.82   10.02   9.94
Sample 3 = 9.78   9.64   9.47
Sample 4 = 9.97   9.81   9.85
Sample 5 = 9.83   9.88   9.68

Mean (of 15 values) = 9.79; variance = 0.0228; s.e of mean = 0.039

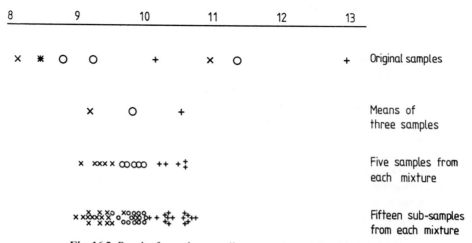

Fig. 16.2 Results from the sampling procedure defined in Fig. 16.1.

16.4 VARIANCE COMPONENTS AND SPLIT PLOT DESIGNS

Providing there is appropriate random sampling at each level, the contribution of each level of variation to the standard error of the final estimate can be predicted. This allows an assessment of the amount of replication that would be desirable at each level of variation. For example, in a field experiment the effects of different treatments on the amounts of nitrogen in the leaves of plants are to be investigated. In each experimental plot there will be variation between plants, between leaves within plants, between sample areas cut from each leaf and finally, between replicate chemical determinations made on each sample area. Writing σ_p^2, σ_l^2, σ_s^2 and σ_d^2 for the variances, called variance components, associated with these different levels of variation with p, l, s and d representing the number of replications at each level, the standard error of the mean amount of nitrogen estimated for a plot can be shown to be

$$\sqrt{\left(\frac{\sigma_p^2}{p} + \frac{\sigma_l^2}{lp} + \frac{\sigma_s^2}{slp} + \frac{\sigma_d^2}{slpd}\right)} \tag{1}$$

Since p, the number of plants, provides replication at all levels, the obvious scheme to use for a fixed number of determinations would be as many plants as possible with one leaf per plant, one sample area per leaf one determination per sample area. In practice, the efforts and costs involved in sampling at the different levels are not equal. Increasing the number of plants is, for example, probably more expensive than increasing the number of leaves sampled per plant. Also, single replication at any level makes it impossible to estimate the variance component at that level and this is often useful information for levels of desirable replication in future experimentation. The variance components are estimated from an hierarchical form of the analysis as shown in Table 16.2. The calculated means squares are set equal to the expected values shown in the Table and the resulting four equations solved to obtain estimates of σ_p^2, σ_l^2, σ_s^2 and σ_d^2. Clearly if any of p, l, d or s are equal to 1 then the corresponding variance component cannot be estimated separately from the variance component at the level above. A frequent compromise that results in a large number of

Table 16.2

Source of Variation	d.f.	Expectation of M.S.
Between plants	$p - 1$	$\sigma_d^2 + d\sigma_s^2 + ds\sigma_l^2 + dsl\sigma_p^2$
Between leaves within plants	$p(1 - 1)$	$\sigma_d^2 + d\sigma_s^2 + ds\sigma_l^2$
Between sample areas within leaves	$pl(s - 1)$	$\sigma_d^2 + d\sigma_s^2$
Between chemical determinations	$pls(d - 1)$	σ_d^2
	$plsd - 1$	

plants but allows the estimation of all variance components, and avoids problems from occasional missing values is to take p as large as possible but set $1 = s = d = 2$.

We have already seen the effects of different levels of variation in assessing treatment differences when discussing split plot experiments (sections 7.6, 7.7). The special feature of split plot designs is that different groups of treatments are deliberately applied to units of different sizes and levels of variation. There are also links with matters to be discussed in the next chapter on sampling from finite populations. The main distinctions are that here we are assuming that we are sampling at each level from a real or hypothetically very large population of units and, for example, that the amount of variation between leaves is the same for all plants.

We have discussed ways of predicting the values of the variance components from previous experiments and of choosing the best values for p, l, s and d to achieve acceptably low values of the standard error (1) for the plot mean. This is only a component, and often a small component, of the standard error that will eventually be used to compare treatment means. The standard errors for comparing treatment means will be derived in the usual way from an analysis of variance of the individual plot values and will include both plot-to-plot variation and any lack of repeatability of treatment effects from block-to-block. Standard errors representing the sampling variation within plots should be made as small as possible but cannot be used to compare treatment means that are bound to have been calculated from measurements from different plots.

Consider again the split plot design where we have two different levels of variation because we have treatments applied to experimental units of two different sizes. The only real justification for using a split plot design is that of practical necessity because of the impossibility of using small plots for some of the treatments. Sometimes the use of split plot designs is advocated for other reasons; it is argued that in a split plot design, the split error mean square has a smaller expectation than the main plot error mean square which includes an extra variance component. Hence, as compared with a fully randomized block design, comparisons of split plot treatments and interactions are likely to be more precise, while comparisons of main plot treatments are likely to be less precise, and that if interest is primarily in the former than the split plot design is preferable. This argument is an oversimplification for two reasons. First, the loss of precision on the main plot comparison is greater than the gain of precision on the split plot comparisons. And second, in interpreting interaction effects by comparing combinations of main plot and split plot treatments, many comparisons involve different main plots and are therefore reduced in precision.

We believe that split plot designs should be avoided except where practically necessary. However, there are many situations where they are necessary and where even more complex designs involving three levels of units may have to be considered. For example, if in an ordinary split plot design it is desired to harvest plants at several different harvest dates, it might be sensible to split

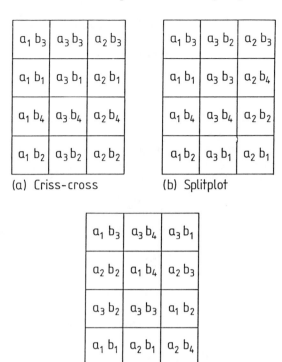

(a) Criss-cross   (b) Splitplot

(c) Fully randomized
factorial

Fig. 16.3 Different designs for a 4 × 3 factorial set of treatments; (a) criss-cross, (b) split plot, (c) fully randomized within each block.

each plot into a number of split plots, each split plot to be harvested at a different harvest date.

Another variation on the split plot theme which is occasionally useful is the criss-cross design in which each block is divided into plots in one direction for one treatment factor, and again into plots in the opposite direction for a second treatment factor. Such a design is shown in Fig. 16.3 with a simple split plot design and the even simpler randomized block for comparison. A criss-cross design would be appropriate with two treatment factors and the levels of each factor needing long thin plots. An example is an experiment comparing different cultivation treatments and different management techniques. Except when this form of practical need leads to the use of a criss-cross design, the design should be avoided, because the randomization procedure and analysis is complex and inefficient. The analysis of variance is split into three parts, a main plot analysis for each main effect and a split plot analysis for the interaction. Each part has its own error mean square and these often have rather few d.f. and consequently give poor precision. If the design is used then the levels of each treatment factor are randomized separately, the two separate randomizations completely determining the allocation of the treatment combinations; hence there can be no independent randomization at the split plot level.

16.5 RANDOMIZATION

We have discussed randomization frequently and have justified it by arguing that any other method of allocating treatments to units is liable to subjective bias. Subjective bias can lead to biases in estimating the effects of treatments and in estimating the standard errors for comparing treatment means. There is a second justification which, although its acceptance can lead to tedious computation, is important and so we shall outline the philosophy of this justification briefly.

When discussing the analysis of experimental data we have made the assumption that the errors in the observations are a random sample from a normally distributed population. The assumption that there is a large population of experimental units of which our set are a random sample is not always easy to justify. For example, if we are feeding cows to investigate the digestibility of different diets, then the population we are considering is presumably a large group of cows, perhaps defined by breed. The experimental animals have to be a random sample of the breed if the conclusions based on the results are to apply to the larger population. In fact the experimental animals are generally a particular group available to the experimenter and not a sample drawn at random from a large population. Also the cows are different from other cows simply from being in an experiment. For both these reasons inferences beyond the set of animals in the experiment cannot be firmly based on a statistical argument that assumes that they are truly a random sample. Similarly, any tightly controlled set of plant growth conditions is extremely difficult to imagine as a random sample of normal environments in which the plants are or will be grown. In summary, the wide applicability of the conclusions from an experiment depends on the assumption that the experimental units are, in some sense, a random sample of units from the wider population.

The device of randomization ensures that the results obtained from a particular experiment are indeed a random sample but only in the sense that a different randomization of treatment allocation would have lead to a different set of results from the same set of experimental units. This allows us to make less useful but totally valid inferences about the effects of the treatments on the units actually used in the experiment. To illustrate this consider the following example.

To compare two different forms of trap for small mammals, four traps of each type are placed in a wood and the numbers of animals caught in each trap over a three-week period recorded. The results are

| Type A | 18 | 23 | 19 | 15 |
| Type B | 11 | 14 | 14 | 12 |

The choice of the four sites for the traps of type A, out of the eight possible sites for traps, was made randomly. Thus the apparent difference between the results for the two different traps must arise either by chance or because type A is more effective, most people would argue that there must be a difference between the trap types because the chance of type A traps getting all four highest results must be very small. This is the simplest form of an argument based purely on randomization. In fact if the eight results (11, 12, 14, 14, 15, 18, 19, 23) are allocated randomly, four to each of two groups, the chance that all four highest values will occur in a single group is 1 in 35 which is a reasonable justification for inferring that type A traps are more effective. The important point, however, is that this argument depends only on the eight traps. When the difference between two groups is less extreme than in the examples above, it is still possible to use the random allocation of treatments to argue the existence of a real effect. Thus if instead of type A traps getting the four highest results they had been first, second, third and fifth largest we should still be fairly convinced of the reality of the difference. But first, second, third and sixth would be less convincing and first, second, fourth and sixth much less convincing. Exact probabilities can, of course, be calculated to make these rather vague statements precise.

Working with the ranks, the order of the results, ignores the actual values of the catches. Thus the difference between 23 and 19 is treated as equivalent to the difference between 19 and 18. We can use the additional information from the numerical values by working with them rather than with the ranks. Thus, the actual numbers of animals caught were 11, 12, 14, 14, 15, 18, 19 and 23. The biggest possible differences that could have appeared between the total catch at two groups of four sites is

$$(18 + 23 + 19 + 15) - (11 + 14 + 14 + 12) = 24$$

which is the observed difference. The next biggest total differences are, in decreasing order:

$$(18 + 23 + 19 + 14) - (11 + 15 + 14 + 12) = 22$$
$$(18 + 23 + 19 + 14) - (11 + 14 + 15 + 12) = 22$$
$$(18 + 23 + 19 + 12) - (11 + 14 + 14 + 15) = 18$$
$$(14 + 23 + 19 + 15) - (11 + 18 + 14 + 12) = 16$$
$$(18 + 23 + 19 + 11) - (15 + 14 + 14 + 12) = 16$$

A reasonable procedure in assessing the significance of a particular observed result is to count how far down the list that result comes and hence what proportion of the possible results show stronger evidence for a real difference. The fact of random allocation allows us to consider the results actually obtained against the background of the results that could have been obtained if a different randomization had occurred and there is in reality no difference in effectiveness between the traps. Note particularly that there is no assumption about the numbers of animals caught being a random sample of the numbers that would have been obtained with a much larger number of traps of each type or over a larger time period. The statistical inferences are solely to the eight traps used and differences observed between the two types of trap over these particularly three weeks. Inferences beyond these boundaries are important but are considered to be non-statistical depending, as they do, on the knowledge of the scientist rather than on the data available.

The significance levels and associated confidence intervals based on randomization alone are generally smaller and larger respectively, than those based on an assumption that the results are a random sample of results from some large population with a defined distribution, such as the Normal distribution, for the characteristic being measured. The greater power and wider validity of the conclusions has to be set against the assumption of random sampling from a large population and, of course, the assumed form of the distribution chosen for the characteristic being measured. We generally choose to make these assumptions but the assumption-free inferences using only the randomization actually applied are often worth calculating to act as a cautionary note when the larger inferences are made but on the basis of the other assumptions.

16.6 MANAGING WITH LIMITED RESOURCES

The resources available to an experimenter are always limited and generally much less than those desired. The experimenter then has to decide whether to revise his objectives by, for example, studying the effects of only two factors by fixing the level of a third factor which was originally also to be varied in the experiment. Because of the 'hidden replication' in factorial experiments (section 7.3), this drastic step will only be necessary if high order interactions between the effects of the factors are likely to be important. Unfortunately, the possibility of large interactions implies that the results obtained when one factor is controlled at a single level cannot be used to infer the effects of the factors varied in the experiment at other levels of the controlled factor. The choice is between some information about the interacting effects of all three factors or more information about the effects of two of the factors but only at a specific level of the third factor. The advice of statisticians between alternatives is often influenced by their views, based on experience, about the likely importance of interaction effects rather than on the relative advantage of gaining knowledge about two or three factors.

Most good experimentation is sequential with each experiment leading to another. The early experiments may involve many factors with very little, if any, replication or even fractional replication (section 7.4). Factors with large main effects or involved in large two-factor interactions are selected for further study in later experiments where more replication is used. Again the choice is being made to study the effects of some factors but at fixed levels of other factors. In this case the factors to be at fixed levels have been chosen on the basis of the results of the earlier experiments. This approach obviously has advantages in experiments aimed at increased understanding of processes and in experiments with the more practical objective of choosing as operating conditions, a combination of the levels of quantitative factors that will optimize the value of some product. The combination of levels may, for example, be a combination of fertilizer application, time of sowing and spacing between plants that will result in maximum yield of a crop. If there is a large second order interaction involving all three factors then a choice of the level of one factor on the basis of previous knowledge with experimentation concentrated on the effects of the other two factors may be preferred to experimentation varying the levels of all three factors.

16.7 FACTORS WITH QUANTITATIVE LEVELS

The choice of levels to investigate the effects of a quantitative factor is a sadly neglected aspect of experimental design. The objectives of an experimenter in considering the way in which yield varies in response to different levels of a factor can range from the simple wish to estimate the rate of increase of yield with increasing level, to the desire to describe the whole pattern of response. Another rather different objective is to determine the level of the factor or, in the case of several factors, the combination of levels which gives the maximum yield or, alternatively, the maximum economic yield. There is a large amount of statistical theory about the optimal choice of levels. Much of this is not relevant to practical experimentation because it is concerned with narrowly defined and largely irrelevant objectives, but the general philosophy which emerges from the theory is often quite simple.

For example, suppose the primary object of an experiment is to estimate the decay of seed germination with storage time. For many types of seed and forms of storage it is strongly believed that the relationship of seed germination percentage with time is linear. If it is intended to investigate the rate of decay over a two-year storage period then statistical theory says that the most efficient use of resources is to estimate the germination rate at the beginning of the period using half the available seeds, and to use the other half to estimate the germination rate at the end of the period. The basis of this theory is the variance of the regression coefficient of y on x, var $(\hat{\beta}) = \sigma^2/S_{xx}$, from which it can be shown that to estimate the rate of decay, β, as precisely as possible, we must take observations as far apart as possible.

For most experimenters this theory is too extreme because even if the relationship is believed to be linear there will be a desire to check the linearity. So in practice a sensible choice of design will involve taking 40% of the observations at the beginning of the two-year period, 40% at the end and the remaining 20% within the time interval, probably mostly at the middle of the period because this is likely to give the most sensitive indication of departure from linearity.

Developing the same philosophy further, if is is believed that the yield response to a quantitative factor is curvilinear with a rise to an optimum yield followed by a decay, then to estimate the simplest form of curved relationship, the quadratic $y = a + bx + cx^2$, the experimenter should use three levels and these should be spread over the whole range of x values for which the curved relationship is believed to hold, with the middle level midway between the two extreme levels. Again the philosophy is simple. The minimum number of levels needed to give information about the quadratic response is three. If we are sure the response is quadratic the use of more than three levels diffuses the information inefficiently. We need to make the range of levels as wide as possible within the range which the quadratic model is valid to achieve clearer definition of the position of the peak. The proportion of observations taken at the middle point will depend on the particular objective, but equal replication of the three levels is never badly wrong. By analogy consider how easy it is to determine the position of the top of a hill if you are some distance away from the hill, but how difficult it is if you are climbing the hill and are near the top.

In general, statistical theory indicates that for a given level of resources it is preferable to use relatively few different levels of a quantitative factor with substantial replication of each level, rather than relatively many levels with less replication. This is appropriate even when the experimenter wishes, in a very general sense, to investigate the whole response function. It might seem that using many different levels would enable the experimenter to detect sudden changes of response wherever they occur. In practice, however, the overall limitation of resources will almost inevitably mean that the replication at each of many different levels will be inadequate to detect anything.

One special area of experimental design is in the determination of the combinations of levels of several factors that will maximize the yield response. To investigate a complex realtionship (i.e. where the response to each factor alone is not linear) we need at least three levels of each factor and the simplest suitable design we have considered previously would be a 3^n factorial. With 3^n factorials the number of treatment combinations is large even for small values of n and increases rapidly with n. Designs using fewer points, omitting some of those of the 3^n structure which give little information about the position of the optimum, and including extra points, have been produced by various statisticians starting with Box and Wilson (1951); Chapter 18 in Mead (1988) gives a good introduction to the subject. The most important ideas behind most of the designs are to concentrate information towards the centre of the design and to equalize the information in all direction about the centre, the concept of rotatability. To illustrate these ideas we show, for two factors in Fig. 16.4, the ordinary 3^2 design and then a central rotatable composite design consisting of a 2^2 design with extra axial and centre points.

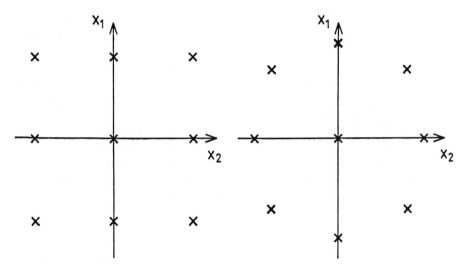

Fig. 16.4 Designs to investigate the effects of two quantitative treatment factors, (a) a 3 × 3 factorial, (b) a control composite rotatable design.

One important point to notice about designs for investigating response surfaces and for optimum seeking is that the analysis of the results from such designs is not primarily an analysis of variance followed by a comparison of treatment means using standard errors. Instead the analysis will normally consist of fitting a regression of yield on the various factor levels, followed by inter-pretation of the fitted regression function. Replication is still important because we still need to know how well the regression function fits, and the standard errors of yield predictions depend on the amount of replication.

16.8 SCREENING AND SELECTION

In the last section we discussed one area of experimentation where we had a specific objective, to find the conditions which maximized yield, and we considered different designs and the subsequent analysis of the data. To illustrate further the need to match particular objectives by appropriate experimental design philosophy, we shall consider briefly two other special cases where the objectives are quite clearly different from the general comparison of a relatively small set of treatments.

Suppose first that we are concerned to select the best r from a total of n new plant varieties. Usually n will be very large and r small and the selection will be made in a number of stages. Assuming that some of the new varieties are almost certain to be higher yielding than existing standard varieties we may decide not to have control varieties in the trial. At each stage, we select those varieties that have performed best, regardless of any significance tests or confidence intervals. Theoretical results (Finney, 1958; Curnow, 1961) suggest that under certain assumptions, a multistage selection programme will be most efficient when the proportion of varieties selected to go forward to the next trial stage is about the same at each stage of selection and the resources allocated to each stage are approximately equal. Thus, if we wish to select three varieties from 250 using three stages the proportion to be selected at each stage, p, will be such that $p^3 = 3/250$. After the first stage we will have $250p$ varieties; after the second $(250p) \times p$; and after the third $250p \times p \times p$ which must be 3. The value for p is 0.23 giving the ideal numbers at each stage as 57 after stage 1 and 13 after stage 2. A reasonable practical programme would therefore be to have one plot for each variety at stage 1, 4 plots for each of about 60 varieties at stage 2, and about 20 plots for each of about 13 varieties at stage 3, the total plots per stage then being roughly equal. Similar considerations apply to the planning of animal breeding programmes and to drug screening programmes.

If, on the other hand, we are conducting a trial to assess a range of new varieties produced by different breeders we will wish to compare these new varieties with the standard varieties being used in the agricultural or horticultural industry. We shall use these standard varieties as controls and then be interested primarily in the comparison of each variety with a control. We should then need to consider whether it would be advantageous to replicate the control variety within each block. This situation may also occur in the latter stages of a selection programme. If only the comparison with the control is important then the theoretical results of section 16.2 demonstrate that the number of replicates of the control in each block should be about equal to the square root of the number of new varieties. Thus if it was decided to bring in a control variety at the third stage of the above selection programme when about 13 varieties remain, each block should contain 3 or 4 plots of the control variety to obtain maximum precision of the comparison with the control.

In screening large numbers of chemicals, drugs for example, to detect those which have a particular effect, it is sometimes known that only a very small proportion of the chemicals will be effective. The simple version of this problem is when the distinction between a unit showing an effect and one showing no effect is unmistakable. If we can ignore the possibility of chemical interactions between chemicals in the same group, the experimental design could then consist of allocating a group of chemicals to each experimental unit, eliminating all chemicals in the group for any unit which shows no effect, and retesting in smaller groups all those chemicals from groups for units which did show an effect. A strategy of testing in groups of 8, and then subsequently in groups of 4, then 2, then singly, is very nearly optimal for a wide range of proportions of effective chemicals. More details of this particular design problem can be obtained from Sobel and Groll (1959). When the difference between an effect and no effect is not so clear cut then we have a more difficult problem and have to use experimental designs with several chemicals applied to each unit but with each chemical appearing more than once.

Block	1	2	3	4	5	6	7	8	9	10
	A	E	I	M	A	B	C	D	A	B
	B	F	J	N	E	F	G	H	F	E
	C	G	K	O	I	J	K	L	K	L
	D	H	L	P	M	N	O	P	O	D
Block	11	12	13	14	15	16	17	18	19	20
	C	D	A	B	C	D	A	B	C	D
	H	G	G	H	E	F	H	G	F	E
	I	J	I	K	J	I	J	I	L	K
	N	M	N	M	P	O	O	P	M	N

Fig. 16.5 Balanced incomplete block design with 16 varieties in blocks of four plots such that each pair of varieties occurs together in a block exactly once.

Table 16.3 Sixteen varieties in 20 blocks of 4

Block	1	2	3	4	5	6	7	8	9	10
	A	E	I	M	A	B	C	D	A	B
	B	F	J	N	E	F	G	H	F	E
	C	G	K	O	I	J	K	L	K	L
	D	H	L	P	M	N	O	P	O	D
Block	11	12	13	14	15	16	17	18	19	20
	C	D	A	B	C	D	A	B	C	D
	H	G	G	H	E	F	H	G	F	E
	I	J	I	K	J	I	J	I	L	K
	N	M	N	M	P	O	O	P	M	N

Special designs are needed for variety trials that compare a large number of varieties. In these trials, unlike factorial experiments, the treatment varieties generally have no structure, each treatment being a separate entity unrelated to any of the others. With a large number of varieties, a randomized complete block design would require very large and therefore heterogeneous blocks to accommodate all the varieties in each block. Such blocking would be ineffective and randomized incomplete block designs, of the form introduced briefly in section 5.7 (see section 7.4) are therefore often used. Where possible, we would choose to use balanced incomplete block designs that arrange that every variety occurs the same number of times in the trial and every pair of varieties occur together in a block the same number of times, generally just once. This degree of balance ensures that all varieties are compared with equal accuracy, that is all comparisons of pairs of varieties have the same standard error. The designs in Figs 5.4 and 5.5 are two examples of balanced designs. A rather larger balanced design for 16 varieties in blocks of 4 plots is shown in Table 16.3.

There are relatively few balanced designs so that it is very unlikely that for a given number of varieties, number of plots per block and given replication,

Block	1	2	3	4	5	6	7	8	9	10
	C	D	E	A	B	A	D	B	E	C
	O	H	J	G	F	H	M	I	G	F
	R	L	M	I	K	J	P	Q	K	N
	T	Q	S	N	P	O	T	R	L	S

Fig. 16.6 Incomplete block design for 20 varieties in ten blocks of four plots, using a α-design, not balanced.

Table 16.4 Twenty varieties blocks in 10 blocks of 4

Block	1	2	3	4	5	6	7	8	9	10
	C	D	E	A	B	A	D	B	E	C
	O	H	J	G	F	H	M	I	G	F
	R	L	M	I	K	J	P	Q	K	N
	T	Q	S	N	P	O	T	R	L	S

a balanced design exists. Practical designs for comparing large numbers of varieties in sensibly small blocks are usually not exactly balanced. There is a large and very useful class of designs, the α-lattice designs, which are nearly balanced in the sense that the precision of estimates of differences between varieties varies only by a few percent. A program ALPHAGEN is available for constructing such designs based on methods described by Patterson and Williams (1976) and Patterson, Williams and Hunter (1978). An example of an α-design for 20 varieties in ten blocks of 4 plots is shown in Table 16.4.

Because blocks and treatments are no longer orthogonal the analysis of incomplete block designs requires the general linear model analysis methods of section 11.2. As mentioned in section 5.7, there is some extra information to be obtained by comparing block totals since only a subset of varieties is contained within each block. The use of such information involves additional problems of estimating the relative weights to be attached to the within and between block estimates of the treatment differences. General programs to analyse incomplete block designs using information from both within and between blocks are now available (REML: 1987).

There have also been new developments in the use of 'nearest neighbour' methods of designing and analysing the results of variety trials. The trial lay-out is designed so that each variety is surrounded in some balanced way by a subset of all the other varieties. The analysis consists of adjusting the yields of each variety plot by the information provided by the yields of the neighbouring plots. This method of reducing the effect of spatial variation on the assessment of each variety has been claimed to be an improvement on the use of blocked designs but there is continuing controversy about the relative merits of nearest neighbour and blocking approaches to design and analysis. Some relevant recent references are Besag and Kempton (1986) and Cullis and Gleeson (1989).

17
Sampling Finite Populations

17.1 EXPERIMENTS AND SAMPLE SURVEYS

The purpose of experiments is to make inferences, as unambiguously as possible, about the effects of treatments applied to experimental units. The units are taken as representative of the population of units to which the treatments may in future be applied. Randomization is used to remove biases in comparing different treatments and to allow a measure to be obtained of the effects of uncontrolled factors that may contribute to the differences observed between treatment means. Blocking is used to control the effects of factors, other than the treatments, that are known to effect the response measured.

The objectives of sample surveys are different, being to allow inferences to be made from a sample about the whole, but always finite, population from which it has been drawn. There is no imposition of treatments. What is required from the sample is a description of certain properties of the population as it naturally exists. With people, we may wish to know the average expenditure of a household on food or the way in which people will vote in the next election. We may know that the characteristic we wish to measure is affected by, for example, socio-economic class and so will choose a sample from the population that takes this into account. To do this we divide the population into strata according to socio-economic class, sample separately from each stratum and then combine the results taking account of the proportions of the population that belong to each of the socio-economic strata. Stratification in sample surveys is clearly related to blocking in experiments, both being ways of controlling unwanted sources of variation.

There are many examples of the use of sampling in descriptive biology. A forester may wish to know the volume of standing timber in a forest. A zoologist may wish to know the amount of food available to a certain species of fish in a stretch of river. A botanist may be interested in the abundance and association of different plant species in an area of woodland. Neither the whole forest, nor the whole river nor the whole woodland can be studied. Sample volumes or areas must be used instead. More specialized sampling procedures based on capture, mark and recapture methods or on line–transect sampling are used to estimate the size of animal populations. In this chapter, we shall not discuss these more specialized procedures but will consider the basic ideas of sampling with special reference to the use of sampling in experiments. In an experiment with plants, the analysis of the growth of plants subjected to different treatment regimes generally requires that, at various times during the experiment, plants be sampled from each experimental plot. This has already been mentioned in section 12.6. In an animal experiment, cost may dictate that post-mortems can be carried out on only a sample of the animals from each treatment group. As already mentioned, the principles of sampling will be discussed in this chapter with reference to experiments but they are of much wider application. Cochran (1977) should be consulted for a thorough discussion of both the general principles and the practice of sampling methods.

In this book we have usually assumed that each sample of data to be analysed can be thought of as a sample of results from a hypothetical population of infinite size consisting of all the possible results for that particular treatment applied to the specified types of plants or animals. This ideal is always unachievable because the number of animals or the area of land available for experimentation is always limited. In fact, the experimenter often uses all the animals or fields that are available. The hope is that the animals or plots of land used are sufficiently representative of the population about which inferences are to be made that the differences between the effects of the treatments measured on the available samples will be good indicators of the differences that would be obtained with the large population that is of real interest.

At the other extreme, in some sampling procedures the sample can be a substantial proportion of the whole population of interest. The forester may be able to estimate the volume of timber in, say, one-third of the trees in the forest. We have already mentioned stratification. Although in some ways equivalent to blocks in experiment, both being a means to control unwanted sources of variation, strata in sampling surveys have an additional feature. They may show different levels of within stratum variation and these differences may be known to some reasonable level of accuracy. The purpose of stratification is that the reduced variability within stratum means that fewer samples are needed to represent each stratum in estimating the overall population mean or total. Indeed, as we shall see, the number of samples required in each stratum will vary according to the size of the stratum, the relevant amount of variation within the stratum and the cost of sampling within each stratum.

17.2 SIMPLE RANDOM SAMPLING

To illustrate the use of simple random sampling we shall discuss the sampling, for the purpose of leaf area measurement, of leaves from an experimental plot. The plot contains N leaves. We shall write the unknown areas of these N leaves as

$$x_1, x_2, \ldots, x_N$$

By using tables of random numbers we can choose a random sample of n

different leaves and measure their leaf areas. These n leaf areas will be written

$$y_1, y_2, \ldots y_n$$

and are, of course, n of the N values $x_1, x_2, \ldots, x_N$ above.

Note that we insist on having n different leaves in our sample. This is inevitable when sampling from an infinite population but must be deliberately arranged when sampling from a finite population. The mean leaf area of the sample

$$\bar{y} = \frac{y_1 + y_2 + \ldots + y_n}{n}$$

will, because it is a random sample, be an unbiased estimate of the population mean

$$\bar{x} = \frac{x_1 + x_2 + \ldots + x_N}{n}$$

The reader may find it helpful to verify this unbiasedness by considering particular small populations. For example consider the following population of size 5

$$x_1 = 3 \qquad x_2 = 5 \qquad x_3 = 4 \qquad x_5 = 6$$

There are ten possible samples of size 3 $[(x_1 \, x_2 \, x_3), (x_1 \, x_2 \, x_3)\ldots]$ and the means of the ten samples of size 3 are

| 4 | 10/3 | 14/3 | 3 | 13/3 | 11/3 | 11/3 | 5 | 13/3 | 4 |

The average value of the ten sample means will be found to be equal to the population mean,

$$\bar{x} = \frac{3 + 5 + 4 + 2 + 6}{5} = 4$$

and this is what is meant by unbiasedness.

Because the population being sampled is finite, and the sample consists of n different elements from that population, the variance of $\bar{y}$ is not, as in previous chapters σ^2/n. Instead the variance can be shown to be

$$\frac{\sigma^2}{n}\left(1 - \frac{n}{N}\right)$$

where

$$\sigma^2 = \frac{\sum_{i=1}^{N}(x_i - \bar{x})^2}{N - 1}$$

is the unknown population variance. The factor $[1 - (n/N)]$ in the variance of

$\bar{y}$ is called the finite population correction. When the sample consists of the whole population, i.e. $n = N$, the sample mean is the population mean and the finite population correction ensures that the variance, which measures the likely inaccuracy of the sample mean as an estimate of the population mean, is zero. At the other extreme, the finite population correction can be ignored if the sample is only a small proportion of the population, i.e. n is small compared with the population size N.

For our artificial example the population variance, σ^2, is

$$\sigma^2 = \frac{(3-4)^2 + (5-4)^2 + (4-4)^2 + (2-4)^2 + (6+4)^2}{4} = 2.5$$

which, substituted in the formula above, tells us that the variance of the mean of a sample size 3 is

$$\frac{\sigma^2}{n}\left(1 - \frac{n}{N}\right) = \frac{2.5}{3}\left(1 - \frac{3}{5}\right) = 0.33$$

In practice, of course, only one sample will be available and the variance of the sample mean will have to be estimated by substituting the sample variance,

$$s^2 = \sum_{i=1}^{n} \frac{(y_i - \bar{y})^2}{n - 1}$$

in place of the unknown population variance σ^2. In our artificial example, we can verify that the sample variance is an unbiased estimate of σ^2 by calculating s^2 for each of the ten possible sample of size 3 as

| 1 | 7/3 | 7/3 | 1 | 7/3 | 13/3 | 7/3 | 1 | 13/3 | 4 |

and noting that the average of these ten numbers is 2.5, the value of σ^2.

In replicated experiments, the standard errors for treatment differences are obtained from the error means square in the analysis of variance. When experimental units are sampled, the sample mean becomes the observation from each unit to be analysed in the analysis of variance. The formula discussed above for the variance of the sample mean, $\bar{y}$, tells us about the accuracy of the sampling process and, in particular, how this is affected by the size of the sample, n. It can be used to decide how large a sample is needed but cannot be used to calculate the standard errors for comparing the treatment means. The standard errors for comparing treatment means must take account of the variability between units and, in blocked experiments, the lack of repeatability of treatment differences from block to block. This is exactly the variability measured by the error mean square in the analysis of variance.

In sampling, the sample mean is often not the most relevant statistic to use in describing the 'yields' from the unit. In the leaf area example, the total leaf area on the plot would be estimated by $N\bar{y}_n$ and this may be a more relevant statistic than the sample mean, $\bar{y}_n$. Any set of single values, one from each unit, can be submitted to the appropriate analysis of variance and standard errors calculated in the usual way.

17.3 STRATIFIED RANDOM SAMPLING

We have already referred in section 17.1 to the possibility of dividing the population to be sampled into homogeneous strata so that a small sample from each stratum will provide a good estimate of the quantity of interest for the whole of that stratum. The means from the different strata can then be combined to give a good estimate of the mean for the whole population.

To illustrate the ideas of stratified random sampling, consider the problems of sampling plants from a large experimental plot in which the plants are grown in four rows. The total 'yield' of the plot is to be estimated from a sample of plants taken from the plot. From a visual inspection, there are clearly differences between the yields of the plants in different rows. The number of plants in row i is N_i and the unknown standard deviation of the yields of the plants in row i is σ_i ($i = 1, 2, 3$ and 4). Because of the differences between rows, the rows will be used as strata. The main question is how many plants should we sample from each row? Having taken samples of size n_1, n_2, n_3 and n_4 from rows $1, 2, 3$ and 4 respectively we can estimate the yields per plant in each row by the sample means $\bar{y}_1, \bar{y}_2, \bar{y}_3$ and $\bar{y}_4$. Knowing the total number of plants in each row, N_i, we can then estimate the total yield of the plot from the weighted average of the sample means

$$N_1\bar{y}_1 + N_2\bar{y}_2 + N_3\bar{y}_3 + N_4\bar{y}_4$$

The variance of $\bar{y}_1$ is, from the previous section,

$$\frac{\sigma_1^2}{n_1}\left(1 - \frac{n_1}{N_1}\right)$$

So the variance of $N_1\bar{y}_1$ is

$$\frac{\sigma_1^2 N_1^2}{n_1}\left(1 - \frac{n_1}{N_1}\right)$$

and the variance of the sum of $N_1\bar{y}_1, N_2\bar{y}_2, N_3\bar{y}_3$ and $N_4\bar{y}_4$, which is the estimate of the total yield of the plot, is

$$\frac{\sigma_1^2 N_1^2}{n_1}\left(1 - \frac{n_1}{N_1}\right) + \frac{\sigma_2^2 N_2^2}{n_2}\left(1 - \frac{n_2}{N_2}\right) + \frac{\sigma_3^2 N_3^2}{n_3}\left(1 - \frac{n_3}{N_3}\right) + \frac{\sigma_4^2 N_4^2}{n_4}\left(1 - \frac{n_4}{N_4}\right)$$

The standard error of the estimate of the total yield of the plot is the square root of this variance. For a given total number of plants to be sampled, the standard error can be shown to be minimised by choosing each sample size, n_1, proportional to $N_i\sigma_i$. The larger the number of plants in the row or the more variable the yields in the row, the larger should be the size of the sample.

As an example of the calculations, consider the following artificial example involving three strata:

i	N_i	σ_i
1	100	2
2	200	4
3	400	1

Practical considerations restrict the total units to be sampled to 140.
Three possible schemes are:

(i) To take 47 units equally from each stratum (one extra unit can probably be managed!);
(ii) To take 20% of the units from each stratum—this is called proportional sampling; or
(iii) To take optimal size samples from each stratum. This last method would have sample sizes proportional to $N_i\sigma_i$ (200:800:400).

The stratum sample sizes for the three methods would be as follows:

i	Equal sampling	Proportional sampling	Optimal sampling
1	47	20	20
2	47	40	80
3	47	80	40
	141	140	140
Standard error of estimated population total	118	126	100

In the optimal sampling scheme, stratum 2 has the largest sample size because it is the most variable stratum, and also the second largest. The final row of the table gives the standard error of the estimated population total, the square root of the formula above for the variance. The advantage, in this example, of optimal sampling is clear. To achieve the accuracy of optimal sampling using equal sampling would have required a total of $(118/100)^2 \times 141 = 196$ units instead of 140. We have, of course, assumed that we know the standard deviations, σ_i, within each stratum. This is rarely, if ever, the case, and usually we will have to estimate the standard deviations. Fortunately the standard error achieved is generally not very sensitive to errors in anticipating the values of the standard deviations within each stratum.

If the cost or effort in sampling is different for the strata then it can be shown that the sample sizes should be inversely proportional to the square root of the costs or efforts involved, as well as to the product of the stratum size and the stratum standard deviation.

In experiments, as explained when we were considering simple random sampling, the formula for the variance of the estimate of the total yield of the plot only measures the sampling precision for the plot. It can be used to decide on the number of plants to be sampled from the plot but is not to be used to calculate standard errors for comparing treatment means. These are calculated in the usual way from an analysis of variance of the estimates of the yield for each plot.

Stratifying populations is clearly only worthwhile if the quantity being analysed shows consistent differences between the various strata. Having discovered such differences we must take note of them. With an experimental plot, using the weighted average of the yields of the plants in the various strata within the plot is only justified if the proportion of the plants in the various strata represent the proportions likely to obtain in the future and if the average or the total yield of the whole plot is the quantity of the direct interest. The existence of difference between strata within plots may suggest that the different treatments could have different effects on the plants of different strata and this may justify an investigation of possible treatment–stratum interactions.

17.4 CLUSTER SAMPLING, MULTISTAGE SAMPLING AND SAMPLING PROPORTIONAL TO SIZE

In estimating the total leaf areas of plots, there may be advantages in terms of cost, of effort and of reduced disturbance to the plants, of sampling whole plants rather than individual leaves. This is called cluster sampling. The leaves on a plant are considered to be a 'cluster'. Cluster sampling can be shown to be less efficient than simple random sampling when the same number of leaves are sampled. However, the convenience of sampling whole plants may allow more leaves to be measured for the same expenditure of effort. The plants must be chosen at random and the total leaf area of each plant averaged over plants to give an estimate of the leaf area per plant. Clearly the total leaf area of the plot is estimated by the estimated leaf area per plant multiplied by the number of plants in the plot. Cluster sampling is only one particular form of multistage sampling. A certain number of plants may be sampled, perhaps using stratified random sampling with the rows of plants forming the strata, and then a random sample of leaves chosen to represent each sampled plant. In cluster sampling, all the leaves on the sampled plants would be measured, but here there are two stages of sampling, first of plants and then of leaves on plants. Because there are certain to be differences between plants as well as between leaves on the sample plant, the variance of the estimate of the leaf area for the whose plot will be minimized, for a fixed total number of leaves examined, when as many plants as possible are sampled and only one leaf per plant is examined. However, the effort involved may be related to the number of plants sampled rather than to the number of leaves. A compromise then has to be made between the number of plants sampled and the number of leaves examined on each plant. However the sampling is arranged, in replicated experiments a single value is obtained for each experimental plot and subjected to the usual analysis of variance procedures.

In our example of sampling whole plants from a field plot, some plants may have many more leaves than others and therefore contribute more to the total leaf area of the plot. In these circumstances a better estimate of the total leaf area is often obtained by sampling 'proportional to size'. Instead of choosing the plants to be sampled at random as in cluster sampling, we choose the plants to be sampled with a probability proportional to the number of leaves they have. This can most easily be arranged by assigning a number to each leaf and choosing a plant if the random number selected is the number assigned to any of its leaves. With this method of sampling, the average area per leaf for each plant is an unbiased estimate of the average area per leaf of the whole population of leaves. The total leaf area of the plot is estimated by multiplying the mean of the average area per leaf for the sampled plants by the total number of leaves on the plot. The accuracy of the estimate depends on the variation from plant to plant of the average area of a leaf. Because the number of leaves per plant is variable, the variation in the average area of a leaf from plant to plant is often less than the variation in the total leaf area from plant to plant. In these circumstances, sampling plants in proportion to the number of leaves they have will be more accurate than the completely random sampling of plants implied by cluster sampling.

In this section we have discussed some of the standard modifications of sampling procedures. This is only the tip of a very large iceberg and, as mentioned earlier, a comprehensive discussion is given in Cochran (1977).

17.5 RATIO AND REGRESSION ESTIMATES

A zoologist wishes to know how many beetles of a particular kind are on an experimental plot following treatment with an insect spray. The beetle lives in the bark of fallen trees. Using quadrats, or sampling areas, of constant size the zoologist measures the number of y of beetles in each quadrat area and also the total area x of bark in that quadrat. The obvious way to estimate the total number of beetles on the plot is to average the number of beetles per unit surface area of bark found in each quadrat and multiply by the total surface area X of the bark on the whole plot. We can represent this estimate by

$$\frac{1}{n}\left(\frac{y_1}{x_1} + \ldots + \frac{y_n}{x_n}\right) X$$

where n is the number of quadrats. This may not be the most accurate estimate. The variance of the number of beetles y in a quadrat may well be proportional to the expected number of beetles in that plot and hence to the surface area, x, available for them (see section 14.5 on the Poisson distribution). The variance of the number of beetles per unit surface area, y/x, will then have a variance inversely proportional to the surface area, x. Thus to obtain an estimate of the number of beetles per surface area over the whole plot, the y/x values should be combined together as a weighted, rather than an unweighted, average with the weights proportional to x. The greater the amount of bark in a plot the more accurate will be the estimate of the number of beetles per unit area of bark, x. The estimate of the total number of beetles on the plot will now be

$$\frac{x_1(y_1/x_1) + x_2(y_2/x_2) + \dots + x_n(y_n/x_n)}{x_1 + x_2 + \dots + x_n}X = \frac{y_1 + y_2 + \dots + y_n}{x_1 + x_2 + \dots + x_n}X$$

which is simply the total number of beetles in all the quadrats divided by the total area of bark in all the quadrats multiplied by the total area of bark on the whole plot, X. This so-called ratio estimate gives more correct weight to the information from each quadrat and, as a result, is often found to be more accurate than the estimate based on the unweighted average of the values of y/x from each quadrat. The following artificial data illustrates this point:

Quadrat	Area of bark (x)	Number of beetles (y)	y/x
1	10	21	2.1
2	20	38	1.9
3	30	66	2.2
4	40	88	2.2
Total	100	213	

Using the four values of y/x, one from each quadrat, we can estimate the number of beetles per unit area of bark by:

$$\frac{2.1 + 1.9 + 2.2 + 2.2}{4} = 2.10$$

If we use the total number of beetles divided by the total area of bark, the number of beetles per unit area of bark is:

$$\frac{y_1 + y_2 + y_3 + y_4}{x_1 + x_2 + x_3 + x_4} = \frac{213}{100} = 2.13$$

The standard error of this last ratio of totals, assuming that the variance of the y values are proportional to the x values, can be shown to be 0.069. This compares with a standard error, with the same assumptions, of 0.078 for the estimate based on averaging the four separate y/x values. With real data, the gains are often considerably greater than this.

The ratio estimate is a good estimate if the variance of each y value is approximately proportional to the x value and if the relationship of y to x is such that y/x is not very variable. The latter condition implies that the relationship of y to x is approximately a straight line through the origin. If a plot of the number of beetles in each quadrat against the surface area of the bark in that quadrat is a straight line not through the origin, $y = a + bx$ say, then the ratios should not be used but a regression line should be fitted to the quadrat data,

$$y = \bar{y} + b(x - \bar{x}),$$

using the least squares methods described in Chapter 9. The total number of beetles can then be estimated by

$$N[\bar{y} + b(\bar{X} - \bar{x})]$$

when N is the total plot area divided by the area of a quadrat, and $\bar{X}$ is the surface area of bark over the whole plot divided by N, i.e. the average area of bark to be expected in a single quadrat. The expression in brackets is the number of beetles to be expected, according to the linear relationship fitted to the data, in a quadrat containing an area of bark, $\bar{X}$.

Note that, for both the ratio and regression estimates, the mean or total values of x for the whole plot is needed. The variable x can be any quantity which is related to the y values either by a linear regression through the origin for the ratio method or by a linear regression not through the origin for the regression method.

The need to know, or to have an estimate of, the mean or total value of x for the whole population has led to a technique known as double sampling. In our beetle example, the total surface area of bark on the whole plot might be estimated from a number of large quadrats and the number of beetles counted on a number of small areas chosen randomly within each large quadrats.

References

Besag, J. and Kempton, R. (1986) Statistical analysis of field experiments using neighbouring plots. *Biometry*, **42**, 231–251.

Box, G. E. P. and Wilson, K. B. (1951) On the experimental attainment of optimum conditions (with discussion). *Journal of the Royal Statistical, Society* **B 13**, 1–45.

Carroll, L. (1871) *Through the Looking Glass*, Macmillan, London.

Causton, D. R. and Venus, J. C. (1981) *The Biometry of Plant Growth*, Arnold, London.

Cochran, W. G. (1977) *Sampling Techniques*, 3rd edn, Wiley, New York.

Collett, D. (1991) *Modelling Binary Data*. Chapman and Hall, London.

Crowder, M. J. and Hand, D. J. (1990) *Analysis of Repeated Measures*, Chapman and Hall, London.

Cullis, B. R. and Gleeson, A.C. (1989) The efficiency of neighbour analyses for replicated variety trials in Australia *Journal of Agricultural Science*, **113**, 233–239.

Curnow, R. N. (1961) optimal programmes for varietal selection. *Journal of the Royal Statistical Society*, **B 23**, 282–318.

Dear, K. B. G., and Mead, R. (1983) The use of bivariate analysis techniques for the presentation, analysis and interpretation of data. *Statistics in Intercropping Technical Report I*. Department of Applied Statistics, University of Reading.

Draper, N. R. and Smith, H. (1981) *Applied Regression Analysis*, 2nd edn, Wiley, New York.

Finney, D. J. (1958) Statistical problems of plant selection. *Bulletin of the International Statistical Institute*, **36**, 242–268.

Finney, D. J. (1971) *Probit Analysis*, 3rd edn, Cambridge University Press, Cambridge, UK.

Fisher, R. A. and Yates, F. (1963) *Statistical Tables for Biological, Agricultural and Medical Research*, 6th edn, Oliver and Boyd, Edinburgh.

Krzanowski, W. J. (1988) Principles of Multivariate Analysis A user's perspective, Oxford.

McCullagh, P. and Nelder, J. A. (1990) *Generalized Linear Models*, 3rd edn, Chapman and Hall, London.

Mead, R. (1988) *The Design of Experiments : Statistical Principles for Practical Application*, Cambridge University Press, Cambridge, UK.

Nelder, J. A. (1961) The fitting of a generalisation of the logistic curve. *Biometrics*, **17**, 89–110.

Patterson, H. D. and Williams, E. R. (1976) A new class of resolvable incomplete block designs. *Biometrika*, **63**, 83–92.

Patterson, H. D., Williams, E. R. and Hunter, E. A. (1978) Block designs for variety trials. *Journal of Agricultural Science*, **90**, 395–400.

Pearce, S. C. and Gilliver, B. (1979) Graphical assesment of intercropping methods. *Journal of Agricultural Science*, **93**, 51–58.

REML (1987) *Program and Manual Scottish Agricultural Statistics Service*, Edinburgh.

Richards F. J. (1959) A flexible growth function for empirical use. *Journal of Experimental Botany*, **10**, 290–300.

Sobel, M. and Groll, P. A. (1959) Group testing to eliminate effectively all defectives in a binomial sample. *Bell. System. Tech. J. 38*, 1179–1252.

Sprent, P. (1969) *Models in regression*, Methuen, London.

Wetherill, G. B. (1977) *Sampling inspection and quality control*, Chapman & Hall.

Appendix

Table A.1 The standardized normal distribution

The distribution tabulated is that of the normal distribution with mean zero and standard deviation 1. For each value of z, the standardized normal deviate, the proportion, P, of the distribution less than z is given. For a normal distribution with mean μ and variance σ^2, the proportion of the distribution less than some particular value, x, is obtained by calculating $z = (x - \mu)/\sigma$ and reading the proportion corresponding to this value of z.

z	P	z	P	z	P	z	P
-4.00	0.00003	-1.50	0.0668	0.00	0.5000	1.55	0.9394
-3.50	0.00023	-1.45	0.0735	0.05	0.5199	1.60	0.9452
-3.00	0.0014	-1.40	0.0808	0.10	0.5398	1.65	0.9505
-2.95	0.0016	-1.35	0.0885	0.15	0.5596	1.70	0.9554
-2.90	0.0019	-1.30	0.0968	0.20	0.5793	1.75	0.9599
-2.85	0.0022	-1.25	0.1056	0.25	0.5987	1.80	0.9641
-2.80	0.0026	-1.20	0.1151	0.30	0.6179	1.85	0.9678
-2.75	0.0030	-1.15	0.1251	0.35	0.6368	1.90	0.9713
-2.70	0.0035	-1.10	0.1357	0.40	0.6554	1.95	0.9744
-2.65	0.0040	-1.05	0.1469	0.45	0.6736	2.00	0.9772
-2.60	0.0047	-1.00	0.1587	0.50	0.6915	2.05	0.9798
-2.55	0.0054	-0.95	0.1711	0.55	0.7088	2.10	0.9821
-2.50	0.0062	-0.90	0.1841	0.60	0.7257	2.15	0.9842
-2.45	0.0071	-0.85	0.1977	0.65	0.7422	2.20	0.9861
-2.40	0.0082	-0.80	0.2119	0.70	0.7580	2.25	0.9878
-2.35	0.0094	-0.75	0.2266	0.75	0.7734	2.30	0.9893
-2.30	0.0107	-0.70	0.2420	0.80	0.7881	2.35	0.9906
-2.25	0.0122	-0.65	0.2578	0.85	0.8023	2.40	0.9918
-2.20	0.0139	-0.60	0.2743	0.90	0.8159	2.45	0.9929
-2.15	0.0158	-0.55	0.2912	0.95	0.8289	2.50	0.9938
-2.10	0.0179	-0.50	0.3085	1.00	0.8413	2.55	0.9946
-2.05	0.0202	-0.45	0.3264	1.05	0.8531	2.60	0.9953
-2.00	0.0228	-0.40	0.3446	1.10	0.8643	2.65	0.9960
-1.95	0.0256	-0.35	0.3632	1.15	0.8749	2.70	0.9965

Table A.1 (*Contd*)

z	P	z	P	z	P	z	P
−1.90	0.0287	−0.30	0.3821	1.20	0.8849	2.75	0.9970
−1.85	0.0322	−0.25	0.4013	1.25	0.8944	2.80	0.9974
−1.80	0.0359	−0.20	0.4207	1.30	0.9032	2.85	0.9978
−1.75	0.0401	−0.15	0.4404	1.35	0.9115	2.90	0.9981
−1.70	0.0446	−0.10	0.4602	1.40	0.9192	2.95	0.9984
−1.65	0.0495	−0.05	0.4801	1.45	0.9265	3.00	0.9986
−1.60	0.0548	0.00	0.5000	1.50	0.9332	3.50	0.99977
−1.55	0.0606					4.00	0.99997

Table A.2 The Student's *t* distribution

This table gives the value of t for which a particular percentage, P, of the Student's t distribution lies outside the range $-t$ to $+t$. These values of t are tabulated for various degrees of freedom.

Degrees of freedom	P							
	50	20	10	5	2	1	0.2	0.1
1	1.00	3.08	6.31	12.7	31.8	63.7	318	637
2	0.82	1.89	2.92	4.30	6.96	9.92	22.3	31.6
3	0.76	1.64	2.35	3.18	4.54	5.84	10.2	12.9
4	0.74	1.53	2.13	2.78	3.75	4.60	7.17	8.61
5	0.73	1.48	2.02	2.57	3.36	4.03	5.89	6.87
6	0.72	1.44	1.94	2.45	3.14	3.71	5.21	5.96
7	0.71	1.42	1.89	2.36	3.00	3.50	4.79	5.41
8	0.71	1.40	1.86	2.31	2.90	3.36	4.50	5.04
9	0.70	1.38	1.83	2.26	2.82	3.25	4.30	4.78
10	0.70	1.37	1.81	2.23	2.76	3.17	4.14	4.59
12	0.70	1.36	1.78	2.18	2.68	3.05	3.93	4.32
15	0.69	1.34	1.75	2.13	2.60	2.95	3.73	4.07
20	0.69	1.32	1.72	2.09	2.53	2.85	3.55	3.85
24	0.68	1.32	1.71	2.06	2.49	2.80	3.47	3.75
30	0.68	1.31	1.70	2.04	2.46	2.75	3.39	3.65
40	0.68	1.30	1.68	2.02	2.42	2.70	3.31	3.55
60	0.68	1.30	1.67	2.00	2.39	2.66	3.32	3.46
∞	0.67	1.28	1.64	1.96	2.33	2.58	3.09	3.29

Table A.3 The F-distribution: 5% points

These tables give the values of F for which a given percentage of the F-distribution is greater than F.

n_2 \ n_1	1	2	3	4	5	6	7	8	10	12	24
2	18.5	19.0	19.2	19.2	19.3	19.3	19.4	19.4	19.4	19.4	19.5
3	10.1	9.55	9.28	9.12	9.01	8.94	8.89	8.85	8.79	8.74	8.64
4	7.71	6.94	6.59	6.39	6.26	6.16	6.09	6.04	5.96	5.91	5.77
5	6.61	5.79	5.41	5.19	5.05	4.95	4.88	4.82	4.74	4.68	4.53
6	5.99	5.14	4.76	4.53	4.39	4.28	4.21	4.15	4.06	4.00	3.84
7	5.59	4.74	4.35	4.12	3.97	3.87	3.79	3.73	3.64	3.57	3.41
8	5.32	4.46	4.07	3.84	3.69	3.58	3.50	3.44	3.35	3.28	3.12
9	5.12	4.26	3.86	3.63	3.48	3.37	3.29	3.23	3.14	3.07	2.90
10	4.96	4.10	3.71	3.48	3.33	3.22	3.14	3.07	2.98	2.91	2.74
12	4.75	3.89	3.49	3.26	3.11	3.00	2.91	2.85	2.75	2.69	2.51
15	4.54	3.68	3.29	3.06	2.90	2.79	2.71	2.64	2.54	2.48	2.29
20	4.35	3.49	3.10	2.87	2.71	2.60	2.51	2.45	2.35	2.28	2.08
24	4.26	3.40	3.01	2.78	2.62	2.51	2.42	2.36	2.25	2.18	1.98
30	4.17	3.32	2.92	2.69	2.53	2.42	2.33	2.27	2.16	2.09	1.89
40	4.08	3.23	2.84	2.61	2.45	2.34	2.25	2.18	2.08	2.00	1.79
60	4.00	3.15	2.76	2.53	2.37	2.25	2.17	2.10	1.99	1.92	1.70

Table A.4 The *F*-distribution: 1% points

n_2	n_1 1	2	3	4	5	6	7	8	10	12	24
2	98.5	99.0	99.2	99.2	99.3	99.3	99.4	99.4	99.4	99.4	99.5
3	34.1	30.8	29.5	28.7	28.2	27.9	27.7	27.5	27.2	27.1	26.6
4	21.2	18.0	16.7	16.0	15.5	15.2	15.0	14.8	14.5	14.4	13.9
5	16.3	13.3	12.1	11.4	11.0	10.7	10.5	10.3	10.1	9.89	9.47
6	13.7	10.98	9.78	9.15	8.75	8.47	8.26	8.10	7.87	7.72	7.31
7	12.3	9.55	8.45	7.85	7.46	7.19	6.99	6.84	6.62	6.47	6.07
8	11.3	8.65	7.59	7.01	6.63	6.37	6.18	6.03	5.81	5.67	5.28
9	10.6	8.02	6.99	6.42	6.06	5.80	5.61	5.47	5.26	5.11	4.73
10	10.0	7.56	6.55	5.99	5.64	5.39	5.20	5.06	4.85	4.71	4.33
12	9.33	6.93	5.95	5.41	5.06	4.82	4.64	4.50	4.30	4.16	3.78
15	8.68	6.36	5.42	4.89	4.56	4.32	4.14	4.00	3.80	3.67	3.29
20	8.10	5.85	4.94	4.43	4.10	3.87	3.70	3.56	3.37	3.23	2.86
24	7.82	5.61	4.72	4.22	3.90	3.67	3.50	3.36	3.17	3.03	2.66
30	7.56	5.39	4.51	4.02	3.70	3.47	3.30	3.17	2.98	2.84	2.47
40	7.31	5.18	4.31	3.83	3.51	3.29	3.12	2.99	2.80	2.66	2.29
60	7.08	4.98	4.13	3.65	3.34	3.12	2.95	2.82	2.63	2.50	2.12

Table A.5 The *F*-distribution: 0.1% points

n_2	n_1 1	2	3	4	5	6	7	8	10	12	24
2	999	999	999	999	999	999	999	999	999	999	1000
3	167	149	141	137	135	1331	132	131	129	128	126
4	74.1	61.3	56.2	53.4	51.7	50.5	49.7	49.0	48.1	47.4	45.8
5	47.2	37.1	33.2	31.1	29.8	28.8	28.2	27.7	26.9	26.4	25.1
6	35.5	27.0	23.7	21.9	20.8	20.0	19.5	19.0	18.4	18.0	16.9
7	29.3	21.7	18.8	17.2	16.2	15.5	15.0	14.6	14.1	13.7	12.7
8	25.4	18.5	15.8	14.4	13.5	12.9	12.4	12.1	11.5	11.2	10.3
9	22.9	16.4	13.9	12.6	11.7	11.1	10.7	10.4	9.87	9.57	8.72
10	21.0	14.9	12.6	11.3	10.5	9.93	9.52	9.20	8.74	8.44	7.64
12	18.6	13.0	10.8	9.63	8.89	8.38	8.00	7.71	7.29	7.00	6.25
15	16.6	11.3	9.34	8.25	7.57	7.09	6.74	6.47	6.08	5.81	5.10
20	14.8	9.95	8.10	7.10	6.46	6.02	5.09	5.44	5.08	4.82	4.15
24	14.0	9.34	7.55	6.59	5.98	5.55	5.23	4.99	4.64	4.39	3.74
30	13.3	8.77	7.05	6.12	5.53	5.12	4.82	4.58	4.24	4.00	3.36
40	12.6	8.25	6.59	5.70	5.13	4.73	4.44	4.21	3.87	3.64	3.01
60	12.0	7.77	6.17	5/31	4.76	4.37	4.09	3.86	3.54	3.32	2.69

Table A.6 The chi-squared distribution

This table gives the values of χ^2 for which a particular percentage, P, of the chi-squared distribution is greater than χ^2. These values of χ^2 are tabulated for various degrees of freedom.

Degrees of freedom	P					
	50	10	5	2.5	1	0.1
1	0.45	2.71	3.84	5.02	6.64	10.8
2	1.39	4.61	5.99	7.38	9.21	13.8
3	2.37	6.25	7.82	9.35	11.3	16.3
4	3.36	7.78	9.49	11.1	13.3	18.5
5	4.35	9.24	11.1	12.8	15.1	20.5
6	5.35	10.6	12.6	14.5	16.8	22.5
7	6.35	12.0	14.1	16.0	18.5	24.3
8	7.34	13.4	15.5	17.5	20.1	26.1
9	8.34	14.7	16.9	19.0	21.7	27.9
10	9.34	16.0	18.3	20.5	23.2	29.6
12	11.3	18.5	21.0	23.3	26.2	32.9
15	14.3	22.3	25.0	27.5	30.6	37.7
20	19.3	28.4	31.4	34.2	37.6	45.3
24	23.3	33.2	36.4	39.4	43.0	51.2
30	29.3	40.3	43.8	47.0	50.9	59.7
40	39.3	51.8	55.8	59.3	63.7	73.4
60	59.3	74.4	79.1	83.3	88.4	99.6

Table A.7 Random numbers

```
10 27   53 96   23 71   50 54   36 23   54 51   50 14   28 02   12 29   88 87
85 90   22 58   52 90   22 76   95 70   02 84   74 69   06 13   98 86   06 50
44 33   29 88   90 49   07 55   69 50   20 27   59 51   97 53   57 04   22 26
47 57   22 52   75 74   53 11   76 11   21 16   12 44   31 89   16 91   47 75
03 20   54 20   70 56   77 59   95 60   19 75   29 94   11 23   59 39   14 47

64 17   18 43   97 37   66 55   86 08   74 50   43 43   23 29   16 24   15 62
91 14   61 71   03 40   15 69   44 46   54 66   35 01   87 61   23 76   36 80
27 71   29 93   52 89   64 78   32 97   65 28   99 82   41 10   97 52   41 91
12 96   17 70   72 76   17 93   38 26   72 96   28 73   27 64   78 16   72 81
54 30   61 13   60 50   61 56   40 20   19 22   30 61   43 89   60 09   82 39

83 32   99 29   30 06   19 71   11 32   69 17   86 34   50 76   37 41   76 54
27 17   25 61   91 76   19 54   99 73   97 21   44 87   39 63   24 22   74 30
40 89   21 88   56 84   11 75   74 88   23 55   48 98   19 48   79 81   92 62
51 66   17 48   29 96   00 83   81 23   58 09   21 39   39 20   83 46   30 75
95 22   63 34   58 91   78 22   50 22   77 21   14 19   58 66   49 25   03 51

93 83   73 70   80 88   71 85   64 44   57 50   19 82   60 77   38 95   93 33
42 02   33 18   33 55   96 66   88 38   16 80   77 51   17 96   49 76   99 28
42 42   13 33   66 00   18 37   58 80   54 32   00 96   25 16   15 37   34 12
66 71   67 54   79 25   64 34   82 15   28 97   88 84   84 51   62 90   17 71
73 05   53 85   63 18   06 47   71 00   32 31   59 72   34 28   70 83   12 90
```

Table A.7 (*Contd*)

02 80	12 24	34 78	22 50	57 02	07 01	13 00	78 80	94 93	14 53
22 89	81 32	32 72	48 92	95 75	88 56	75 53	79 17	53 81	54 17
94 45	64 84	17 28	06 57	71 96	81 36	37 65	42 62	43 84	45 23
10 30	05 07	21 34	59 18	85 95	21 87	73 16	78 37	15 98	16 66
73 39	21 94	01 84	28 20	50 35	57 82	88 13	52 53	76 73	68 22
47 91	87 36	45 69	03 01	24 25	13 64	42 74	36 67	77 67	00 92
39 24	26 77	62 37	82 46	93 96	82 75	75 16	95 05	30 68	83 02
77 29	09 12	41 77	29 57	34 89	94 95	45 70	59 85	38 04	04 80
04 78	20 07	17 15	68 12	38 26	01 90	68 30	83 80	19 89	98 65
83 81	53 08	09 23	22 61	99 41	27 90	35 43	07 09	62 26	45 83
97 67	74 54	96 14	63 28	98 11	18 33	82 60	90 41	33 11	77 59
52 80	26 89	13 38	70 08	73 22	64 70	83 44	49 24	20 93	12 59
80 69	43 27	33 56	39 88	73 31	24 44	87 33	08 21	40 06	77 91
00 48	24 08	73 92	37 19	69 87	91 79	86 27	47 91	31 70	53 52
14 91	97 37	53 40	46 26	29 25	96 42	57 22	94 34	59 71	23 59
50 62	28 51	94 10	15 18	06 02	39 94	13 91	54 50	60 27	26 68
17 59	53 08	58 06	80 00	75 71	95 13	76 91	24 55	34 09	97 12
73 17	99 45	85 28	63 17	99 31	24 62	75 82	78 89	27 59	18 62
37 95	74 96	25 44	95 66	42 02	31 48	82 21	76 87	86 75	07 95
76 95	18 76	76 28	18 60	44 92	76 09	46 96	39 37	27 12	30 44

Index